THE
POWER
OF
PLAGUES

THE
POWER
OF PLAGUES

Irwin W. Sherman

Department of Biology

University of California

Riverside, California

ASM
PRESS

Washington, D.C.

Address editorial correspondence to ASM Press, 1752 N St. NW, Washington, DC 20036-2904, USA

Send orders to ASM Press, P.O. Box 605, Herndon, VA 20172, USA
Phone: (800) 546-2416 or (703) 661-1593
Fax (703) 661-1501
E-mail: books@asmusa.org
Online: estore.asm.org

Library of Congress Cataloging-in-Publication Data

Sherman, Irwin W.
  The power of plagues / Irwin W. Sherman.
     p. ; cm.
  Includes bibliographical references and index.
  ISBN-13: 978-1-55581-356-7 (alk. paper)
  ISBN-10: 1-55581-356-9 (alk. paper)
   1. Communicable diseases—History. 2. Epidemics—History.
3. Diseases and history. I. American Society for Microbiology.
II. Title
   [DNLM: 1. Communicable Diseases—history. 2. Communicable Disease
Conrol—history. 3. Disease Outbreaks—history. WC 11.1 S553p 2006]

RA643.S55 2006
614.4—dc22

2005031975

10 9 8 7 6 5 4 3 2 1

Cover photo: The Four Horseman of the Apocalypse by Daniel Stolpe. Hand-pulled lithograph, 21 × 48 inches, 1995. www.nativeimagesgallery.com.

Contents

Preface   *vii*

1. The Nature of Plagues   *1*
2. Plagues, the Price of Being Sedentary   *23*
3. Six Plagues of Antiquity   *43*
4. An Ancient Plague, the Black Death   *67*
5. A Modern Plague, AIDS   *89*
6. Typhus, A Fever Plague   *117*
7. Malaria, Another Fever Plague   *135*
8. King Cholera   *159*
9. Smallpox, The Spotted Plague   *191*
10. Preventing Plagues   *211*
11. The Plague Protectors   *231*
12. The Great Pox Syphilis   *255*
13. The People's Plague: Tuberculosis   *275*
14. Leprosy, the Striking Hand of God   *303*
15. Six Plagues of Africa   *313*
16. Plagues without Germs   *355*
17. Plagues On Order   *383*

Appendix. Cells and Viruses   *401*
General Works on Disease and History   *405*
Notes   *407*

Index   *419*

Preface

Plagues, the historian Asa Briggs observed, "are a dramatic unfolding of events; they are stories of discovery, reaction, conflict, illness and resolution. They test "the efficiency and resilience of local and administrative structures" and "expose the relentlessly political, social and moral shortcomings . . . rumors, suspicions, and at times violent . . . conflicts." This book was written to make the science of epidemic diseases—plagues—accessible and understandable. It is a guide through the maze of contagious diseases, their past importance, the means by which we came to understand them, and how they may affect our future.

In 1976 I purchased *Plagues and Peoples* by the University of Chicago historian, William H. McNeill. That book changed my way of thinking about infectious diseases. McNeill wrote that while carrying out research for a new book, *The Rise of the West: A History of the Human Community*, he was struck by a perplexing question: how was Hernan Cortez, with an army of less than 600 men, able to conquer the millions of subjects of the Aztec empire? Discarding notions about the superiority of the Spaniards because of their guns and horses, McNeill proposed that an epidemic of smallpox broke out among the Aztecs and, because of its selective lethality, had a profound psychological influence on the Indians. "There could be no doubt about which side of the struggle enjoyed divine favor . . . Little wonder then that the Indians accepted Christianity and submitted to Spanish control so meekly . . . The lopsided impact of infectious disease upon Amerindian populations offered a key to understanding the ease of Spanish conquest of America—not only militarily, but culturally as well." McNeill went on to hypothesize that the course of history was influenced by human encounters with infectious diseases, especially when an unfamiliar infection attacked a population for the first time. If this were indeed true, he asked, then why did historians neglect the significance of epidemic disease? He answered: it was overlooked because the older records of deaths from disease were so imperfect that the scale and significance were lost. Further, modern day historians, conditioned by their own experiences with epidemic infections in which immunity tended to limit the severity of the outbreak, tended to consider older accounts of massive die-offs as gross exaggerations. In other words, there was (with rare exceptions such as the Black Death) a failure of historians to appreciate the differences in outcome "between the outbreak of a disease in an experienced population and the ravages of the same infection when loosed on a community lacking acquired immunities."

McNeill went on to write that the assumption made by many historians was that infections were always the same as those experienced by Europeans before the advent of modern medicine and therefore historical treatises mentioned epidemics not at all or only casually. Further, because epidemic diseases appeared to have an unpredictable quality and because they lacked the decisiveness of wars or natural catastrophes such as floods and volcanic eruptions, historians were left with an uncomfortable feeling. Disease, with its lack of predictability, did not seem to fit into a larger picture of human history and so it was of little interest to the serious professional historian. McNeill's aim was "to bring the history of infectious diseases into the realm of historical explanation by showing how varying patterns of disease circulation have affected human affairs in ancient as well as modern times."

A decade later the world was faced with the ever-expanding AIDS epidemic, and most of us were suffering from historical amnesia with regard to disease. I soon realized that a new book on disease and history could help to bridge the gap in the understanding of plagues and their impact on lives past, present, and future. McNeill's objective was to place infectious disease in a historical context. Mine was also to provide this, but as a biologist I also wanted to describe the nature and evolution of diseases and then to show how the past could prepare us for future encounters with infectious diseases. In 1986, most of us were unfamiliar with typhus, diphtheria, whooping cough, malaria, sleeping sickness, or snail fever. If we had heard about sexually transmitted diseases other than AIDS, such as syphilis and herpes, we didn't know much about the different kinds of parasites that were responsible. Most people under the age of 25 lacked a smallpox vaccination scar because the disease had been eradicated. Indeed, it was a commonly held belief that all diseases in the developed world had been eliminated by vaccines or soon would be. If not, we assured ourselves, we could rely on new drugs such as antibiotics to cure every new disease. We were convinced that our water was safe to drink and our food could be eaten with little fear of contracting a disease. We were sure that the transmitters of disease—mosquitoes, flies, lice, and fleas—could easily be eliminated by insecticides, and if by chance an individual did become infected, we were confident that that person could be treated quickly and effectively and thus disease spread would be curbed.

These views were to change rather quickly: drug-resistant tuberculosis emerged as a worldwide threat, and there were outbreaks of Ebola, hantavirus pulmonary syndrome, and SARS (severe acute respiratory syndrome).

In the summer of 1999, an outbreak of West Nile virus (WNV) caused illness in 62 people and 7 died. This took place in New York City, not Africa. By 2003, WNV, spread by infected birds, was present throughout the United States. A particularly lethal bird influenza—one that killed millions of animals in a dozen Asian countries—caused alarm in 2004. This virus, which rarely infects humans, spread rapidly in the human population and killed 42 people. If this "new" flu had been able to be transmitted from person-to-person by coughing, sneezing, or even a handshake, then a natural bioterrorist attack would have occurred. Luckily, that did not happen, but the prediction is this: in the foreseeable future it will occur and there will be no place for us to hide. (Chapter 16 discusses WNV and the manner by which "new" flu strains emerge.) Clearly, in a very short time all of us began to appreciate that epidemic disease was not a thing of our past.

This book has been conceived as a conversation about how we came to understand the nature of severe outbreaks of epidemic disease—plagues, if you will. It tells about scientists and how they discovered the causes and developed controls for infectious diseases, as well as the ways in which plagues and culture interact to shape values, traditions, and the institutions of Western civilization. It also provides a status report—where we are today with epidemic diseases—and offers possible solutions to the coming plagues such as malaria, tuberculosis, cholera, AIDS, mad cow disease, influenza, and West Nile.

I have not taken a chronological approach in the examination of the plagues that have afflicted humankind. Chapters have been written so that they are more or less independent, and as a result they need not be read in a proscribed sequence. However, in some instances readers may attain a somewhat better understanding when the chapters on principles and protection (chapters 1, 2, 10, and 11) are read early on. Some readers will be disappointed that their "favorite" plague has not been included in these pages. To have added many more epidemic diseases would have made the book much too long and encyclopedic—something I wanted to avoid. Rather, the particular plagues included have been selected for their value in teaching us important lessons. The style of *The Power of Plagues* is such that readers without any background in the sciences can easily understand its message. This book is intended to promote an understanding of infectious disease agents by a sober and scientific analysis and is not a collection of horror stories to provoke fear and loathing. Learning about how infectious diseases have shaped our past has proven to be an exciting and enlightening experience for me. My hope is that readers of this book will also find that to be true.

Figure 1.1   *Woman with Child*, by Pablo Picasso, 1903. © 2005 Estate of Pablo Picasso/Artists Rights Society (ARS), New York.

Chapter 1

The Nature of Plagues

Disease can be a personal affair. Peter Turner, a World War II veteran, was a commander of the Pennsylvania division of the American Legion. In the summer of 1976, Turner, a tall, well-built 65-year-old, decked out in full military regalia, attended the American Legion convention in Philadelphia. As a commander, Turner stayed at the Bellevue-Stratford Hotel, headquarters for the meeting. Two days after the convention, Turner fell ill with a high fever, chills, headache, and muscle aches and pains. He dismissed the symptoms as nothing more serious than a "summer cold." His diagnosis proved to be wrong. A few days later, he had a dry cough, chest pains, shortness of breath, vomiting, and diarrhea. Within a week, his lungs filled with fluid and pus, and he experienced confusion, disorientation, hallucinations, and loss of memory. Of 221 legionnaires who became ill, Commander Turner and 33 others died from pneumonia. The size and severity of the outbreak, called Legionnaires' disease, quickly gained public attention, and federal, state, and local health authorities launched an extensive investigation to determine the cause of this "new" disease. There was widespread fear that Legionnaires' disease was an early warning of an epidemic. Although no person-to-person spread could be documented, few people attended the funerals or visited with the families of the deceased veterans.

Statistical studies of Legionnaires' disease revealed that all who became ill had spent a significantly longer period of time in the lobby of the Bellevue-Stratford Hotel than those who remained healthy. Air was implicated as the probable pathway of spread of the disease, and the most popular theory was that infection resulted from aspiration of bacteria (called *Legionella*) in aerosolized water from either cooling towers or evaporative condensers. Unlike infections caused by inhalation, aspiration is produced by choking. Secretions in the mouth get past the choking reflex and, instead of going into the esophagus and stomach, mistakenly enter the lungs. Protective mechanisms that normally prevent aspiration are defective in older people, smokers, and those with lung disease. The Legionnaires were near-perfect candidates for contracting the disease. Since the Philadelphia outbreak, there have been numerous reports of Legionnaires' disease in the general population: 11,000 documented cases annually in the United States and estimates as high as 100,000, with a fatality rate of 15%. These outbreaks have been traced to water heaters, whirlpool baths, respiratory therapy equipment, and ultrasonic misters used in grocery stores.

A few years later, another "new" disease appeared. Mary Benton, a graduate student and English composition teaching assistant at UCLA, knew something was amiss as she prepared for Monday's class. She had spent the previous day happily celebrating her 24th birthday, but by evening she was doubling over in pain every time she went to the bathroom. Mary, who was previously healthy and active, figured she had an infection or was suffering from overeating; she became concerned as her symptoms worsened. By the time she saw her physician, she had nausea, chills, diarrhea, headache, and a sore throat. Her temperature was 104.7°F, her heart rate was 178 beats/minute, and she had a red rash, initially on her thighs but becoming diffuse over her face, abdomen, and arms. Her blood pressure had fallen to 84/50, she had conjunctivitis in both eyes, and her chest X ray was normal, but a pelvic examination revealed a brownish discharge. Despite administration of antibiotics, oxygen, and intravenous fluids, her condition deteriorated over the next 48 hours. She died of multiorgan failure—low blood pressure, hepatitis, renal insufficiency, and internal blood clots. Laboratory tests provided clues to the cause of death. Cultures made from her blood, urine, and stools were negative, but the vaginal sample contained the bacterium *Staphylococcus aureus*. The "new" disease that had felled Mary Benton was named "toxic shock syndrome," or TSS. The source of Mary's infection, and the possibility that it might be spread through the population as a sexually transmitted disease (STD), raised many concerns. For the next 10 years, TSS continued to appear among previously healthy young women residing in several states. As with Mary Benton, each case began with vomiting and high fever followed by lightheadedness and fainting, the throat felt sore, and the muscles ached. A day later, a sunburn-like rash appeared, and the eyes became bloodshot. Within 3 to 4 days, victims suffered confusion, fatigue, weakness, thirst, and a rapid pulse; the skin became cool and moist; and breathing became rapid. This was followed by a sudden drop in blood pressure; if it remained low enough for a long enough period, circulatory collapse produced shock.

TSS is a gender-specific disease. From 1979 to 1996, it affected 5,296 women, median age 22, with a peak death rate of 4%. However, TSS was not an STD. Ultimately, it was linked to the use of certain types of tampons, especially those containing cross-linked carboxymethyl cellulose with polyester foam, which provided a favorable environment for the toxin-producing *S. aureus*. Elevated vaginal temperature and neutral pH, both of which occur during menses, were enhanced by the use of these superabsorbent tampons.

In addition, tampons obstruct the flow of menstrual blood and may cause reflux of blood and bacteria into the vagina. By the late 1980s, these tampon brands were removed from the market, and the number of deaths from TSS declined dramatically.

The effects of disease at the personal level can be tragic (Fig. 1.1), but when illness occurs in many people, it may produce another emotion—fear—for now the disease might spread rapidly, causing death as well as inflaming the popular imagination. The 2003 outbreak of SARS (severe acute respiratory syndrome) had all the scary elements of a plague—panic, curtailed travel and commerce, and economic collapse. It began in February 2003 when a 64-year-old Chinese physician who was working in a hospital in Guandong Province in southern China traveled to Hong Kong to attend a wedding and became ill. He had a fever, a dry cough, a sore throat, and a headache. Unconcerned, he felt well enough to sightsee and shop with his brother-in-law in Hong Kong; however, during that day his condition worsened, and he had difficulty in breathing. Seeking medical attention at a nearby hospital, he was taken immediately to the intensive care unit and given antibiotics, anti-inflammatory drugs, and oxygen. This was to no avail, and several hours later he suffered respiratory failure and died. The brother-in-law, who was in contact with him for only 10 hours, suffered from the same symptoms 3 days later and was hospitalized. Again, all measures failed, and he died 3 weeks after being hospitalized.

Laboratory tests for the physician (patient 1) and his brother-in-law (patient 2) were negative for Legionnaires' disease, tuberculosis, and influenza. A third SARS case occurred in a female nurse who had seen the physician in the intensive care unit, and the fourth case was a 72-year-old Chinese-Canadian businessman who had returned to Hong Kong for a family reunion. He stayed overnight in the same hotel and on the same floor as the physician. (He would ultimately carry SARS to Canada when he returned home.) Patient 5 was the nurse who attended the brother-in-law, and patients 6, 7, 8, and 9 were either visitors to the hospital or nurses who had attended patient 4. Patient 10 shared the same hospital room with patient 4 for 5 days. In less than a month, 10 patients had SARS, six of whom (3, 4, 6, 8, 9, and 10) survived and four of whom (1, 2, 5 and 7) died. Over the next 4 months, the SARS survivors sowed the seeds of infection that led to more than 8,000 cases and 800 deaths in 27 countries, representing every continent.

Despite the recognition that sickness, such as SARS, Legionnaires' disease, and TSS, may appear suddenly and with disastrous consequences, more often than not, little notice has been taken of the ways that disease can shape history. The influence of disease on history has often been neglected because there appeared to be few hard-and-fast lessons to be learned from a reading of the past. Sickness seemed to have no apparent impact, except for catastrophic epidemics such as the bubonic plague, or it was outside our experience. We live in an age when diseases appear to have minimal effects—we are immunized as children, we treat illness with effective drugs and antibiotics, and we are well nourished. Thus, our impressions of how disease can affect human affairs have been blunted. This, however, is an illusion: the sudden appearances of SARS, Legionnaires' disease, TSS, and acquired immunodeficiency syndrome (AIDS) are simply the most recent examples of how disease can affect society. Our world is much more vulnerable than it was in the past. New and old diseases can erupt and spread throughout the world more quickly because of the increased and rapid movements of people and goods. Efficiencies in transportation allow people to travel to many more places, and almost nowhere is inaccessible. Today, few habitats are truly isolated or untouched by humans or our domesticated animals. We can move far and wide across the globe, and the vectors of disease can also travel great distances. Aided by fast-moving ships, trains, and planes, they introduce previously remote diseases into our midst (such as West Nile virus infection, SARS, influenza, and mad cow disease). New diseases may be related to advances in technology: TSS resulted from the introduction of "improved" menstrual tampons that favored the growth of a lethal microbe, and Legionnaires' disease was the result of the growth and spread of another deadly "germ" through the hotel's air-conditioning system.

This book chronicles the recurrent eruptions of plagues that marked the past, influence the present, and surely threaten our future. The particular occurrence of a severe and debilitating outbreak of disease may be unanticipated and unforeseen, but despite the lack of predictability, there is a certainty: dangerous "new" diseases will occur.

Living Off Others

The "germs" that caused SARS, Legionnaires' disease, and TSS are parasites. To more fully appreciate the nature of these and other diseases and how they may be controlled, it helps to know a little more about parasites. No one likes to be called a parasite. The word suggests, at least to some, a repugnant alien creature that insinuates itself into us and cannot be shaken loose. Nothing could be further from the truth. Within the range of all that lives, some entities are unable to survive on their own and require another living being for their nourishment. These life-dependent entities that "feed at the table of the rich" are called parasites, from the Latin word *parasitus*, meaning "food." The business they practice, parasitism, is neither disgusting nor unusual. It is simply a means to an end: obtaining the resources needed for their growth and reproduction. We do the same—eating and breathing—in order to survive.

Parasitism is the intimate association of two different kinds of organisms (species) wherein one benefits (the parasite) at the expense of the other (the host), and as a consequence of this, parasites often harm their hosts. The harm inflicted, with observable consequences, such as those seen in Commander Peter Turner and Mary Benton and those patients afflicted with SARS, is called "disease," literally, "without comfort." Though parasites can be described by the one thing they are best known for—causing harm—they come in many different guises (Fig. 1.2 and 1.3). Some may be composed of

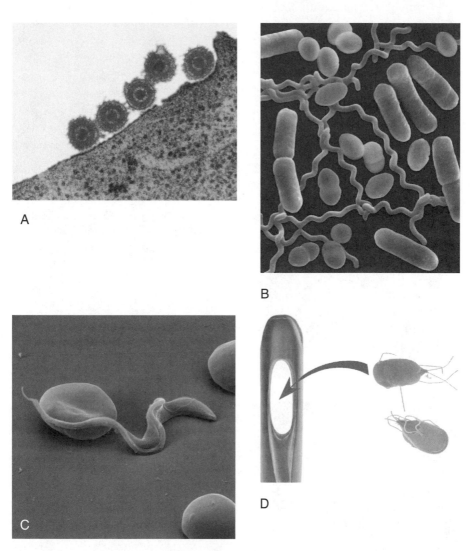

Figure 1.2 A catalog of microparasites. (A) Cold sore virus, as seen with the transmission electron microscope. (B) Three kinds of bacteria—spherical (coccus), rod shaped (bacillus), and corkscrew (spirillum or spirochete)—as seen with the scanning electron microscope. (C) Trypanosome (ribbon-like organism) among red blood cells, as seen by the scanning electron microscope. (D) Using a scanning electron microscope it can be seen that *Giardia* easily fits through the eye of a needle. (Panels A, B, and D courtesy of Dennis Kunkel Microscopy, Inc. Panel C courtesy of Eye of Science/Photo Researchers, Inc. © 2005 Photo Researchers, Inc.)

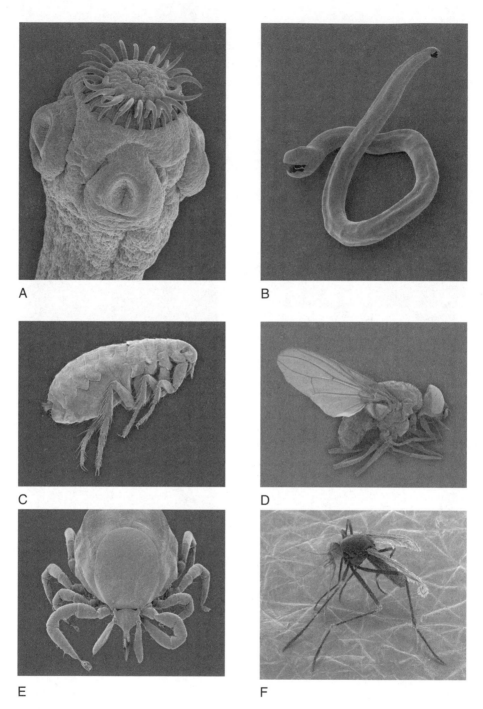

Figure 1.3 A catalog of macroparasites. (A) The head of a tapeworm (a flatworm); (B) the hookworm (a roundworm); (C) flea; (D) fly; (E) tick; and (F) mosquito. All seen through the scanning electron microscope. (Courtesy of Dennis Kunkel Microscopy, Inc.)

a fragment of genetic material wrapped in protein, such as a virus. Others (bacteria, fungi, protozoa) consist of a single cell (see "Cells and Viruses" in the Appendix), and some are made up of many cells (roundworms, flatworms, mosquitoes, flies, ticks). Some parasites, such as tapeworms, hookworms, the malaria parasite (*Plasmodium falciparum*), and the human immunodeficiency virus (HIV), live inside the body, whereas others (ticks and chiggers) live on the surface. Parasites are invariably smaller in mass than their host. Consider the sizes of the malaria parasite (a microparasite) and hookworm (a macroparasite). Both produce anemia or, as one advertisement for an iron supplement called it, "tired blood." A malaria parasite lives within a red blood cell that is 1/5,000 of an inch in diameter. If only 10% of your blood cells were infected, the total mass of the malaria parasites would not occupy a thimble, yet in a few days they could destroy enough of your red blood cells that the acute effects of blood loss could lead to death. In effect, you could die from an internal hemorrhage. Although the "vampire of the American South," the blood-sucking, threadlike hookworm, is only 0.5 inches in length and 0.05 inches in girth, if your intestine harbored 50 worms, you would lose a cupful of blood a day. Yet the entire mass of worms would weigh less than 5 hairs on your head.

Some parasites have complex life cycles and may have several hosts. In malaria, the hosts are mosquitoes and humans; in blood fluke disease, the "curse of the Pharaohs," the hosts are humans and snails; and in sleeping sickness, the hosts are tsetse flies, game animals, and humans. All parasites—whether they are large or small—cause harm to their host, though not all kill their host outright. This is because resistance may develop in any population of hosts, and not every potential host will be infected—some individuals may be immune or not susceptible because of a genetic abnormality or the absence of some critical dietary factor (e.g., vitamin deficiency).

To succeed in a hostile world where individual hosts are distinct and separate from one another, parasites need to disperse their offspring or infective stages to reach new hosts. To meet this requirement they produce lots of offspring, thereby increasing the odds that some will reach new hosts. It is a matter of numbers: more offspring will have a greater probability of reaching a host and setting up an infection. In this way, the parasite enhances its chances for survival. Three cases will illustrate this: the malaria parasite, the red blood cell-destroying hookworms, and the white blood cell killer HIV.

When a malaria-infected mosquito feeds, it injects with its saliva perhaps a dozen of the thousands of parasites that are present in its salivary glands. Each malaria parasite invades a liver cell and, after a week, each produces up to 10,000 offspring; in turn, every one of these offspring infects a red blood cell. Within the infected red blood cell, a malaria parasite produces 10 to 20 additional infective forms to continue the destructive process. In little more than 2 weeks, a person infected by a single malaria parasite will be infected with more than 100,000 parasites, and 2 days later the blood will contain millions of malaria parasites.

Hookworms (Fig. 1.3B) live attached to the lining of the small intestine, which they pierce with their razor-sharp teeth, allowing them to suck blood, as would a leech. Each female hookworm—no bigger than an eyelash—can live within the intestine for more than 10 years, producing each day more than 10,000 eggs. In her lifetime, this "Countess Dracula" can produce more than 36 million microscopic eggs.

The AIDS-causing virus, HIV, is a spherical particle so small that if 250,000 were lined up they would hardly be an inch in length. However, each virus has an incredible capacity to reproduce itself. After it invades a specific kind of white blood cell (the T-helper lymphocyte), where it replicates, a million viruses will be produced in a few short days. To gain some appreciation of the high reproductive capacity of this virus, we might think of the infecting HIV as a person standing on a barren stretch of beach; if we were to return to this beach a few days later, we would find it jammed and overcrowded with millions—a population explosion.

Any environment other than a living host is inimical to the health and welfare of the parasite. Some parasites have got around this with resistant stages, such as spores, eggs, or cysts, that enable them to move from one host to another in a fashion akin to "island hopping." Hookworms, tapeworms, blood flukes, and pinworms have eggs that are able to survive outside the body; the microscopic cysts of the roundworm *Trichinella* are able to resist the ordinarily lethal effects of the acids in our stomach to cause trichinosis; and we are all too familiar with the possibility of a bioterrorist attack from anthrax spores, which spread by inhalation of "anthrax dust." The movement of a parasite from host to host—whether by direct or indirect means—is called transmission. When the transmission of parasites involves living organisms such as flies, mosquitoes, ticks, fleas, lice, or snails, these "animate intermediaries" are called vectors. Transmission by a vector may be mechanical (e.g., the bite wound of a mosquito or fly) or developmental (e.g., parasites that grow and reproduce in snails, as in blood fluke disease, or in mosquitoes, as in malaria and yellow fever). Transmission of a parasite may also occur through contamination of eating utensils, drinking cups, food, needles, bedclothes, towels, or clothing or in droplet secretions. In the 1976 outbreak of Legionnaires' disease in Philadelphia, transmission was not from person to person but through a fine mist of water in the air-conditioning system, whereas in the case of SARS (and influenza), transmission is from person to person via droplet secretions from the nose and mouth.

Parasites and their free-living relatives come in a variety of sizes, shapes, and kinds (species). Bacteria, 1 to 5 μm in size, are prokaryotes (see "Cells and Viruses" in the Appendix) that can be free living or parasitic. They may assume several body forms—rods (bacilli), spheres (cocci), or spiral. Protozoa, 5 to 15 μm in size, are one-celled eukaryotes (see "Cells and Viruses" in the Appendix) that can lead an independent existence (such as the freshwater *Amoeba* sp.) or be parasitic (such as the *Entamoeba* sp. that causes amebic dysentery or the corkscrew-shaped trypanosomes that cause African sleeping sickness). Bacteria and protozoa are too small to be seen with the unaided eye. The technological advance—the microscope—perfected in the 1600s allowed for their discovery, and so they are called microparasites (Fig. 1.2 illustrates some parasites that cannot be seen without the aid of a microscope). The ultimate microparasite is a virus. Viruses are smaller than bacteria and cannot be seen with the light microscope but only with the electron microscope, which can magnify objects more than 10,000 times. Although a virus's genetic code (in the form of either RNA or DNA) contains all the information needed for assembling a new virus, it lacks that which is necessary for reproduction. Therefore, for a virus to reproduce, it must enter a living cell and use the cellular machinery to replicate itself. Because viruses are not completely independent, they are not alive, and yet they can be killed if their DNA or RNA is destroyed. Viruses—the agents of SARS, AIDS, and the flu—are neither cells nor organisms. Microparasites reproduce within their hosts and are sometimes referred to as infectious microbes or, more commonly, "germs." Larger parasites that can be seen without the use of a microscope are referred to as macroparasites; they are composed of many cells. Those that most often cause disease in humans or domestic animals are the roundworms, such as the hookworm; the flatworms, such as the tapeworm and the blood fluke; the blood-sucking insects, such as mosquitoes, flies, and lice; or the arachnids, such as ticks (Fig. 1.3 illustrates parasites that can be seen with the human eye). Macroparasites do not multiply within an infected individual (except in the case of larval stages in the intermediate hosts); instead, they produce infective stages that usually pass out of the body of one host before transmission to another.

"What's in a name? That which we call a rose by any other name would smell as sweet." When William Shakespeare penned these lines in *Romeo and Juliet*, he gave value to substance over name calling. However, being able to tell one microbe from another is more than having a proper name for a germ—it can have practical value. Imagine you have just returned from a trip and now suffer with a fever, headache, and joint pains; worst of all, you have a severe case of diarrhea. What a mess you are! When you see your physician, she tells you that your distress could be due to an infection with *Salmonella*, *Giardia*, *Entamoeba*, or the influenza or SARS virus. Prescribing an antibiotic for a disease caused by a virus would do you no good, but in the case of "food poisoning" caused by *Salmonella*, a bacterium, a course of antibiotic therapy might restore you to health. On the other hand, if your clinical symptoms are due to the presence of protozoan parasites such as *Giardia* or *Entamoeba*, they would not respond to antibiotics; other drugs would have to be prescribed to cure you. Therefore, determining the kind of parasite (or parasites) you harbor will do more than provide the name of the offender; it will allow for selective treatment of your illness.

Plagues and Parasites

In antiquity, all disease outbreaks, irrespective of their cause, were called plagues; the word "plague" comes from the Latin *plaga* meaning "to strike a blow that wounds." When a parasite invades a host, it establishes an infection and wounds the body (Fig. 1.4). Individuals who are infected and can spread the disease to others (such as SARS patient 4) are said to be contagious or infectious. Initially, Legionnaires' disease and TSS were thought to be contagious. However, despite the obvious clinical signs of coughing, nausea, vomiting, and diarrhea, a person-to-person transmissible agent was not found. In short, the victims of TSS and Legionnaires' disease were not infectious, in contrast to patients with influenza, SARS, and the common cold, who display a similar array of symptoms. Influenza and SARS are different kinds of upper respiratory diseases: the flu is contagious 24 hours before

Figure 1.4   *The Plague of Ashod* by Nicolas Poussin (1594–1665). Poussin's painting is probably that of bubonic plague since rats are shown on the plinth. (Courtesy of Corbis.)

symptoms appear, has a short (2- to 4-day) incubation period, and infrequently requires hospitalization, whereas with SARS there is a longer (3- to 10-day) incubation period, the individual is infectious only after symptoms appear, and hospitalization is required.

Infectiousness, however, may persist even after disease symptoms have disappeared; such infectious but asymptomatic individuals are called carriers. The most famous of these carriers was the woman called "Typhoid Mary," an Irish immigrant to the United States whose real name was Mary Mallon. In 1906 she began working as a cook for a wealthy New York banker, Charles Henry Warren, and his family. The Warren family rented their large house in Oyster Bay, Long Island, from a Mr. George Thompson. That summer, 6 of 11 people in the house came down with typhoid fever (caused by the "germ" *Salmonella enterica* serovar Typhi), including Mrs. Warren, two daughters, two maids, and a gardener. Mr. Thompson, fearing he would be unable to rent his "diseased house" to others, hired George Soper, a sanitary engineer, to find the source of the epidemic. Soper's investigation soon led him to Mary Mallon, who had been hired as a cook just 3 weeks before the outbreak of typhoid in the Warren household. Mary had remained with the Warrens for only a month and had already taken another position when Soper found her. On June 15, 1907, Soper published his findings in the *Journal of the American Medical Association*: Mary was a healthy carrier of typhoid germs. Although she was unaffected by the disease (which causes headache, loss of energy, diarrhea, high fever, and, in 10% of cases, death) she still could spread it. When Soper confronted Mary and told her she was spreading death and disease through her cooking, she responded by seizing a carving fork, rushing at him, and driving Soper off. Soper, however, was undaunted, and he convinced the New York City Health Department that Mary was a threat to the public's health. She was forcibly carried off to an isolation cottage at Riverside Hospital on North Brother Island in the Bronx. There, her feces were examined and found to contain the typhoid bacteria. Mary remained at the hospital, without her consent, for 3 years and was then allowed to go free as long as she remained in contact with the Health Department and did not engage in food preparation. She disappeared from Health Department view for a time but then took employment as a cook at the Sloane Maternity Hospital under an assumed name, Mrs. Brown. During this time she spread typhoid to 25 doctors, nurses, and staff, two of whom died. She was sent again to North Brother Island, where she lived the rest of her life, 23 years, alone in a one-room cottage. During her career as a cook, "Typhoid Mary" probably caused many more than the well-documented cases, and she surely caused more than three deaths. Mary Mallon was not the only human carrier of typhoid. In 1938 when she died, the New York City Health Department noted that there were 237 others living under their observation. However, she was the only one kept isolated for years, and one historian has ascribed this to prejudice toward the Irish and a noncompliant woman who could not accept that unseen and unfelt "bugs" could infect others. Mary Mallon told a newspaper, "I have never had typhoid in my life and have always been healthy. Why should I be banished like a leper and compelled to live in solitary confinement . . .?"

Forecasting Storms, Predicting Plagues

In October 1991, meteorologists predicted "the perfect storm." As a result, the public was warned of an impending storm of epic proportions that would send high winds and Atlantic Ocean waves crashing on the East Coast of the United States, from Boston, Mass., to Cape Hatteras, S.C. The storm was created by three factors: a collision between a high-pressure system, a low-pressure system, and the remnants of a dying hurricane. Understanding weather factors allows for better tracking and predicting of storms; similarly, recognizing the elements required for a parasite to spread in a population allows for better forecasting of the course a disease may take. Three factors are required for a parasite to spread from host to host: there must be infectious individuals, there must be susceptible individuals, and there must be a means for transmission between the two. Transmission may be by indirect contact involving vectors such as mosquitoes (in malaria and yellow fever), flies (in sleeping sickness and river blindness), or ticks (in Lyme disease and Nantucket fever), or it may be by direct contact, as it is with measles, influenza, SARS, and tuberculosis, whereby it is influenced by population density.

In the past, the sudden increase in the number of individuals in a population affected by a disease was called a plague. Today we frequently refer to such a disease outbreak as an epidemic, a word that comes from the Greek words *epi* ("among") and *demos* ("the people"). Weather forecasters use satellites, Doppler radar, and measures of barometric pressure, wind velocity and direction, water and land surface temperatures, humidity, and cloud formations to produce statistical models that predict "the perfect storm" and lesser weather disturbances. The predictions of meteorologists are heard daily: there's a 20% chance of rain; the snow level will be down to 4,000 feet; road conditions will be hazardous; or tomorrow will be sunny and mild. Epidemiologists are disease forecasters who study the occurrence, spread, and control of a disease in a population by using statistical data and mathematical modeling to identify the causes and modes of disease transmission, to predict the likelihood of an epidemic, to identify the risk factors, and to help plan control programs such as quarantine and vaccination. When TSS broke out, epidemiological studies linked the syndrome to the use of tampons, principally Rely Tampons, in menstruating women. The recommendation was that the illness could be controlled by removal of such tampons from the market. Acting on this advice, Procter and Gamble stopped marketing Rely Tampons, and cases of TSS virtually disappeared.

For an infection to persist in a population, each infected individual, on average, must transmit the infection to at least one other individual. The number of individuals each infected person infects at the beginning of an epidemic is given by R_0; this is the basic reproductive ratio of the disease or, more simply, the multiplier of the disease. The multiplier helps to predict how fast a disease will spread through the population.

The value for R_0 can be visualized by considering the children's playground game of touch tag. In this game, one person is chosen to be "it," and the object of the game is for that player to touch another, who in turn also becomes "it." From then on, each person touched helps to tag others. If no other player is tagged, the game is over, but if more than one other player becomes "it," then the number of touch taggers multiplies. Thus, if the infected individual ("it") successfully touches another (transmits), then the number of diseased individuals (touch taggers) multiplies. In this example, the value for R_0 is the number of touch taggers that results from their being in contact with "it."

The longer a person is infectious, and the greater the number of contacts that the infectious individual has with those who are uninfected, the greater the value of R_0 and the faster the disease will spread. An increase in the population size or in the rate of transmission increases R_0, whereas an increase in parasite mortality or a decrease in transmission will reduce the spread of disease in a population. Thus, a change that increases the value of R_0 tends to increase the proportion of hosts infected (prevalence) as well as the burden (incidence) of a disease. Usually, as the size of the host population increases, so do disease prevalence and incidence.

If the value for R_0 is larger than 1, the "seeds" of the infection (i.e., the transmission stages) will lead to an ever-expanding spread of the disease—an epidemic or a plague—but in time, as the pool of susceptible individuals is consumed (like fuel in a fire), the epidemic may eventually burn itself out, leaving the population to await a slow replenishment of new susceptible hosts (providing additional fuel) through birth or immigration. Then a new epidemic may be triggered by the introduction of a new parasite or mutation, or there may be a slow oscillation in the number of infections, eventually leading to a persistent low level of disease. However, if R_0 is less than 1, each infection produces fewer than one transmission stage and the parasite cannot establish itself.

The economic costs of the outbreak of SARS in 2003 were nearly $100 billion as a result of decreased travel and decreased investment in Southeast Asia. The University of California at Berkeley was so concerned about this epidemic that it put a ban on Asian students planning to enroll for the summer session. The question raised at the outset was: How long will the SARS outbreak last? Calculating the value of R_0 provided an answer. Analysis of ~200 cases during the first 10 weeks of the epidemic gave an R_0 value of 3.0, meaning that a single infectious case of SARS would infect about three others if control measures were not instituted. This value suggested that there was a low to moderate rate of transmissibility and that hospitalization would block the spread of SARS. The prediction was borne out: transmission rates fell as a result of reductions in population contact rates and improved hospital infection control as well as more rapid hospitalization of suspected but asymptomatic individuals. By July 2003 the R_0 value was much smaller than 1, and the ban on Asian students enrolling at the Berkeley campus of the University of California was lifted.

Epidemiologists know that host population density is critical in determining whether a parasite can become established and persist. The threshold value for disease establishment can be obtained by finding the population density for which $R_0 = 1$. In general, the size of the population needed to maintain an infection varies inversely with the transmission efficiency and directly with the death rate (virulence). Thus, virulent parasites, that is, those causing an increased number of deaths, require larger populations to be sustained, whereas parasites with reduced virulence may persist in smaller populations.

Measles, caused by a virus, provides an almost ideal pattern for studying the spread of a disease in a community. The virus is transmitted through the air as a fine mist released through coughing, sneezing, and talking. The virus-laden droplets reach the cells of the upper respiratory tract (nose and throat) and the eyes and then move on to the lower respiratory tract (lungs and bronchi). After infection, the virus multiplies for 2 to 4 days at these sites and then spreads to the lymph nodes, where another round of multiplication occurs. The released viruses invade white blood cells and are carried to all parts of the body, using the bloodstream as a waterway. During this time, the infected individual shows no signs of disease. But after an incubation period of 8 to 12 days, there is fever, weakness, loss of appetite, coughing, a runny nose, and tearing of the eyes. Virus replication is now in high gear. Up to this point, the individual probably believes that his or her suffering is a result of a cold or influenza, but when a telltale rash appears—first on the ears and forehead and then spreading over the face, neck, trunk, and to the feet—it is clearly neither influenza nor a common cold. Once a measles infection has begun, there is no treatment to halt the spread of the virus in the body.

Measles (along with mumps, whooping cough, smallpox, and chicken pox) passes from one host to another without any intermediary; recovery from a single exposure produces lifelong immunity. As a consequence, measles and the other diseases commonly afflict children and are therefore called "childhood diseases." Although measles has been eradicated in the United States because of childhood immunization, it can be responsible for a death rate of about 30% in less developed countries. It is one of the 10 most frequent causes of death in the world today. One of the reasons that measles may disappear from a community is immunity that may be the result of natural recovery from an infection or immunization.

Hurricanes are areas of low pressure that form over oceans in the tropics. (Over the Pacific Ocean and north of the equator, such storms are called typhoons, and in the South Pacific and Indian Oceans they are referred to as tropical cyclones.) Meteorologists classify these storms into five categories depending on the wind speed, barometric pressure, storm surge height, and potential for damage. A category 1 hurricane has winds of 74 to 95 mph and may cause damage to shrubbery and mobile homes, whereas a category 5 hurricane has winds of greater than 155 mph and has the potential to do extensive damage to roofs on buildings; massive evacuation of residents may be recommended. This classification informs us of the strength of these tropical storms and helps predict their development. However, as valuable as this classification can be, it is not a precise indicator of all the possible aspects, e.g., the location of landfall, extent of damage to property, number of deaths, economic losses, and when it will decline in force. Like hurricanes, epidemics or plagues can be classified into three types with their own stages of development (Fig. 1.5), and each may produce a very different set of outcomes.

Measles is highly infectious and moves rapidly through a population in epidemic waves, coming at regular intervals. Let us consider the pattern of measles in a population. The persistence of measles depends on the critical community size and is defined as the smallest population without any temporary absence of disease. This type I epidemic pattern for measles is found in a population of about 300,000 to 500,000 people, with about 7,000 to 10,000 cases of measles per year. In a type I epidemic, the population is large and the pattern shows a regular series of outbreaks (peaks), but the disease never completely disappears—that is, it is endemic—and cases persist. This is because the number of susceptible individuals is large enough for the chain of transmission of infection to remain unbroken, and so R_0 is greater than 1.

In a type II epidemic, the peaks of infection are discontinuous, but there is a regular pattern of occurrence of cases. But there is no endemicity; there are temporary absences of the disease, and the value of R_0 is less than 1. This pattern occurs because there are not enough susceptible individuals to maintain the chain of virus transmission. Measles shows this pattern in a population of about 10,000 to 100,000. Finally, a type III epidemic occurs in even smaller communities, those with less than 10,000 people. Here, the pattern of an increased number of cases occurs at irregular intervals, and there are long periods when there is no disease. Because of the small size and/or the remoteness of these communities, the chain of transmission of infection is interrupted, infections have to be reintroduced, and R_0 is a number much smaller than 1. In contrast, when an infectious disease becomes worldwide or widespread (and R_0 is a large number), then the epidemic is called a pandemic. Examples of pandemic diseases include SARS, influenza, bubonic plague, cholera, and AIDS.

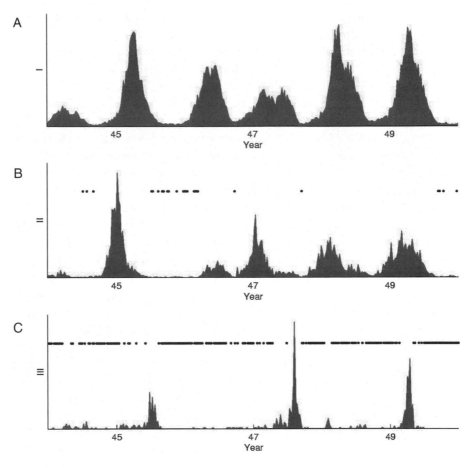

Figure 1.5 Types of epidemics. The plots show the incidence or numbers of cases of measles for three different sizes of populations from 1942 to 1952. (A) In type I, the peaks in the number of cases are regular, and there are always cases (endemicity) and no fadeouts. The population size is 3.4 million. (B) In type II, there are regular outbreaks (peaks in the number of cases), no endemicity, and fadeouts (shown by dots). This occurs with a population size of 300,000. (C) In type III, with a population size of 10,000, there are irregular outbreaks with long fadeouts occurring between the peaks. (Courtesy of Matthew Keeling.)

The spread of infection from an infected individual through the community can be thought of as a process of diffusion whereby the motions of the individuals are random and movement is from a higher concentration to a lower one. Therefore, factors affecting its spread include the size of the population, those communal activities that bring susceptible individuals in contact with infectious individuals, the countermeasures used (e.g., quarantine, hospitalization, immunization), and seasonal patterns. For example, in temperate northern climes, measles spreads most frequently in the winter months because people tend to be confined indoors. However, in Iceland, where the spring thaw is followed by a harvest, there are also summer peaks because of communal activities on the farms. As noted earlier, the spread of SARS in 2003 was controlled by quarantine and hospitalization; winter flu epidemics can usually be limited by immunization.

Epidemiologists have as one of their goals the formulation of a testable theory to project the course of future epidemics. It is possible to calculate the critical rate of sexual partner exchange that will allow an STD to spread through a population, i.e., when R_0 is greater than 1. For HIV, with a duration of infectiousness of 0.5 years and a transmission probability of 0.2, the partner exchange value is 10 new partners per year. For other STDs, such as untreated syphilis and gonorrhea, that have somewhat higher transmission probabilities, the values are 7 and 3, respectively. However, despite the development of mathematical equations, predicting the spread of an epidemic can be as uncertain as forecasting a hurricane, a blizzard, or a tornado. Indeed, making predictions early in a disease outbreak by fitting simple curves can be misleading, because it generally ignores interventions that reduce the contact rate and the probability of transmission. For SARS, fitting an exponential curve to data from Hong Kong that were obtained between 21 February and 3 April 2003 predicted 71,583 cases 60 days later, but using a linear plot, 2,410 cases were predicted. In fact, by 30 May 2003, according to the World Health Organization, there were more than 8,200 cases worldwide and over 800 deaths. By 5 July 2003, a headline in the *New York Times* declared, "SARS contained, with no more cases in the last 20 days."

Other uncertainties in predictability may involve changes in travel patterns with increased contact and risk. Sociological changes may also affect the spread of disease. For example, schoolchildren may influence the spread of measles, as occurred in Iceland when villages grew into towns and cities. Quarantine of infected individuals has also been used as a control measure, e.g., with SARS. Generally speaking, quarantine is ineffective, and more often than not it is put in place to reassure the concerned citizens that steps at control are being taken. However, as noted above, there are other interventions that do affect the spread of disease by reducing the number of susceptible individuals. One of the more effective measures is immunization.

To block parasite transmission, a sufficient number of individuals in the population must be immunized such that the value for R_0 is less than 1. For measles, R_0 is approximately 15. With this multiplier, measles will spread explosively; indeed, with multiplication every 2 weeks and without any effective control (such as immunization), millions could become infected in a few months. It has been estimated that to eliminate measles (and whooping cough), ~95% of children under the age of 2 must be immunized; for mumps and rubella to be eliminated, the percentages are 90 and 85%, respectively. However, elimination of endemic malaria in Africa, where R_0 is 50 to 100, would require 99%

coverage with a lifelong vaccine given at 3 months of age. So if transmission is intense (i.e., R_0 is a large number), mass immunization must take place at the earliest age feasible, and the later the average age of vaccination, the less likely it is that transmission will be blocked. Thus, for disease elimination, not everyone in the population must be immunized, but it is necessary to reduce the number of susceptible individuals below a critical point (called herd immunity). When successful, immunization may convert a type I epidemic to type II, and then convert the type II epidemic to type III—as happened in the United States and Great Britain with measles, mumps, and rubella.

The Evolution of Plagues

"A recurrent problem for all parasites . . . is how to get from one host to another in a world in which such hosts are never contiguous entities," wrote the historian William H. McNeill. He went on: "Prolonged interaction between human host and infectious organism, carried on across many generations and among suitably numerous populations on each side, creates a pattern of mutual adaptation to survive. A disease organism that kills its host quickly creates a crisis for itself since a new host must somehow be found often enough and soon enough, to keep its chain of generations going." Based on this, it would seem obvious that the longer the host lives, the greater the possibility for the parasite to grow, reproduce, and disperse its infective stages to new hosts. Therefore, the conventional wisdom is that the most successful parasites are those that cause the least harm to the host, and that over time there is a tendency for virulent parasites to become benign.

At first glance, it would appear that the progress of myxomatosis in Australia supports this idea. The story of myxomatosis begins in 1839 when the Austin family migrated from England to Australia. Over time, they became rich from sheep farming. To reestablish their English environment, the Austins imported furniture, goods, and a variety of animals. In 1859, a ship came from England to Australia with rabbits. Since the rabbits had no natural predators in Australia, they multiplied rapidly, destroying plants and native animals. The Austins began to wage war on the rabbits. By 1865, more than 20,000 rabbits were killed on the Austin estate. Still, the rabbits continued to spread, traveling as much as 70 miles per year. Control measures such as fences, barbed wire, ditches, and the like did not work. So myxomatosis, a viral disease of wild rabbits from South America and lethal to domestic rabbits, was introduced into Australia in the 1950s to act as a biological control agent. In 1950, 99% of the rabbits died of myxomatosis. Several years later, the virus killed only 90% of the rabbits, and it declined in lethality with subsequent outbreaks. It was also found that the viruses from the later epidemics were less virulent than the earlier forms and that these less virulent forms were much better at being transmitted by mosquitoes, the vector for the myxoma virus. Therefore, the rabbits lived longer, and the number of rabbits infected with milder disease was higher. It was concluded that the virus had evolved toward benign coexistence with the rabbit host.

William McNeill, impressed by the results of the introduction of the myxoma virus into Australia, wrote,

> from an ecological point of view . . . many of the most lethal disease-causing organisms are poorly adjusted to their role as parasites . . . and are in the early stages of biological adaptation to their human host; though one must not assume that prolonged co-existence necessarily leads toward mutual harmlessness. Through a process of mutual accommodation between host and parasite . . . they arrive at a mutually tolerable arrangement . . . [and based on myxomatosis] . . . some 120–150 years are needed for a human population to stabilize their response to drastic new infections.

There is, however, reason to question McNeill's conclusions. A recent reexamination of myxomatosis in Australia shows that the mortality of the rabbits, after the decrease in virulence of the virus and the increase in rabbit resistance was calculated, was comparable to the mortality of humans in most vector-borne diseases, such as malaria. In other words, the virus was hardly becoming benign. Further, the decrease in virulence observed over the first 10 years of the study did not continue, but reversed. It appears that myxomatosis is not an example of benign evolution.

An alternative to the contention that parasites evolve toward a harmless state is that natural selection favors an intermediate level of virulence. This intermediate level is the result of a trade-off between parasite transmission and parasite-induced death. Since the value for R_0 increases with the transmission rate as well as the duration of the host's infectiousness, an increase in transmission would reduce the duration of infection, and selection may favor intermediate virulence. Furthermore, because R_0 depends directly on the density of susceptible hosts in the population if the number of susceptible individuals is great, then a parasite may benefit from an increased rate of transmission even if it kills the host sooner and prevents transmission at a later time. However, if susceptible hosts are not abundant, then the parasite that causes less harm to the host (i.e., is less virulent) may be favored since that would allow the host to live longer, thereby providing more time for the production of transmission stages. The hypothesis that virulence is always favored when hosts are plentiful and is reduced when hosts are less plentiful neglects the fact that a feedback exists in the host-parasite interaction: a change in parasite virulence impacts the density of the host population, which in turn alters the pressures of natural selection on the parasite population, and so on. Thus, although parasite virulence tends to decline over evolutionary time, it never becomes entirely benign; in the process, the parasite population becomes more efficient in regulating the size of the susceptible host population.

The view that parasites evolve toward becoming benign suggests that parasites are inefficient if they reproduce so extensively that they leave behind millions of progeny in an ill or dead host. Indeed, some have contended that enhanced virulence is the mark of an ill-adapted parasite or one recently acquired by the host. This is not true. The number of parasite progeny lost is not of evolutionary significance; rather, it is the number of offspring that pass on their genes to succeeding generations that determines evolutionary success. Natural selection does not favor the best outcome for the greatest number of individuals over the greatest amount of time, but instead favors those characteristics that increase the passing on of a specific set of genes.

Consider a particular species of weed that is growing in your garden. The production of 1,000 seeds that yield only 100 new weed plants might be considered wasteful in terms of seed death and the amount of energy the weed put into seed production, but if the surviving seeds ultimately yield more weed plants in succeeding generations, then that weed species is more efficient in terms of evolutionary success. Parasites are like weeds. They have a high biotic potential, and those that leave the greatest number of offspring in succeeding generations are the winners, evolutionarily speaking. Evolutionary fitness, be it for a parasite, a human, a bird, or a bee, is a measure of the success of the individual in passing on its genes to future generations through survival and reproduction. When the fitness of the host is reduced by a parasite, there is harm, illness, and an increased tendency toward death. Host resistance is the counterbalance to virulence or the degree of harm imposed on the host by the presence of the parasite. If host resistance is lowered, a disease may be more pathogenic, although the parasite's inherent virulence may be unchanged. Thus, how negatively a host will be affected, i.e., how severe or how pathogenic the disease will be, is determined by two components: virulence and host resistance. Virulence is not so much a matter of a particular mutation as it is how that mutation is filtered through the process of natural selection; it is through natural selection that the final outcome may be a lethal outbreak or a mild disease—and, of course, when a new plague emerges, R_0 must be a number greater than 1.

Since parasite survival requires reaching and infecting new hosts, effective dispersal mechanisms may require that the host become sick: sneezing, coughing, and diarrhea may assist in parasite transmission. The conventional wisdom is that it takes a prolonged period of time for virulence to evolve; however, the evolution of parasite virulence need not take years, as in the case of the myxoma virus, but may be quite rapid, on the order of months. The basis for this is that a parasite may go through hundreds of generations during the single lifetime of its host. Then, too, because of competition between different parasites living in a single host, it might be advantageous for one kind of parasite to multiply as rapidly as it can before the host dies from the other infectious species. Succinctly put, the victorious parasite is the one that most ruthlessly exploits the pool of resources (food) provided by the host and produces more offspring, thus increasing its chances to reach and infect new hosts.

If parasite dispersal depends on the mobility of the host as well as host survival, then severe damage inflicted on the host by enhanced virulence could endanger the life of the parasite. Consider, for example, the common cold. It would be very much in the interest of the cold virus to avoid making you very sick, since the sicker you become, the more likely you are to stay at home and in bed; this would reduce the number of contacts you would have with other potential hosts, thereby reducing the opportunities for virus transmission by direct contact. Similarly, the development of diarrhea in a person with cholera or *Salmonella* (which causes "food poisoning") facilitates the dispersal of intestinal microbes via feces-contaminated water and food; in the absence of diarrhea, parasite transmission would be reduced.

AIDS is a consequence of an increase in the virulence of HIV. The enhancement of HIV virulence is believed to have resulted from accelerated transmission rates due to changes in human sexual behavior: the increased numbers of sexual partners was so effective in spreading the virus that human survival became less important than survival of the parasite. As the various kinds of plagues are considered in greater detail in subsequent chapters, recognition of the evolutionary basis for virulence may suggest strategies for public health programs. Thus, clean water may favor a reduction in the virulence of waterborne intestinal organisms (such as the causative agent of cholera), and clean needle exchange and condom use would both reduce transmission and lessen HIV virulence. However, some contend that this indirect mechanism may be too weak and too slow to substantially reduce virulence; a better approach could be direct selection by targeting the virulence factor itself. For example, immunization that produces immunity against the toxin produced by the diphtheria microbe also results in a decline in virulence. Future efforts will determine which strategy is the better means for effective "germ" control to improve the public health.

Figure 2.1   Hollywood's view of *Australopithecus* as seen in the movie *2001: A Space Odyssey*. © Turner Entertainment Co. Licensed by Warner Bros. Entertainment Inc. All Rights Reserved.

Chapter 2

Plagues, the Price of Being Sedentary

In Stanley Kubrick's classic film *2001: A Space Odyssey*, Richard Strauss's music (Thus Spake Zarathustra, op. 30) provides a haunting and frightening background to the sequence of scenes that represent the dawn of humanity. The sun rises on a barren African savannah. A band of squat, hairy ape-men appear; they eat grass. Though herds of tapirs graze close by, the ape-men ignore them because the means and the tools necessary to attack or kill the tapirs have not yet been developed. These ape-men are vegetarians who forage for roots and edible plants. At the dawn of the second day, the ape-men are seen huddled around a water hole; the landscape is littered with bones. The leader of the group picks up a bone and smashes the skeleton of an antelope, and then the bone is used to kill a tapir (Fig. 2.1). Shortly thereafter, the raw pieces of tapir flesh are eaten and shared by other hairy apelike creatures, members of the clan. At the dawn of the third day, the meat-eating, tool-using man-apes drive off a neighboring band of apelike creatures. Bone tools used for killing animal prey are now used to threaten and drive off rival tribes. In slow motion, accompanied by the slowly building tones of Strauss's music, the leader of the man-apes flings his weapon, a fragmented piece of bone, into the air. It spins upward, twisting and turning, end over end. There is a jump cut of 4 million years into the future, and the bone dissolves into a white, orbiting space satellite. Kubrick's science fiction film has been described as a countdown to tomorrow, a visual masterpiece, and a compelling drama of human evolution. Absent from the film is an examination of how the enlightened roving bands of early apelike humans settled down and become increasingly disease ridden. Here is that part of the story.

Becoming Human, Becoming Parasitized

It is now generally accepted that Africa was the cradle of humanity. The earliest evidence of hominids, that is, animals ancestral to modern humans and

23

not closely related to other monkeys and apes, is found in Africa. The evidence for this comes from unearthed bones and teeth (fossils). The fossil record shows that one of our oldest ancestors lived in Africa about 4.2 million years ago. These early hominids, which split from the ape lineage (and were discovered in Kenya in 1994), are named *Australopithecus anamensis.* Based on the structure of the teeth and the position of the opening where the spinal cord enters the skull, they are apelike humans, not apes. Our ancestor *Australopithecus* spent time in trees and behaved similarly to chimpanzees, or so we believe, since fossils provide no record of behavior. Whether *A. anamensis* walked on two feet is uncertain, but evidence for erect, upright posture in *Australopithecus* comes from bones discovered in Ethiopia and Tanzania that are 3.8 to 3.0 million years old, from a species named *A. afarensis.* One of these finds, a small female, was named "Lucy" by Donald C. Johanson of the University of California, Berkeley (Fig. 2.2). The limb structure and the way the hip joint and pelvis articulate make it clear that Lucy walked on two legs. This was dramatically shown by Mary Leakey and her team, who discovered three sets of fossilized footprints left in wet volcanic ash some 3.6 million years ago (Fig. 2.2A). *A. afarensis* weighed about 75 lb and was not very "brainy"; his or her brain was no larger than that of contemporary African great apes. When *A. afarensis* descended from the trees and stood upright with two feet firmly planted on the ground, it not only affected posture, but it dramatically changed lifestyles and diets, and disease patterns also began to change.

Descent from the trees to the ground placed the australopithecines into a new environment, an ecological niche that was very different from the forest canopy. This freed them from some diseases but allowed for the acquisition of new ones. For example, in the treetops, australopithecines would have been bitten by mosquitoes that carried parasites acquired from other animals living in the canopy, but at ground level they would be exposed to other airborne blood suckers, such as ticks and flies, or they would come in contact with different food sources and contaminated water. Their teeth were small and underdeveloped, as in modern human beings (Fig. 2.2B), and the canines, highly developed in existing ape species, were small like ours. Based on their teeth, these australopithecines probably chewed fruits, seeds, pods, roots, and tubers. Since no stone tools were found associated with the fossils, it is believed *A. afarensis* did not make or use durable tools or understand the use of fire. They were opportunistic scavengers or vegetarians. The lifespan of an australopithecine has been estimated to have been between 18 and 23 years.

Beginning about 3 million years ago, the climate in Africa changed from tropical warm and wet to more temperate cool and dry. As a consequence, the dense woodlands were replaced by more open grassy habitats, savannahs. This climate change presented a challenging environment for the woodland-dwelling australopithecines. Although we do not know whether the climate

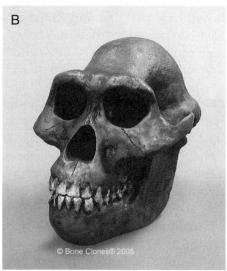

Figure 2.2 *Australopithecus.* (A) Reconstruction of Mr. & Mrs. Lucy. (B) Skull of Lucy. (Panel A courtesy of Ken Mowbray, American Museum of Natural History. Panel B © 2005 Bone Clones.)

change triggered it, at about this same time, ∼2.5 to 1.8 million years ago, there appeared in the fossil record several different kinds (species) of hominids, with two or three coexisting species in eastern and southern Africa. One of these species was *Australopithecus boisei*, a small-brained vegetarian, and the other was *Homo habilis*. The name *Homo habilis*, or "handy man," is based on the fact that altered stones and animal remains have been found with the fossil bones. *H. habilis* was more than a scavenger and a gatherer. *H. habilis* was also a hunter who made and used stone tools: simple stone flakes, scrapers and "choppers" that were chipped from a larger stone. (These stone tools, first found in Africa's Olduvai Gorge, are called Oldowan tools). Tool fashioning suggests a great leap in human intelligence and begins the technological changes that would forever mark *Homo* sp. as a toolmaker and a tool user. *H. habilis* used the flake tools to cut up the carcasses of the animals that were killed; these were transported to a home base where the meat was fed upon. *H. habilis*, with a somewhat larger brain, was "smarter" than *A. afarensis*, but the fossil finds tell us nothing of the numbers of individuals, whether there was division of labor among males and females, or anything about their behavior. However, we speculate that there were 50 to 60 individuals in a group living in an area of 200 to 600 square miles. We imagine that *H. habilis* lived at the edge of shallow lakes and in crude rock shelters.

There is no fossil record of the parasites that afflicted *H. habilis*, since their soft bodies have disintegrated over time, but we do know that with meat eating came an increase in parasitism. As these nomadic hunters encountered new prey, they met new parasites and new vectors of parasites. The result was zoonoses, that is, animal infections transmitted to humans. What were these zoonotic infections? We surmise that the parasites of *H. habilis* were those acquired from the wild animals that they killed and scavenged. The butchered meat might have had parasites such as the anthrax and tetanus bacteria, the roundworm that causes trichinosis, and a variety of intestinal tapeworms (see Fig. 1.3A). *H. habilis* would probably be bitten by mosquitoes, ticks, mites, and tsetse flies and probably also had head lice. *H. habilis* also may have suffered from viral diseases such as the mosquito-transmitted yellow fever, as well as non-vector-borne viruses that cause hepatitis, herpes, and colds, and he may have had a spirochete infection such as yaws (see Fig. 1.2B). It is doubtful, but *H. habilis* could also have been infected with the parasites that cause sleeping sickness, malaria, and leprosy. They certainly must have been infected with filaria, pinworms, and blood flukes (see Fig. 3.2E) but probably did not have typhus, mumps, measles, influenza, tuberculosis, cholera, chicken pox, diphtheria, or gonorrhea. When *H. habilis* roamed the African savannah, the human population was quite small (about 100,000), and we expect that human-to-human transmission rates of parasites were low.

Roughly 1.6 million years ago (or ~1.8 million years ago in Africa), *H. habilis* was replaced by *Homo erectus* (meaning "erect man"). (In Africa, *H. erectus* is equivalent to fossils that have been named *Homo ergaster*.) *H. erectus* was close to modern humans in body size and, based on the skull capacity, was somewhat larger-brained than *H. habilis*—but still with barely half the capacity of modern humans. *H. erectus* had smaller cheek teeth, suggesting that they were omnivores, had smaller faces, developed a culture characterized by living in caves, and hunted game animals using bifacial flake stone stools fashioned into "hand axes." This stone tool technology (called Acheulean tools) allowed *H. erectus* to more completely process the harder parts of animals and plants by grinding, crushing, splitting, and cutting them up before eating. As such, these stone tools represented a technological advance and served as extensions of the hands and teeth to break down food before digestion. *H. erectus* was able to start fires and made use of fire to cook the food. The *H. erectus* population was now somewhat less than a million. According to the "long journey" hypothesis, about 2 million years ago the *H. erectus* populations began to move out of Africa via the Middle East, but climate and geography prevented them from turning west, so they took a more southerly route into China and Indonesia. Then they turned north and moved west across the more central parts of Europe and Asia. The earliest fossil remains of *H. erectus* were found in Indonesia (Java) by Eugene Dubois in 1891 and so were named Java Man; 2 decades later, when Davidson Black found similar fossils in caves in China, they became known as Peking Man. The cave sites in China are about 500,000 years old, and the last of them were abandoned about 230,000 years ago.

When the populations of *H. erectus* left Africa, some of their parasites went with them—but only those that could be transmitted directly from person to person. Those vectors that remained restricted to Africa, such as the species of mosquitoes, snails, and flies that transmit diseases such as filariasis, blood fluke, and sleeping sickness, respectively—would not follow the migratory path. Indeed, even today they remain diseases that are characteristic of Africa. But as *H. erectus* encountered new environments with new kinds of animals, they were subjected to sources of new parasites; and with an increase in the number of humans living in more restricted geographical environments, the probability of large-scale infections was enhanced.

Tool making and tool use, as well as human cooperation, made hunting possible. Together, they contributed to further increases in the size of the human population and, over time, *H. erectus* evolved into humans closely resembling us. A half million years ago, the human population of Africa and those in Europe and Asia began to diverge from one another. Some 200,000 years ago, the fossil record shows individuals who were larger brained; those with a more graceful face, found in southwestern Europe and dated to 40,000 years ago, were called Cro-Magnon man. In a fit of hubris Carolus Linnaeus gave them the scientific name *Homo sapiens*, literally "wise man." *H. sapiens* not only used the stone technology of *H. erectus* but also made tools from bone and antlers. They made artistic carvings and cave paintings and records on bone and stone, played music on simple wind instruments, adorned themselves with jewelry, and buried their dead in ritual ceremonies; their living sites were highly organized and stratified, and they hunted and fished in groups. The intermittent technological advances in tool making seen with *H. habilis* and *H. erectus* were constantly refined. Clearly, they were our immediate ancestors. At this time the human population numbered about a million individuals.

In Western Europe, human skeletons were found first in the Neander Valley of Germany in 1856; they were called Neanderthals. Subsequently, Neanderthal fossils were found in the Middle East and parts of western Asia. They date from 190,000 to 29,000 years ago. Some archeologists have classified them as a separate species, *Homo neanderthalensis*. The Neanderthals, who have frequently been depicted as brutish cavemen (Fig. 2.3), had a brain size slightly greater than ours, and they left evidence showing that they cared for the sick and performed ritual burials; however, their stone tools were cruder than those of Cro-Magnon man. Within a few thousand years the Cro-Magnons, with their superior weapons and other advanced cultural practices, had completely displaced the Neanderthals. In Africa there are skeletal remains that are more modern than those of Neanderthals, dating back 200,000 years. Thus, nearly a quarter of a million years ago populations of *H. sapiens* lived in Africa, Asia and Europe. Their migration into the Americas took place about 12,000 to 35,000 years ago when they crossed the Bering Straits' land bridge from Asia into the Americas. Forty thousand years ago, humans moved into Australia. Thus, over the past 5 million years new hominid species have emerged, coexisted, competed, and colonized new environments; in some instances they succeeded, whereas in others they became extinct. The fossil record is of necessity incomplete, and much will be learned from future anthropologic digs, but what is certain is that we did not arrive by a straight-line descent from the apes. We are not the single topmost limb in the hominid evolutionary tree, but simply one of its many branches.

Hunter-gatherers, unable to preserve and store fruits, vegetables, and meat, were forced to roam over large distances in search of wild edible plants, to hunt down game animals, and to find sources of drinking water. Moving from place to place, these nomadic bands were not surrounded by heaps of rotting meat or feces, and exposure to parasite-infested waters was limited. Though the hunter-gatherers did come together in groups, the size of their populations was small, so diseases of crowds requiring human-to-human transmission were absent. Based on what we know about modern hunter-gatherer societies, like those in present-day New Guinea, the Australian aborigines, and the Kalahari bushmen, we believe that our hunter-gatherer ancestors were a relatively healthy lot. Gradually, however, conditions would change as the size of human populations increased and people adopted sedentary habits, living for extended periods of time in permanent or semipermanent settlements. Thus, over time, the incidence of human disease dramatically increased.

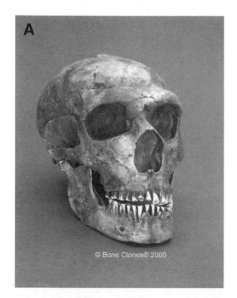

Figure 2.3 Interpretations of *Homo neanderthalis* based on fossil skull and bones. The skull (A) was used in the 1920s by Frederick Blaschke to reconstruct Neanderthal man (B), in which Neanderthal man was a brutish figure. Although later discredited, the stooped and apelike reconstruction was used in a 1953 B-grade movie (C) to create a monsterlike Neanderthal man. (A) © 2005 Bone Clones, Inc. (B) Photo by Tom McHugh/Photo Researchers, Inc. © 2005 Photo Researchers, Inc. (C) Courtesy of Popcorn Posters.

The Road to Plagues: More Humans, More Disease

Today, we speak of the problems associated with the "population bomb"—the unbridled growth of humans—that threatens our very existence. This growth in human populations cannot be calculated with any certainty until the middle of the 18th century, but we can make some educated guesses. Three hundred thousand years ago, the human population was 1 million; 25,000 years ago, that number had grown to 3 million, and 10,000 years ago, it is estimated to have been 5 million. By AD 1, it was 300 million. This phenomenal growth spurt coincides with the initiation of agriculture and the domestication of animals, which is generally dated to 8000 BC. Between 8000 BC and AD 1750, the population of the world increased 160 times to 800 million. The human population was increasing, and so too was the problem of overcrowding. For example, in 8000 BC, population density was 0.2 people/square mile, but by 4000 BC it was 4 people/square mile.

What is the basis for this growth in the human population? The English clergyman Thomas Malthus (1766–1834) wrote *An Essay on the Principle of Population* in 1798 in which he stated that a population that is unchecked increases in geometric fashion. Malthus assumed that there would be a uniform rate of doubling, and this is of course naive, because it leads to impossibly large numbers. (By way of example, if you doubled a penny every day for a month, the final amount would be over $1 billion). It has been said that explosions are not made by force alone, but by a force that exceeds restraint. As Malthus correctly observed, there are factors that will eventually bring population growth to a halt; for example, restraint could result from the fact that the food supply increases only arithmetically. The consequences of unrestrained population growth, in Malthus's words, would lead to "misery and vice" or, in today's vernacular, starvation, disease, and war. These would tend to act as "natural restraints" on population growth. Thus, the Malthusian model suggested that a natural population has an optimal density.

If we make a graph plotting the human population on an arithmetic scale from 500,000 years ago to the present, the resulting curve suggests that the population remained close to the baseline from the remote past to about 500 years ago and then surged abruptly as a result of the scientific-industrial revolution (Fig. 2.4A). More instructive, however, is to plot the same data for a longer time period using a logarithmic scale, since this allows for more of the data points to be placed in a smaller space. This log-log plot (Fig. 2.4B) reveals that the human population moved upward in a stepwise fashion and that there were three surges: one corresponding to the tool-making revolution, one to the agricultural revolution, and one to the scientific-industrial revolution. What were the checks on human growth rates that limited population size so that at equilibrium (the flat part of the logarithmic "curve") there was a zero rate of change and the number of deaths equaled the num-

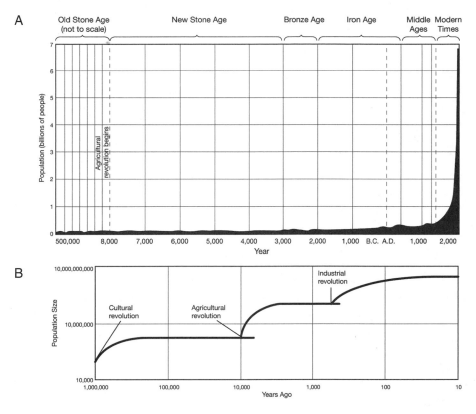

Figure 2.4 (A) Growth of the human population for the last 500,000 years. If the Old Stone Age (Paleolithic) were in scale it would reach 18 feet to the left. (B) Log-log plot of the human population over the last million years.

ber of births? Two kinds of checks occurred to set the upper limit (or the set point) for population growth: external or environmental factors (including limited food, space, or other resources) and self-regulating factors (such as fewer births, deliberate killing of offspring, or an increased death rate due to accidents or more virulent parasites). For Malthus, disease and warfare as well as "moral restraint" (birth control) acted as natural restraints, the Four Horsemen of the Apocalypse—Disease, Famine, War, and the Pale Rider, Death. Indeed, it has been estimated that before the introduction of agriculture, the earth could have supported a population of between 5 million and 10 million people who were engaged in hunting and gathering. Agriculture changed the environmental restraint so that the set point or upper limit of population size was increased.

The Effect of Agriculture

Human history took off 50,000 years ago in what Jared Diamond, Professor of Geography and Physiology at the UCLA School of Medicine, called "The Great Leap Forward." Fifty thousand years ago, *H. sapiens* used standardized stone tools for cutting, scraping, and grinding, and fashioned pieces of bone into fishhooks and spears, needles, awls, harpoons, and eventually bows and arrows. These tools could also be used as weapons, and now humans could hunt down and kill their animal prey at a distance. These early humans not only used the meat of animals for their nourishment, but also began to clothe themselves in the skins of these animals. Through the invention of rope, it was possible to make snares and nets so that birds and fishes could also become part of their diet. All this attests to the fact that between 100,000 and 50,000 years ago there was a significant change in the cognitive capacity of the human brain without a significant change in its size. Coupled with this change in brain organization was the anatomical improvement of the voice box: now not only could humans speak, they could begin to develop language.

For 99% of human existence we were hunter-gatherers, so why, some 10,000 years ago, did we settle down to become farmers? This change from hunting and gathering to farming has been termed the agricultural revolution, the time when humans domesticated plants and animals and exerted control over food production. Although the term "revolution" would seem to indicate that it appeared suddenly and dramatically, this was certainly not the case. The human control of food production was not discovered or invented; nor was it a conscious choice made by our ancestors "to farm or not to farm," since there were no farmers to serve as role models. No, domestication of plants and animals evolved as a consequence of human choice made without any awareness of its long-term consequences.

Development of techniques and practices for agriculture and animal husbandry progressed step by step in sequential fashion. It did not take place over a short time, and not all the wild animals and wild plants that would eventually be domesticated in a particular region were domesticated at the same time. Indeed, it probably took thousands of years to shift the human diet from wild foods alone to both wild and cultivated foods. The reason for this time lag is that food production evolved as a result of the accumulation of many separate choices, and there were trade-offs, especially in the allocation of time and effort.

Consider for the moment that you are a hunter-gatherer who has accumulated enough wisdom and technology to set up a small garden. Some of the choices you would be faced with are: which plants should I grow? how much time should I spend planting instead of hunting or scavenging? what are the benefits of tending the garden over going out to hunt and gather wild plants? Perhaps your most important consideration might be: of the two, hunting or gardening, which will save me from starvation in the future? In his book *Guns, Germs, and Steel*, Jared Diamond speculated that "All other things being equal, people seek to maximize their return of calories, protein or other specific food categories by foraging in a way that yields the most return with the greatest certainty in the least time for the least effort. Simultaneously they seek to minimize the risk of starving. . . . One suggested function of the first gardeners 11,000 years ago was to provide a reliable reserve larder as insurance in case wild food supplies failed." Although the relative importance of the factors that contributed to the shift from hunting and gathering to farming still remain controversial, one thing is certain: once there was a shift from nomadic hunting and gathering to more sedentary food production, there could be no turning back.

Coincident with the climate change by the end of the Pleistocene (~11,000 years ago), a time when the glaciers had receded and the climate had become milder and drier, many large mammals had become extinct in Europe, Asia, and Africa. This led to a decline in the abundance of wild game, and hence the life of the hunter-gatherer became more precarious: to obtain the same amount of food as in the past required the expenditure of greater amounts of time and energy on the part of the hunters and gatherers. The reduction in the number and kind of game animals was coupled with a change in climate that favored the availability of plants with potential for domestication, particularly the cereal grains. Now there were larger and larger areas with wild cereals, and these could be harvested with little difficulty by using stone and bone tools fashioned into sickles with flint blades. In addition to the technologies for harvesting cereals such as wild barley and wheat, the technologies needed for processing and storing these grains came into being. These technologies, which appear crude and simple by today's standards, allowed the first unconscious steps of plant domestication to take place. Tools included baskets to carry the grain from the field to the home base, mortars and pestles to remove the husks and to pulverize the grain, a technique of heating the grain to allow storage without sprouting, and the construction of underground storage pits, some of which were made waterproof by plastering. Coincident with farming, the density of the human population increased. It is not clear whether the rise in density of human populations led to domestication of plants and animals or vice versa, but it is certainly true that with the availability of more and more calories it was possible to feed more and more people. Professor Jared Diamond in his book *Guns, Germs, and Steel* observed that the adoption of food production was an autocatalytic process, a positive feedback, whereby a gradual rise in population densities required that people obtain more food, and in turn those who took the steps to produce it were rewarded. Once farming began, people could become more and more sedentary—they settled down. In turn, birth spacing could be shortened, producing more people, requiring still more food, and so on.

Farming populations became better nourished thanks to an increase in the availability of the number of edible calories per square mile, and eventually farmers replaced the nomadic groups of hunter-gatherers by converting them to engage in the practice of farming or by displacing them by the sheer force of numbers.

The life of nomadic hunter-gatherers was such that population levels were well below the maximum limit that would be imposed by their reproductive biology and the availability of food. Then what limited their increase? The inability of the hunter-gatherer mother to carry more than a single child along with her normal baggage, coupled with her inability to nurse more than one child at a time, limited the practical interval between births to 4 years. It is likely that hunter-gatherers effectively spaced their children's births by means of lactation amenorrhea, sexual abstinence, infanticide, and spontaneous abortion. In contrast, once humans settled down, they were freed from the encumbrances imposed on the hunters and gatherers who had to carry their children around, so now they could have as many children as they could bear and raise. Consequently, the birth interval for the farmer was reduced to 2 years. Agriculture also encouraged higher birth rates because additional children provided cheap labor. Further, farming had another advantage over hunting and gathering: more calories could be produced per unit land area and time expended. While 200 square miles could support 50 to 60 hunter-gatherers, this same land area could support more than 10,000 farmers. The higher birth rate of the food producers, together with their ability to feed more people per square mile, allowed them to achieve much higher population densities than those who were engaged in hunting and gathering.

Once agriculture and animal husbandry yielded a surplus of food that could be stored, there was a need for some members of the sedentary population to guard it. This was of course impractical for the nomadic hunter-gatherers, since they would have to both hunt and gather and at the same time protect their bounty from others. But the availability of surplus stored food allowed members of the settled population to specialize: some became guards, and others were armed and served as soldiers who could group together to steal food from others. Perhaps this foreshadowed the origins of war.

Furthermore, when food production did not require every member of the settled community to directly work the land, it became possible for some members of the group to engage in other activities. Religion has been described as the "opiate of the masses," and recent psychological research suggests that a specific region of the brain contains the neurological seat of the human religious impulse. Once humans formed agriculturally based societies, those who chose to participate in such group activity may have been at a competitive advantage over those who did not. The latter likely became outcasts. Hence, the agricultural revolution may have provided selective advantages to those who were in a controlled group, and superstition and organized religious practices were effective means of control by promoting group cohesion. It may not be accidental that the first known and highly organized religions arose coincident with the agricultural revolution.

Some of the surplus food could be used to feed those "who provide religious justification for wars of conquest, artisans such as metalworkers who develop swords, guns and other technologies; and scribes, who preserve far more information than can be remembered accurately." In time, political stratification would develop: heading the settled community would be the elite consisting of hereditary chiefs (or kings) and bureaucrats. Under the appropriate circumstances, these complex political units that governed "the settled" could also be mustered into formidable armies of conquest. Stored food and the land upon which it was grown became valuable resources that could be taxed, and surpluses could be traded for other goods; thus, commerce and banking began to emerge. With larger populations, family and inheritance schemes resulted, class structures with elaborate religious practices emerged, and writing was invented. Through agriculture and its prospect for increased food production, there was a population expansion that favored technological advances, as well as the development of cities (urbanization) and the rise of civilizations.

Another consequence of humans settling down was an increase in the amount of human disease. Agriculture by itself did not create new infections; it simply accentuated those that were already present, or it converted an occasional event into a major health hazard. This was largely due to the fact that transmission of infectious agents becomes easier as individuals are crowded together; the practice of using human excrement ("night soil") or animal feces (manure) as fertilizer allows for the transmission of infective stages; finally, the closer association with domestic animals allows for their diseases to be transmitted to humans.

The Lethal Gifts of Agriculture

Permanent settlements developed independently in several parts of the world, including the Middle East, China, and the Americas. But those that have been best studied are found in the Fertile Crescent, a region bounded by the Tigris and Euphrates rivers and curving around the Mediterranean and the Nile Valley to include Syria, Lebanon, Israel, Egypt, Turkey, Jordan, Iran, and Iraq. The oldest known village, just outside present-day Jericho in Israel, may have sprung up around a shrine used by roving bands of hunters and gatherers. By 10,500 years ago, it had evolved into a small farming village. At first, this settlement and others like it were simply collections of villages on the banks of the natural streams, but soon they were able to spread out via networks of irrigation canals. The surplus of food and the practice of irrigation contributed to larger and larger concentrations of people, allowing some people to quit farming and to become full-time artisans, priests, or members of other professions. Meanwhile, the farmers who provided the food for these ever-enlarging villages continued to live on their outskirts. About 5,500 years ago, Uruk, the first undisputed city—a place where farmers do not live—was established in Mesopotamia (present-day Iraq).

We have little precise information about the parasitic diseases that afflicted our ancestors more than 10,000 years ago. To be sure, they had their parasites, but we believe their impact may not have been quite as severe as in the period that stretches from the present back to the time of the earliest cities. The reason for this is that the pattern and the impact of disease depend on several factors: the population density; the character and quality of the water supply, food, and shelter; the frequency of contact among individuals; human contact with animals; and the climate. However, once human populations were concentrated into larger and larger communities and their numbers increased, then the potential for infectious disease organisms to be transmitted was enhanced, and thus disease could (and did) affect the size and well-being of the population. Farming and domestication also reduced the biological variability of animals and crop plants, leading to purebred strains and making local populations of plants and animals more and more uniform. As a consequence, any upset in the balance could decimate an entire population whether it was a crop, a flock, or a herd. (A modern example of this is the 1845–1851 potato famine in Ireland.) Agriculture necessitated artificial flooding of land that was naturally devoid of adequate amounts of water, enabling longer growing seasons; it also required tilling or plowing and replenishing the soil with fertilizer. The cheapest and most available fertilizers are human waste and animal feces. Intensive agricultural practices require protection of crops from weeds and pests by using some kind of control measures. Tilling the soil necessitates some kind of work force: animal power, human power, or machines. Disruptions in the availability of any of these could lead to disaster. With purebred strains (monocultures), disruption becomes that much easier. Living in villages or cities—that is, in permanent settlements—involves the risk of parasite invasion. Those infected with intestinal parasites can more easily transmit disease to others through their feces, and when the water supply becomes contaminated—either through the use of night soil or through the use of feces-contaminated streams—the spread of disease can be great indeed. A single contaminated water source serving a large population can be a much greater threat than several sources supplying smaller bands of hunters and gatherers. Thus, irrigation practices created a favorable environment for parasites: moisture was abundant, and there was a liquid medium in which parasites and/or their cysts and eggs could persist. Parasite transmission would take place when humans used the water for drinking, bathing, washing of clothes, or waste disposal.

The disruptive effects of an epidemic disease are more than simply the loss of individual lives. Often, the survivors are demoralized, they lose faith in inherited customs, and, if the epidemic affects those of working age, it can lead to a material as well as a spiritual decline. As a consequence, the cohesion of the community may break down, making it susceptible to invasion from neighbors. Once disease is widespread in an agricultural community, it can produce a listless and debilitated peasantry that is handicapped for sustaining work in the fields, digging irrigation canals, resisting military attack, or throwing off alien political domination. All of this may lead to economic exploitation.

The smaller population size of hunters and gatherers makes it seem probable that person-to-person "civilized" infectious diseases such as measles, influenza, smallpox, and polio could not have established themselves because they are density-dependent diseases requiring a critical number of individuals for transmission. Although there is no hard literary or archeological evidence, it does seem reasonable to suggest, as William H. McNeill did, that "The major civilized regions of the Old World each developed its own peculiar mix of infectious, person-to-person diseases between the time when cities first arose (about 3000 BC) and about 500 BC. Such diseases and disease-resistant populations were biologically dangerous to neighbors unaccustomed to so formidable an array of infections. This fact made territorial expansion of civilized populations much easier than would otherwise have been the case."

The Accident That Caused Societal Differences

Imagine for a moment that you are living in the year 1492 and you have just graduated from the University of Padua in Italy with a degree in medicine. Before you are able to set up your practice, you receive a letter from an old friend, Giuseppe Diamonte, who writes from Spain that King Ferdinand and Queen Isabella are about to provide funds for the discovery of a new route to India; the expedition is to be under the command of a fellow Italian, Christopher Columbus, and he is in need of a naturalist to collect plants and animals and to act as the ship's doctor. Giuseppe writes that he has recommended you for the position. You are enthusiastic, board a ship in Venice, land in Barcelona, and travel by horseback to Palos, Spain. The journey across the Atlantic Ocean begins on 3 August. Upon arrival in "India" (but in fact the present-day Dominican Republic) on 12 October, you are astounded to be greeted by a band of near-naked "Indians" who have paddled their canoes to greet the ship and its crew. No iron tools or ocean-going vessels can be seen, the village consists of a scattering of huts, and there is little that could be considered a city. The natives have no writing; what agriculture that exists is on an entirely different scale from what you were familiar with in Europe. Six weeks earlier you left iron tools and weapons, agriculture, ocean-going ships, large cities, horses, carts and carriages, writing, money, banking, painting, sculpture, cathedrals, palaces, buildings of brick and stone, established religion and ritual, and music. You are perplexed. You ask yourself: why has the rate of technological and political development been so much faster in Europe and Asia than it has in the Americas (or, for that matter, Africa or Australia)? In short, why were the Americas technologically a few thousand years behind Europe? Why were stone tools, comparable to those used by Europeans and Asians 10,000 years earlier, being used?

Fast forward to the future. In *Guns, Germs, and Steel*, Jared Diamond argues persuasively that it was not biological differences but geography that was the decisive element. It wasn't differences in the braininess or genetics of the human populations, but the plant and animal resources available on a particular continent—an accident of geography—that made the difference. Diamond believes that the fortuitous accident began in the Fertile Crescent, which contained a suitable array of plants and animals, called "founders," that were the basis for domestication. What were the founder plants? Those locally available in southwest Asia and the Fertile Crescent that would not serve as the basis for domestication were plants with a large amount of indigestible material, such as bark, or that were poisonous, or low in nutritional value, or tedious to prepare and gather. The desirable attributes of plants that make them suitable for domestication include a larger proportion of edible parts (large seeds) and a smaller proportion of woody, inedible parts; easy to harvest en masse (with a sickle); seasonal; easy to grind, sow, and store; high in yield; and high in calories. Plants with these characteristics fall into four categories: grasses (wheat, barley, oats, millet, and rice), legumes (peas and beans), fruit and nut trees (olives, figs, dates, pomegranates, grapes, apples, pears, cherries), and fiber crops (flax, hemp, and cotton).

Of the 148 big, wild, terrestrial, plant-eating animals (herbivores)—those suitable for domestication—only 14 were able to serve as founders. What were these founder animals that could be domesticated? In Europe and Asia, about 4000 BC, it was the "big five"—sheep, goats, pigs, cows, and horses. In East Asia, cows were replaced by yaks, water buffalos, and gaurs. In contrast, the Americas had mountain sheep and goats, llamas, bison, peccaries, and tapirs; Australia had kangaroos; and Africa had zebras, buffalos, giraffes, gazelles, antelopes, elephants, and rhinoceroses—but all of these were unsuitable for domestication. Domestication involves more than taming; it requires a special set of animal characteristics: social species that occupy territories, and animals that are herbivores. A domesticated animal species must also have the right reflexes—it must be predictable and not panic easily, and it must not be ferocious or nasty in disposition. It must grow quickly, and it must be able to breed in captivity. The appropriate domesticated animals were cows, goats, horses, sheep, donkeys, yaks, and camels.

Once these animals were domesticated, what benefits did they provide? Food in the form of meat and milk, clothing and fiber from wool and hides, manure for use as fertilizer, and animal power for land transport of goods and people, as well as for plowing fields. Indeed, before there were domesticated beasts of burden, the only means for moving goods and people across the land was on another person's back! Domesticated horses, goats, camels, and cows were hitched to wagons to move humans and their possessions, and reindeer and dogs were used to pull sleds across the snow. Horse-drawn chariots revolutionized warfare, and after the invention of saddles and stirrups it became possible for marauding Huns on horseback to strike fear into the legions of Rome.

There was a downside to animal domestication. Domesticated animals could be the source of human disease. As human populations settled down, they created heaps of waste—middens of animal bones, garbage, and feces—that served as breeding grounds for and a source of microparasites; they also attracted insects that were vectors of disease, as well as wild birds and rodents carrying their own parasites and potential sources of human disease. With each domesticated species came possible human exposure to new disease agents—parasites. For example, the number of diseases acquired from domestic animals (zoonotically) has been estimated as follows: dogs, 65; cattle, 45; sheep and goats, 46; pigs, 42; horses, 35; rats, 32; and poultry, 26. Specifically, the human measles virus has its counterpart in the distemper virus of dogs and rinderpest in cattle. Smallpox has its closest relatives in the virus of cows and poxviruses in pigs and fowl, and human tuberculosis is a cousin of bovine tuberculosis. More recent examples of the jump from one animal species to another include HIV, which was a chimpanzee virus that became humanized; monkey pox, which is transmitted to humans by the bite of pet prairie dogs; and SARS (severe acute respiratory syndrome), which originated in civet cats.

With the clearing of forests, the planting of crops, and the destruction of wild game animals, new ecological niches were created for insects and scavenging rodents. Mosquitoes and flies that once fed on game animals now found a new source of blood: humans. These "bloodsuckers" could act as vectors for malaria, yellow fever, and African sleeping sickness. Ditches, irrigated fields, and pottery vessels also served as breeding grounds for insects and snails, facilitating the transmission of blood fluke disease, yellow fever, malaria, elephantiasis, and river blindness.

The crowd diseases of humans, such as smallpox, measles, pertussis (whooping cough), tuberculosis, and influenza, were initially derived from very similar ancestral infections of domesticated animals. At first, those who hunted, farmed, and domesticated animals fell prey to the parasites they acquired; some died, but in time, resistance to these new diseases developed. When such a partially immune people came in contact with others who had no such protection, a devastating epidemic could occur. It was these contagious diseases (caused by a wide variety of worms and "germs") that would ultimately play a decisive role in the European conquests of Native Americans, Africans, and Pacific Islanders; determine the outcomes of wars; loom large in the economic growth and prosperity of nations; and contribute to slavery and colonialism.

Figure 3.1 *Plague in an Ancient City* (detail), circa 1652–1654 by Michael Sweerts (1624–1664). Museum number AC1997.10.1. Courtesy of the Los Angeles County Museum of Art, gift of The Ahmanson Foundation. Photograph © 2005 Museum Associates/LACMA.

Chapter 3

Six Plagues of Antiquity

As humans changed their lifestyles, their relationship with infectious diseases came to be altered. For 2 million years these human populations consisted of small groups of hunter-gatherers with limited contact with other such groups, and there were no domesticated animals. Such a population structure, with little or no exposure to new sources of infection and with minimal survival and transmission of parasites, led to a situation in which epidemic diseases were virtually nonexistent. Indeed, only those diseases with very high transmission rates that induced little or no immunity, as well as macroparasitic diseases that did not involve vectors for transmission and sexually transmitted diseases, were able to establish themselves in the groups of hunter-gatherers. Although some vector-borne diseases such as malaria and yellow fever may have been present at this stage of human history, it was only after human populations settled down and adopted an agricultural life, or continued a nomadic existence that depended on the husbandry of large herds of animals, that conditions favored the emergence of epidemic diseases (plagues). Historically, plagues (Fig. 3.1) came to be recorded only in our recent past, in a time when we became farmers.

By 8000 BC the human population was settled in villages—first in the valleys of the Tigris and Euphrates Rivers in Mesopotamia, and then along the Nile in Egypt, the Indus in India, and the Yellow River in China. Agriculture provided increased amounts of food for the people, but it also contributed to the conditions that would result in a decline in human health. It was the agricultural revolution, with the cultivation of crops and animal husbandry, that provided the driving force for the growth of cities (urbanization). Urban life also enhanced the transmission of certain diseases through the air and water, by direct contact, and by vectors such as snails, mosquitoes, and flies. The diseases of antiquity (5000 BC to AD 700) were characterized by parasites with long-lived transmission stages (e.g., eggs) as well as those involving person-to-person contact. Thus, most diseases became established only when a persistent small number of infectious individuals could be maintained, i.e., when the disease became endemic; this required populations greater than a few hundred thousand.

The Pharaohs' Plague
A look back

Assyrian and Babylonian literature, as well as the Egyptian papyrus from Kahun written in about 1900 BC, describes a disease that causes blood to appear in the urine (hematuria). Near the Louvre Museum in Paris, there is a stone from ancient Egypt that reads, "Anyone who moves this boundary stone will be covered with bloody urine." Hematuria was described by the father of Arabian medicine, Avicenna (AD 980–1037) in his *Canon Medicine*, but the condition, called *aaa*, was recognized much earlier. It is mentioned in the Ebers papyrus, dated ~1500 BC and named after Georg Ebers, who in 1862 found the paper in a tomb in Thebes, Egypt. There is even a hieroglyphic sign showing a penis dripping fluid, and this too may be blood (Fig. 3.2A). Such a sign was not considered to be connected with disease, but to be a mark of puberty in the male child. Many remedies for this disease are described in the Ebers papyrus, suggesting that this condition was widespread. Furthermore, in a relief of the tomb of Ptah-Hetep I and Mehou of the VIth Dynasty at Sakarrah, there are figures of fishermen and bargemen with enlarged abdomens, surely representing the pathology of chronic snail fever or blood fluke disease. In 1910 Marc Armand Ruffer (1859–1917) examined several Egyptian mummies from the XXth dynasty (1200 to 1000 BC) and found the calcified eggs of the blood fluke in the kidneys of several mummies (Fig. 3.2B). Fossil snails capable of transmitting blood fluke disease have been found in the well water of Jericho. It has been hypothesized that the water was infested with infected snails, resulting in a high level of disease. Too debilitated by disease to defend the city or repair the decaying walls, the people of Jericho were easily defeated by Joshua's army. Joshua, unaware of the cause of the disease that contributed to his success but wanting to prevent its spread, destroyed Jericho and proclaimed a curse on anyone who would rebuild it. The city remained deserted for 500 years. Centuries of recurring drought destroyed the snails, and thus the city has remained free of disease to this day. All of this suggests that snail fever has existed in tropical and subtropical parts of the world, but especially in Egypt, since ancient times.

Figure 3.2   The blood fluke *Schistosoma*, causative agent of the Pharaoh's Plague. (A) Hieroglyphic; (B) calcified egg from a mummy; (C) *Schistosoma haematobium* egg, as seen with a light microscope; (D) *Schistosoma mansoni* egg with miracidium inside; (E) adults in copula, as seen with the scanning electron microscope (from David Halton); (F) ciliated miracidium, as seen with the scanning electron microscope (courtesy of Vaughan Southgate); and (G) cercaria, as seen by scanning electron microscopy (from David Halton). mw, male worm; fw, female worm; gc, gynecophoric canal.

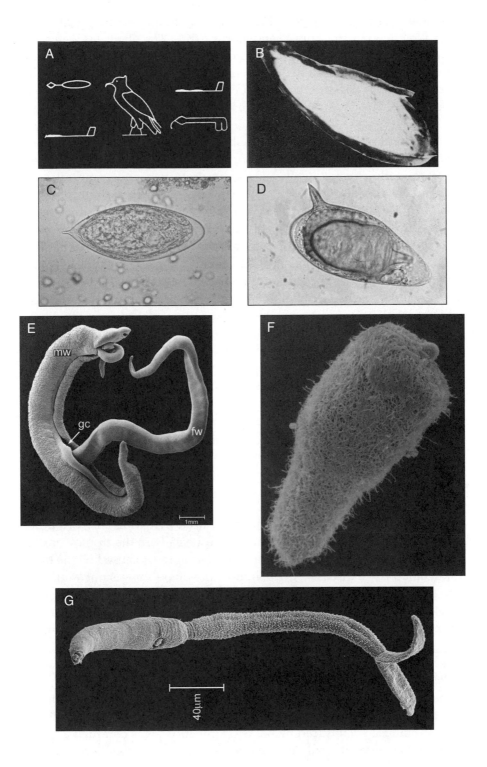

Soon after food production began in the Fertile Crescent (~8000 BC), it spread to other parts of Eurasia and North Africa. The plants (as well as the domesticated animals) that formed the basis for agriculture in the valleys between the Tigris and Euphrates Rivers were also cultivated in the Nile Valley of Egypt; this triggered the rise of Egyptian civilization. But these same agricultural practices, sustained by the Nile, also sowed the seeds of Egypt's decline.

The Land of the Pharaohs flourished for 27 centuries, and its accomplishments, even today, are truly impressive. When Herodotus, the Greek historian, made a tour of Egypt in 400 BC, he wrote of "wonders more in number than those of any other land." And he went on, "when the Nile inundates the land all of Egypt becomes a sea and only the towns remain above water. Anyone, traveling from Naucratis to Memphis sails right alongside the pyramids and when the waters recede they leave behind a layer of fertile silt—'black land'—the Egyptians call it, to distinguish it from the sterile 'red land' of the deserts. Egypt is the gift of the river."

In some ways, nature favored Egypt. Unlike Mesopotamia, which stood on an open plain and was unprotected from marauding tribes, the deserts that bordered the Nile discouraged invasion; thus, the people lived in relative security. The villages shared the river and merged into cities. To tap the bounty of the Nile required the cooperation and organization of the people, with social and political structures developing therefrom. All power was invested in the pharaohs, who were both kings and gods. Below the pharaoh was a vast bureaucracy that rested on the shoulders of the workers and the peasantry. However, Egypt's people, who built enduring stone monuments for their pharaohs, were racked with a debilitating disease, snail fever. Although medical science began in Egypt, the doctors and surgeons could not keep this disease at bay. The reason for this was that the early civilizations of Egypt and those of the Fertile Crescent (Sumer, Assyria, and Babylon) were based on agriculture, which required irrigation and/or natural flooding by the rivers. Irrigation farming, especially in the tropics, created conditions favorable for the transmission of snail fever caused by the blood fluke. Blood fluke disease—the plague of the pharaohs—is not a fatal disease as malaria and yellow fever are; however, it is a corrosive disease. Although there may have been a time when the natural flooding of the Nile made snail fever a seasonal problem, once there was irrigation it became an all-year problem: infections could be acquired from the standing water in the irrigation channels. Consequently, as Professor William H. McNeill wrote in *Plagues and Peoples*, "there was a listless and debilitated peasantry handicapped . . . for the . . . demanding task of resisting military attack or throwing off alien political domination and economic exploitation. Lassitude and chronic malaise . . . induced by parasitic infections was conducive to successful invasion by the only kind of large-bodied predators human beings have to fear: their own kind, armed and organized for war and political conquest." McNeill also suggested that the rule of the pharaohs may have been due to the power of the snail and the blood fluke and malaria—the classic plagues of Egypt—which debilitated the populace.

And so it was that snail fever did its work. By 660 BC, Egypt became subject to internal political dissension and attack by their iron-armed neighbors (the Assyrians), and their civilization, based on agriculture and copper weapons, began to collapse. The Persians overran Egypt in 525 BC. The cause of snail fever, the disease that set the Egyptian civilization on its inexorable downward spiral, was unknown to the ancient Egyptians because the transmission stages of the parasite (eggs, miracidia, and cercariae) are microscopic; in addition, the adult worms themselves are tiny and live within the small blood vessels, so they were unnoticed for thousands of years.

Search for the destroyer

Blood fluke disease, also known as snail fever or endemic hematuria, involves feces or urine, water, snails, and a flatworm. The first Europeans to experience the disease on any scale appear to have been the soldiers of Napoleon's army during the invasion of Egypt (1799 to 1801). The symptom of the disease—bloody urine—was rife among the soldiers, and Baron Jean Larey, a military surgeon, noted its high frequency in the men; however, he believed that the excessive heat during the long marches was the cause. The connection between hematuria and a parasite was not recognized until 1851, when Theodor Bilharz, a German physician working in Egypt, made a startling discovery while carrying out an autopsy on a young man: worms were found in the blood vessels, a location never before encountered (Fig. 3.2E). He named the worm *Distomum* (meaning "two mouths") *haematobium* (from the Greek words *haema* ["blood"] and *bios* ["to live in"]). In 1858 the name was changed to *Schistosoma* (from the Greek words *schisto* ["split"] and *soma* ["body"]). Today, blood fluke disease is called schistosomiasis or bilharzia, the latter in honor of Bilharz's discovery. (During World War I the British soldiers found it easier to call the disease "Bill Harris.")

In 1851, Bilharz reported seeing microscopic eggs with a pointed spine in the female worm (Fig. 3.2C); in the following year, he observed these eggs in the bladder, and within the egg he observed a small, motile embryo. He also found that the eggs would hatch to release a small, ciliated larva (Fig. 3.2F) that swam around for about an hour and then disintegrated. This work was confirmed in 1863 when John Harley, a London physician, examined a patient with hematuria who had previously lived in the Cape of Good Hope in South Africa. Examining the blood-tinged urine in a drop of water under the microscope, he found schistosome eggs, several of which hatched to give progeny that swam by using their cilia (Fig. 3.2F). But there remained a puzzle: how was the infection transmitted? Bilharz and others were aware that flukes closely related to *Schistosoma* had intermediate stages in snails, but when Harley examined snails from a region where schistosomiasis was prevalent, he found no evidence of larval stages. Despite this failure, the suspicion remained that humans acquired the infection either by eating infected snails or by drinking water containing the ciliated larvae, called miracidia. In 1870 Spencer Cobbold, working in London, obtained eggs from a young girl living in the Cape of Good Hope and found that, although the eggs would not hatch in urine, they did so in fresh or brackish water. Then, in about 1904, Japanese physicians found that a related blood fluke, named *Schistosoma japonicum*, could also infect humans, but this species had eggs without a spine. In 1905 Patrick Manson discovered another type of schistosome egg, one with a spine on its side (Fig. 3.2D); this was in the feces of an Englishman who had lived in the West Indies but had never visited Africa; it was duly named *Schistosoma mansoni*. Now there were three known species of human-infecting blood flukes.

The life cycle and mode of transmission of the schistosome to a human was first demonstrated between 1908 and 1910 in Japan. Fujinama and Nakamura found that, when the tails of mice were immersed in water from rice fields known to have a high incidence of bilharzia, they became infected with *S. japonicum*. Shortly thereafter, it was possible to show that the miracidia were able to penetrate freshwater snails in the rice paddies, and Ogata found that a tailed larva (called a cercaria) emerged from infected snails and could directly penetrate the skin of mice (Fig. 3.2G). This suggested that species other than *S. japonicum* might have a similar life cycle. At the outbreak of World War I, the British became concerned about the potential deleterious effects of schistosomiasis on their troops in Egypt. In 1915 the British War Office sent Robert Leiper to Cairo "to investigate bilharzia . . . and advise as to preventive measures to be adopted." Leiper collected freshwater snails, identified them, and determined whether they were infected, either by allowing the snails to release cercariae or by dissecting the snails to find other larval stages (called sporocysts). Within weeks, he and his team identified the snails *Bulinus* and *Biomphalaria* as the vectors. (Because the snail vector is critical to transmission, schistosomiasis is called "snail fever.") Leiper went on to show that the skin of mice could be directly invaded by the cercariae by placing the tails of the mice in cercaria-infested water. This suggested that the infection was acquired by bathing in infested water. But could the infection also be acquired by ingestion? Because Leiper was able to show that when cercariae were placed in dilute hydrochloric acid (similar in concentration to that found in the human stomach) they were killed, this route of infection seemed most unlikely. Leiper was also able to show that the adult *S. mansoni* and *S. haematobium* were different from each other and that cercariae that hatched from *Biomphalaria* produced eggs with lateral spines, whereas those from *Bulinus* produced eggs with a terminal spine. The pathology of the two species was also found to differ: *S. mansoni* remained in the liver and laid its eggs there, whereas *S. haematobium* early in its development left the liver for the veins surrounding the bladder. Thus, 65 years after the discovery of the adult worm by Bilharz, the life cycle of snail fever was finally known: on reaching fresh water, the discharged eggs release a swimming larva, the miracidium. Miracidia are short-lived, but if they encounter a suitable snail they penetrate the soft tissues (usually the foot), migrate to the liver, and change in form (sporocyst); for 6 to 7 weeks, by asexual reproduction, the numbers of parasites increase. During this time, the snail sheds thousands of fork-tailed cercariae, which can swim and directly penetrate human skin, and in 5 to 8 weeks they develop into adult worms.

Snail fever, the disease
Schistosomes differ from other flukes (Trematodes) in that the sexes are separate and they inhabit the blood vessels. The adult worms are ~10 mm in length, and the stouter males have a groove running lengthwise, called the gynecophoric canal, where the female normally resides (Fig. 3.2E). It is this groove in the male that is the basis for the worm's generic name *Schistosoma*, meaning "split body." Both males and females have two suckers at the head-end of the worm, and the more anterior one surrounds the mouth. (Bilharz mistakenly took the two suckers for two mouths and thus he called the worm *Distomum*, "two mouths.") The schistosome adults, in copula, live in blood vessels (veins) close to the bladder and small intestine. Mating occurs in the gynecophoric canal, and then the paired worms move "upstream" into smaller veins, where the female worm deposits the fertilized eggs. The pathology of schistosomiasis is due not to the adult worms themselves but to the eggs. Each day, hundreds of embryo-containing eggs move across the walls of the veins into the bladder or intestine, aided by the host's inflammatory response, and in the process, eggs become enclosed in a small tumor called a granuloma. It is passage of eggs through the bladder wall that results in bleeding and gives the telltale sign of hematuria. Once in the bladder or intestine, the egg becomes freed of the granuloma and is eliminated from the body, either with the urine or in the feces.

More than two-thirds of the eggs, however, fail to work their way out of the body and are washed back in the veins; by means of the bloodstream, they scatter throughout the body, where they accumulate in various organs. Accumulation is greatest in the liver and spleen. The piling up of eggs blocks the normal blood flow, and this leads to tissue death. The egg also acts as an irritating foreign substance that the body attempts to wall off by surrounding it with a fibrous capsule. The egg-laden liver eventually becomes filled with scar tissue. In the bladder blood fluke, *S. haematobium*, the scarred areas block the migration of eggs through the lower bowel tissues, and more eggs are swept back into other sites. The earliest signs of infection occur within 1 to 2 months and are fever, chills, sweating, headache, and cough. Six months to a year later, the accumulation of eggs produces organ enlargement, especially the liver and spleen; the enlarged and cirrhotic liver causes the abdomen to become bloated, appetite diminishes, blood loss leads to anemia, and there is dysentery (Fig. 3.3).

Schistosomiasis is an arithmetic disease: the severity of its symptoms and cumulative damage are directly related to the number of worms present, and the latter depends on the degree of exposure. In heavy cases there may be hundreds of worms, and the adults may live 20 to 30 years. Clearly, with time and increased invasion by cercariae, a person becomes more and more debilitated. Yet, over the centuries the adult inhabitants of areas with endemic disease, such as Africa, developed some measure of immunity, largely as a result of continuous exposure. Europeans and Americans, with no such immunity, suffer more severe symptoms as a result of higher worm burdens.

Where snail fever is found

Wars and human migrations carried the blood flukes of the East African lakes to the Nile River, and from there they were distributed along the trade routes to most of the continents of the world. Although in 1902 Manson believed schistosomiasis to be a disease unique to Africa, he had to revise his thinking when he discovered an Englishman, who had resided in the West Indies but had never been to Africa, passing eggs with a lateral spine. In 1908, Piraja da Silva, living in Bahia, Brazil, wrote that the schistosome common in the Americas was probably introduced from Africa by West African slaves, beginning as early as 1550. Indeed, Bahia was one of the ports of entry for the African slaves; more recently, it has been suggested that under the Dutch (1630–1654), Recife may also have been an important slave entry point.

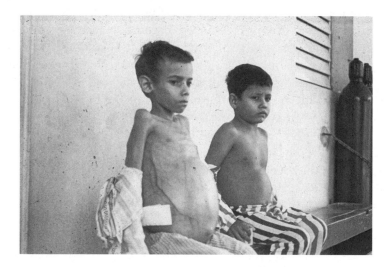

Figure 3.3   Two young boys infected with blood flukes.

Although snails native to the Caribbean and South America have been found to be effective vectors (and different from the snail species in Africa), snails introduced from Africa have also been important in transmission.

Schistosomiasis has not been eliminated. At present there are an estimated 200 million people infected with schistosomes, resulting in more than a million deaths annually. *S. japonicum* is found in Southeast Asia and the western Pacific as well as China, the Philippines, and Indonesia. *S. haematobium* and *S. mansoni* are both found in 43 countries in Africa, but *S. mansoni* is also found in the Americas (Brazil, Suriname, Venezuela, and the Caribbean).

People who come to the freshwater pools to work, bathe, drink, wash clothes, and swim may also use the water for elimination of their body wastes. Individuals may be infected and reinfected almost daily as they paddle through the cercaria-infested waters that they have come to use as their outdoor toilets. Schistosomiasis remains one of Africa's greatest tragedies, with an estimated 50 million cases. The highest incidence occurs in children. Between 1934 and 1957, Ghana reported an annual occurrence of 70% in children, Nigeria reported over 80%, and Zaire reported 60%. These numbers have hardly changed since then. In some instances, technology has expanded the number of cases. Indeed, every new irrigation scheme and each new dam may pose a new threat.

The Aswan High Dam of Egypt, begun in 1960 with Soviet financing and engineering and requiring 30,000 Egyptians toiling around the clock, was completed in 1971. The High Dam, by controlling the level of water in Lake Nasser, has brought electricity to many parts of Egypt as well as making it possible to cultivate four crops per year through year-round irrigation. However, the dam has also created conditions favorable for the schistosome-carrying snails. Before High Dam construction, there was already perennial irrigation in the Nile Delta and the prevalence of schistosomiasis was 60%, whereas in the 500 miles of river between Cairo and Aswan, when there was annual flooding, the prevalence was 5%. Some 4 years after the dam was completed, the average prevalence of schistosomiasis between Cairo and Aswan increased sevenfold (35%; range, 19 to 75%).

Schistosomiasis is generally a disease associated with agriculture, but it has also been a military problem ever since the days of Napoleon. During World War II, when U.S. troops stormed ashore on the Pacific island of Leyte in October 1944, they were unaware that in addition to being attacked by Japanese bullets, they were also being invaded by the cercariae of *S. japonicum*. By January 1945, the first cases were diagnosed; in the end, 1,700 men were put out of action at a cost of 300,000 fighting man-days and $3 million. Five years later, 50,000 Chinese communist soldiers prepared for an invasion of Taiwan, but schistosomiasis became so widespread among the troops that the campaign was abandoned and the island was retained by Chiang Kai-shek's forces.

Where did schistosomiasis originate? It probably first occurred in animals living in the rainforests and lakes of East Africa and then spread, together with its vector snails, along the Nile and out into the Middle East and Asia via the trade routes. Blood flukes occur in birds and mammals other than humans. Indeed, "swimmer's itch" or "cercarial dermatitis" is found in lakes and along the seashore in Michigan, Minnesota, Wisconsin, New Jersey, and New England, as well as in other parts of the world, and is caused by cercariae (Fig. 3.2G) that normally infect aquatic birds and mammals. The skin rash and pustules are the result of their failure to continue their migration past human skin.

Snail fever today

Diagnosis of schistosomiasis is made by examining stools and urine under the light microscope and finding eggs. Sometimes this is supplemented by biopsy and immunological methods. Preventive measures include education of the population in means to prevent transmission, treatment of infected persons, and control of the snail vector using molluscicides. In some cases (e.g., growing rice), avoidance of contact may be impossible. However, human exposure can be reduced by providing a safe water supply for bathing and washing as well as sanitary disposal of human wastes. Other measures may be lining irrigation channels with cement to discourage snails, intermittently irrigating rice paddies to disrupt the life cycle, or storing water away from snails for 2 to 3 days, a time that exceeds the survival time of the miracidia.

The earliest treatment for the disease was developed in 1918 and required intravenous administration of an antimony compound (tartar emetic). In 1929, intramuscular injections of another antimony compound, stibophen, were used, but the cure rates were not as good as those with tartar emetic. However, both drugs showed severe toxic reactions and sometimes resulted in death. Later, an oral drug, niridazole, was introduced (1964), but it wasn't until the 1970s that a truly effective drug with low toxicity was developed: praziquantel (trade name Biltricide). There is no preventive vaccine or drug for this disease.

The development of new drugs for the treatment of schistosomiasis can be a long and expensive undertaking. Further, determination of drug efficacy and safety may require extensive animal and human testing. One such heroic effort is worthy of mention. Dr. C. Barlow, an American physician, volunteered for a chemotherapy trial and exposed his abdomen to 224 cercariae; when cercarial dermatitis developed, he knew he had been infected. He came down with severe schistosomiasis, which required many intravenous injections of tartar emetic for cure. In his old age Barlow wrote: "even today I shudder every time I see a hypodermic needle."

Today, with hindsight, it is easy to understand why those living in ancient Egypt were unable to control schistosomiasis, but why, 2000 years later, are we still failing? There are certainly economic constraints—such as the cost of molluscicides and drugs, as well as the necessary infrastructure for providing clean water and sanitary disposal of wastes—but in the final analysis it is the habits of the human population that are of critical importance to elimination of this disease. Since there is no animal reservoir, humans are required for the perpetuation of the disease. As long as infected individuals continue to urinate and defecate in the same waters where the vector snail lives and to expose their bare skin, there will be blood fluke disease.

The Plague of Athens

The valleys of the Nile and Tigris-Euphrates spawned the civilizations of Egypt and Mesopotamia. The flat fertile valleys made it easier to subject the populations to a single ruler and to oversee individuals so that each had a prescribed role in the society. As a consequence, there emerged in the Fertile Crescent and Egypt a unified system of kingdoms. The different geography of Greece gave rise to another kind of sociopolitical system. Greece, the southernmost extremity of the Balkan landmass, is composed of limestone mountains separated by deep valleys and is cut almost in two by the narrow Corinthian Gulf. The southwestern peninsula was called Peloponnesia, and the northeastern part was called Attica. East of the mainland are the islands of the Aegean Sea; to the south is Crete. Because the small populations were separated from one another by mountains and the sea, central control in Greece was impossible. Individuals were not compelled to become specialists but rather masters in a range of accomplishments of hand and mind. Greece, smaller in area than Florida, was never able to support more than a few million inhabitants, but it has played an enormous role in the history of Western civilization. We see some of this today in their monuments, sculptures, paintings, and writings, and we speak of the "glory that was Greece." Indeed, the prestige of the Greeks in the arts and their ideas on medicine, astronomy, and geography were accepted with unquestioning faith until the 17th century, when a new scientific spirit of experiment and inquiry came into being.

The land and the climate of Greece were unsuitable for farming grains; as a result, the economy rested on the large-scale movement of goods by ship. The Greeks planted vines and olive trees and produced wine and oil, and these were exchanged for grains and other, less valuable commodities. Indeed, an acre of land with vines and olive trees could yield a quantity of wine and oil that could be exchanged for an amount of grain requiring many acres for its production. The outlying communities on the Mediterranean and Black Seas that provided these less valuable commodities were called by the Greeks "barbarians," from the Greek word *barbarous* ("foreign" or "uncivilized"). It was the barbarian societies that provided grain, metals, timber, and slaves in exchange for oil and wine and so contributed to the emergence of the Greek civilization.

The Greeks had an unshakable belief in the worth of the individual. While to the east there were absolute monarchies, the Greeks evolved a democratic society in which each individual was respected and counted. What developed were states consisting of a city and its surrounding lands whose inhabitants, citizens (literally "those living in the city"), were valued. (Not all individuals, however, were citizens; slaves and women did not have the same rights.) The city-states of Greece, insulated from the barbarians, consisted of urban centers dedicated to commercial transactions, with limited local farming; as they prospered, the population grew. The inhabitants of the city-states were a healthy population but not immune to the diseases that were endemic in the Middle East; this was to prove decisive in the Peloponnesian War.

Before the Greeks of historical times (~750 BC), a civilization had flourished in Mycenae (1600 to 1200 BC), and before that in Crete (the Minoans). Later, the Dorians invaded from the north, destroying the Mycenaean cities and society; this ushered in what for 450 years (1200 BC to 750 BC) has been called the Dark Ages. At the close of the Dark Ages, there emerged two powerful city-states that were essentially military garrisons governed by a commander and his captains. These were Athens (in Attica) and Sparta (on the Peloponnesus). Sparta was settled by the Dorians, and Athens gave refuge to Mycenaeans. Each city-state represented opposing philosophies: stern military discipline in Sparta versus intellectual and political freedom in Athens. As the population of Athens grew too large for the limited space of Attica, the Athenians began sailing out of their port of Piraeus to colonize the Aegean islands and the western coast of Asia Minor (in what is now Turkey). These Greek colonies were called Ionia.

Belief in liberty and freedom made Greek city-states resist domination by others, and since each state had its own habits, rules, and government, the loyalties of the citizens were to a particular city-state, the polis. (People engaged in the civic life of the polis give us the term "politics.") As a consequence, war between Athens and Sparta was inevitable. The Peloponnesian war began between these two city-states in 431 BC, and it lasted for 27 years. The cause of the war was probably economic, although we cannot be certain of this. What is known, however, is that its outcome was determined by disease.

The Greeks, apparently against all odds, managed to defeat the numerically far superior Persian forces in two battles, one in 490 BC (at Marathon), and the other in 480 BC in the great naval battle of Salamis. This led a number of Greek city-states to join together with Athens in a sea league, both to punish the Persians and to obtain recompense for the cost of the war. However, in time, Athens turned this league into an instrument of its own imperial power, appropriating the funds of the league for the creation of monuments of imperial splendor (notably, the Parthenon). This naturally provided a focal point for the jealousies and rivalries of the various city-states, especially Sparta.

Corinth was a commercial and colonial power and an ally of Sparta. Because of this, its interests were in competition with those of Athens. Athens began to interfere in the affairs of Corinth; Corinth naturally objected and threatened Athens, so in retaliation, Athens began an embargo of Corinth and other city-states on the Corinthian isthmus. This crippling embargo caused Corinth to urge Sparta to declare war on Athens. The Athenians had a great fleet but a poorly trained army, whereas the Spartans were an effective military power on land but not at sea, because they lacked a fleet. Pericles, the leader of the Athenians, decided to rely mainly on Athenian naval supremacy. His strategy was to bring all the people in Attica into the city, abandon the outlying countryside to destruction by the Spartans, and rely upon the navy to supply the city with food and other necessities that would be carried through the fortified corridor from the port of Piraeus into the city itself. The hope was that Sparta would eventually be worn out and frustrated. Sparta did invade Attica from the north, and the Athenians gathered themselves and remained secure within the walled fortifications of their city. But when the Spartans destroyed the olive and grape orchards in the outlying countryside, the source of Athens' wealth came into jeopardy. Further, the large numbers of peasants from the countryside who sought refuge within Athens resulted in the city becoming overcrowded. In 430 BC disaster struck. An epidemic that started in Ethiopia moved into Egypt and from there was brought by ship to Piraeus. The epidemic raged for about 2 years and killed about one-fourth of the Athenians, including Pericles, in 429 BC.

In 1994 and 1995, a mass grave was uncovered prior to construction of a subway station just outside Athens' ancient cemetery. There were some 90 skeletons, ten belonging to children; the grave may have contained as many as 150 people. The skeletons in the graves were placed helter-skelter with no soil between them, and the bodies were placed in the pit within a day or two, suggesting burial in a state of panic. The grave was dated to between 430 and 426 BC. It is believed these are the remains of Athenians killed by the plague. The historian Thucydides, himself a survivor of the plague, wrote,

> The bodies of the dying were heaped one on top of the other, and half-dead creatures could be seen staggering about in the streets or flocking around the fountains in their desire for water. . . . The catastrophe was so overwhelming that men, not knowing what would happen next to them, became indifferent to every rule of religion or of law. All the funeral ceremonies which used to be observed were now disorganized, and they buried the dead as best they could. Many people, lacking the necessary means of burial because so many deaths had already occurred in their households, adopted the most shameless methods. They would arrive first at a funeral pyre that had been made by others, put their own dead upon it and set it alight; or, finding another pyre, they would throw the corpse that they were carrying on top of the other one and go away. . . . Seeing how quick and abrupt were the changes of fortune which came to the rich who

suddenly died and to those who had previously been penniless . . . , people began openly to venture on acts of self indulgence which before they used to keep in the dark. . . . No fear of god or law of man had a restraining influence. As for the gods, it seemed to be the same thing whether one worshiped them or not, when one saw the good and bad dying indiscriminately.

What was this devastating plague? Despite Thucydides' detailed description, the precise identity of the disease is not known. It was clearly not the bubonic plague, because the characteristic symptom of the bubo (swelling of the lymph nodes in the region of the groin and armpits) is not found in Thucydides' description. Other suggested candidates are measles, typhus, Ebola, mumps, and even toxic shock syndrome. The case for typhus seems strongest, from the standpoint of both epidemiology—the age group is similar—and symptoms. Typhus is characterized by fever, pustules, and a rash of the extremities; it is known as a "doctors' disease" from its frequent incidence among caregivers. The disease confers immunity. During World War I, patients with typhus reportedly jumped into water tanks to alleviate extreme thirst. But the fit is not exact. The rash in Thucydides' description does not precisely match that of typhus, nor does the state of mental confusion.

The plague of Athens demoralized the citizenry, destroyed the fighting power of the Athenian navy, and prevented the launching of an attack against Sparta. Though the war dragged on for many more years, the spirit of Athens had already been broken; by 404 BC, defeat was complete. Sparta deprived Athens of her navy, and her land defenses were razed. The plague of Athens clearly changed the course of history.

The Roman Fever

Of the antiquity of human malaria, there is no doubt. Enlarged spleens, presumably due to malaria, have been found in Egyptian mummies more than 3,000 years old, and the Ebers papyrus mentions fevers. More recently, evidence of malaria has been detected in lung and skin samples from mummies dating from 3204 to 1304 BC. Clay tablets from the library of Ashurbanipal, king of Assyria (668–627 BC), mention enlarged spleens, headaches, and periodic chills and fever, indicating that more than 4,000 years ago the region between the Tigris and Euphrates Rivers was already malarious. Malaria probably came to Europe from Africa via the Nile Valley or resulted from closer contact between Europeans and the people of Asia Minor. Early Greek poems from the end of the sixth century BC describe intermittent fevers, and Homer's *The Iliad* (800 BC) mentions malaria, as do the writings of Aristophanes (ca. 448–380 BC), Plato (ca. 428–347 BC), and Sophocles (ca. 496–406 BC). The Greek physician Hippocrates (ca. 470–380 BC) discussed in *Of the Epidemics* the two kinds of malaria, one with recurrent fevers every third day (benign tertian), and another with fevers on the fourth day (quartan), which are today called *Plasmodium vivax* and *Plasmodium malariae*, respectively. He also noted that those living near marshes had enlarged spleens (Fig. 3.4), but he never speculated as to the relationship. Indeed, Hippocrates believed that the intermittent fevers were the result of an imbalance in the body's fluids (bile, blood, and phlegm) brought about by drinking stagnant marsh water. Hippocrates recognized that at harvest time, when Sirius (the dog star) was dominant in the night sky, fever and misery would soon follow. From the historical descriptions, malaria was obviously present in Greece but probably had little influence on military campaigns. Only once in his account of the Peloponnesian War did Thucydides refer to an illness suggestive of malaria that affected the Athenian army besieging Syracuse while encamped on marshy grounds.

Hippocrates did not describe the deadly malignant tertian malaria (*Plasmodium falciparum*), so we suspect that it did not exist or was rare. It has been speculated that although this kind of malaria was periodically brought into southern Europe from North Africa and Asia Minor, such infections were infrequent owing to the absence of a suitable mosquito vector in the Mediterranean. With greater agricultural activities, including deforestation and soil erosion, conditions arose that favored the establishment of a suitable habitat on the shores of Europe for several Asian and North African species of *Anopheles* mosquitoes. Over time, mosquito vector competence increased. The highly virulent *P. falciparum* came to be established in Europe by the second century AD, and from that time onward it plagued the Romans.

Figure 3.4   The physician Jason, a contemporary of Hippocrates, palpating the spleen (from a funerary urn in The British Museum, London).

The great age of Greek expansion lasted 200 years (750 to 550 BC). Greek colonization took place along the shores of the Aegean and Black Seas and west into Sicily and southern Italy. By 750 BC, colonies had been established on the west coast of Italy as far north as the bay of Naples. The Ionian Greeks from Asia Minor sailed and traded further west, along the coast of the Mediterranean, reaching present-day Spain. The Greeks also traded with inland villages as they moved up the Rhone River into Gaul and even as far north as England and Ireland. It was the practice of the Greeks to keep to themselves. Because they remained apart from the indigenous peoples, their colonies were extensions of the homeland and one of the means by which Greek civilization was spread to other parts of the world. The flourishing trade with the colonies also became the means by which infectious diseases were transmitted to the naive populations in distant lands.

About 2000 BC, the Latins—a group of Indo-European origin, possibly with forebears in central Asia, to which the Romans belonged—migrated first into central Europe and then to the northernmost part of Italy. By 1000 BC they had settled on the ~700-square-mile volcanic Latium plain bounded on the north by the Tiber River. The soil was rocky but fertile, and the Latins prospered as farmers. Then, in ~800 BC, the Etruscans coming out of Asia Minor landed on the coast north of the Tiber, from whence they moved inland, and by 600 BC they dominated all of Italy from the Alps in the north to Salerno in the south. There they were halted by the already established Greek colonies. The Etruscans, a highly civilized people who were traders and merchants, brought to the Romans their first contact with the eastern Mediterranean. The Romans made allies of or subdued the other tribes of the Latium plain, including the Etruscans. By the beginning of the fourth century BC, Rome was the leading city in central Italy. In time, that city became an empire that would last 500 years. By 350 BC, the Romans had moved southward, reaching the Greek settlements at the foot of the peninsula, and in 275 BC, when the Greeks were defeated, Rome became the master of the entire Italian peninsula. Later, there would be other conflicts and other victories over Carthage (264 to 241 BC, 218 to 201 BC, and 149 to 146 BC), the Macedonian Empire of Alexander (197 BC), the Seleucid Empire in Syria (190 BC), and the Ptolemaic Empire in Egypt (31 BC). By 55 BC, the Romans had invaded Britain. In 27 BC, the Roman Empire was established when Octavian assumed the title of Augustus and became the first Roman emperor.

Rome lived off its imports. Cargo shipped from the provinces was unloaded at the seaport at Ostia and carried up the Tiber River to the city. The Roman Empire, with its center in Rome, developed an ever-extended series of colonies, and by the year AD 100, there was a vast trade network that included India, China, and the northern parts of Africa and the Middle East. The regular movement of goods and people to and from Rome also made for the spread of infections. Thus, over time the chances of the Mediterranean population's contracting an unfamiliar infection became greater and greater.

There is no evidence that malaria was a public health problem in Italy among the ancient Etruscans, but there is clear evidence of malaria being devastating to the Roman Empire. Indeed, in ancient Rome, temples were dedicated to the goddess Febris, who is described as an old hag with a prominent belly and swollen veins. The medical literature of the time also contains accurate descriptions of malaria, and there are references to marshes as the source of the disease. The disease was so prevalent in the marshland of the Roman Campagna, near Ostia, that the condition was called the "Roman fever"; eventually, it was given the Italian name *mal'aria,* literally "bad air," because this recurrent fever occurring during the sickly summer season was believed to be due to vapors emanating from the marshes. For almost 2,000 years, Rome was the home of the Roman fever.

Although malaria was uneven in its distribution, and its endemicity fluctuated cyclically in the Roman Empire, epidemics of malaria occurred in Rome and the Campagna every 5 to 8 years. The Pontine marshes southwest of Rome were a lethal source of malaria and were described this way: "the Pontine . . . creates fear and horror. Before entering it you cover your neck and face well before the swarms of large bloodsucking insects are waiting for you in this great heat of summer, between the shade of the leaves, like animals thinking intently about their prey . . . here you find a green zone, putrid, nauseating where thousands of insects move around, where thousands of horrible marsh plants grow under a suffocating sun." Not even the Romans, who were able to conquer most of the Western world, were able to master the Pontine marshes. It has been estimated that even if malaria occurred in only one-sixth of the Campagna, it could devastate the agricultural economy. Indeed, it required the efforts of the fascist dictator Benito Mussolini in the late 1920s to drain and fill these swamps, making them habitable and productive agriculturally. Colonies on the coast of Italy also failed because of malaria, and in some districts of Rome where the urban population may have reached 750,000 to a million people, the death rate could be quite high. In some particularly unhealthy places, life expectancy was only 20 years of age, whereas in places where malaria was absent, life expectancy could be as high as 40 or 50. In the Roman Empire, malaria was a disease of children, but severe illness was also found among immigrants without acquired immunity. After the establishment of Christianity, the Roman fever plagued pilgrims visiting the Holy City. Some have claimed that foreign invaders of Rome—particularly the French and the Germans—were more effectively repelled by the deadly fevers of the Pontine marshes and the Campagna than by any manmade weapons. Indeed, Alaric died from malaria during the siege of Rome in the summer of AD 410, and Attila the Hun's failure to march on Rome in AD 452 was partly due to the threat of this disease as well as famine in Italy. And in AD 1155, the army of Frederick Barbarossa was so decimated by malaria at Rome that they were forced to retreat across the Alps.

Plagues and the Rise of Christianity

Plagues of various sorts were not unfamiliar in Roman history. An epidemic—probably smallpox—struck the city in AD 65. It was brought, or so we believe, by Roman troops who had been campaigning in Mesopotamia. Mortality was heavy, and in some areas, half the population died. A new round of smallpox infections began in AD 251 to 266; it was reported that 5,000 people a day died at its height. By this time, measles may also have become established among Mediterranean populations. No accurate estimates of population losses can be given, but they must have been high. It was plagues such as these that strengthened the early Christian Church.

The early Romans owed their loyalty to their pagan gods. Their religion was one of form and ritual. In essence, the Romans had a contractual relationship with their gods: if you do something for me, I will do something for you. At first, their religion was animism: gods represented the spirits in water, rocks, fire, trees, beasts, sun, moon, stars, and lightning. The spirits were amoral, and they either helped or harmed the worshippers according to the manner in which they were treated. In effect, the role of religion was to appease the multitude of gods so that the worshippers would receive some benefit. From the Etruscans, the Romans borrowed elaborate religious ceremonies and gods with a human form that could be represented as painted images or carved statues. It was also the Etruscans who gave the Romans their earliest contact with the gods and goddesses of Greece, many of whom were absorbed into the Roman religion. Cults developed to worship specific gods at specific times (called holy days or, later, holidays) and those who presided over the cult rituals were called priests; however, these individuals were neither moral nor spiritual. In time, the emperor himself became a god, and so loyalty had to be sworn to him.

Then in the first century AD, there appeared a new kind of religion, preached by Jesus of Nazareth (the Christ) and his disciples. Although Jesus preached for less than 3 years in what is now Palestine, his many disciples traveled throughout the Roman Empire spreading the word of the Christian religion. Jesus' preaching took the form of parables and miraculous healings; he encouraged the poor and the oppressed and spoke of forgiveness, detachment from wealth and property, and special care for the outcasts and sick of society. Because the Roman Empire consisted of many cities, Christianity became an urban movement. (Indeed, because of this, those living outside the city were called rustic or, in Latin, *paganus*, from which the word "pagan" [non-Christian] is derived.) Jesus gathered around him a community of followers who regarded themselves as God's people, and they went forth establishing a missionary movement. The early Christians had a moral ideal: they separated themselves from pagan idolatry and espoused universal salvation. Because Jesus' disciples preached the coming of a new king, it appeared to the Romans that there might be a revolution in the making. At the outset, however, the Romans simply regarded Jesus as a minor political rebel whose followers could be used as convenient scapegoats. As a result, the Christians were blamed for all types of disasters, including plagues, inflation, fires, and even barbarian incursions. The reasons for this were many: the Christians did not worship the emperor, they did not observe the pagan ritual acts, they insisted that they alone possessed God's truth, and Christ's teachings were critical of the established order. Christ's omnipotence was also believed to be demonstrable by those who had survived a debilitating or deadly disease.

The rise and consolidation of Christianity may have also been affected by disease. The expectations of the poor Romans were that with Christ's second coming they would be freed from their rich masters. Christianity, unlike paganism, preached care of the sick as a recognized religious duty. Those who were nursed back to health felt gratitude and commitment to the faith, and this served to strengthen Christian churches at a time when other institutions were failing. Another positive feature of Christianity was that the teaching of the faith made life meaningful even in the face of sudden death, since it was perceived as a release from an individual's suffering. The capacity of Christian doctrine to cope with the psychic shock of epidemic disease made it attractive for the populations of the Roman Empire. Paganism, on the other hand, was less effective in dealing with the randomness of death. In time, the Romans came to accept the Christian view. Rome became the headquarters of Christianity, and in AD 337, with the conversion of Emperor Constantine, Christianity became the Church of the Empire.

The Antonine Plague

The Roman Empire expanded its frontiers until AD 161, but from that time onward, its defenses began to crumble. Toward the end of the first century AD, a warlike people riding on horseback from Mongolia—the Huns—began a westward movement. The Huns brought new infections to the Roman Empire, but other infections rebuffed them, such as the Roman fever. Marcus Aurelius Antoninus was born in AD 121, and throughout his reign as emperor (AD 140 to 180), he was engaged in defensive wars on the northern and eastern borders of the Roman Empire. In AD 164, the Roman legions under the command of Avidus Claudius were sent to Mesopotamia to repel an invasion by the Parthians, and in this they succeeded. But the troops returned with a devastating plague that spread throughout the countryside and reached Rome by AD 166. This epidemic of Antoninus, called the Antonine plague, spread to other parts of Europe, causing so many deaths that cartloads of bodies were removed from Rome and other cities. In Rome, Emperor Antoninus made administrative reforms, and concerned himself with famine and plague in the empire. He was a writer (*Meditations*) and a devotee of stoicism, and he ruthlessly persecuted the Christians, believing them to be a threat to imperial power. In AD 161, when the Huns had reached the northeast border of Italy, Antoninus was forced to contemplate battle, but fear and disorganization delayed a direct confrontation with the Germanic tribes on the Rhine-Danube frontier until AD 169. When he and his legions moved into the northern frontier with the objective of securing the empire's northwesterly boundaries (as far as the Vistula River), a plague broke out among the troops; it raged until AD 180 and affected not only the Roman legions but also the Huns. As the plague ravaged his army Antoninus elected to retreat to Rome, but he was never to reach his destination. In Vienna, on the seventh day of his illness, 17 May AD 180, he died from this plague. The plague returned again in AD 189 and, though it was less widespread than the first epidemic, at its peak there were more than 2,000 deaths a day in Rome.

The Antonine plague is also associated with the physician Galen of Pergamum (AD 130–ca. 200), whose ideas dominated medicine until the 16th century. Galen's hero was Hippocrates (470–380 BC), who had laid down the principles of medicine in Greece. Galen was first appointed surgeon to the gladiators in Asia Minor and then moved to Rome, where he practiced medicine. Though Galen was a skilled anatomist, an experimentalist, and a searcher for new drugs, when faced with the plague he fled Rome. He was, however, recalled by Emperor Antoninus to Rome, where he died. But before his death, he left a description of the plague's symptoms: high fever, inflammation of the mouth and throat, thirst, diarrhea, and a telltale sign: pustules on the skin that appeared after 9 days. Even today, precisely what this plague was remains a mystery, but most historians suspect that this was the first record of a smallpox epidemic. Some believe that smallpox moved into the Roman Empire either with the legions returning from Mesopotamia or with the Huns who carried it from Mongolia and then on to Rome.

The Cyprian Plague

In 250, Cyprian, the Christian archbishop of Carthage, described a disease that appeared to have originated in Ethiopia, moved into Egypt, and eventually come to the Roman colonies of North Africa: vomiting, diarrhea, gangrene of the hands and feet, a burning fever, and a sore throat. There were no pustules. The Roman colonies were the breadbasket of the empire, and this plague seemed to coincide with an invasion of locusts that had destroyed the crops. Did the plague have anything to do with famine, or was it ergotism—a convulsive disorder resulting from ingestion of rye contaminated with the parasitic fungus ergot, which contains LSD. (Ergot was hypothesized to explain the hallucinations associated with witchcraft trials in Salem, Mass., in 1692.) We do not know for certain, but the Cyprian plague did become a pandemic, advancing quickly through direct person-to-person contact as well as by contaminated clothing. Mortality is said to have been high: the number of deaths exceeded the number of survivors. The plague of Cyprian lasted 16 years. It caused panic among the people, and those who fled to the surrounding countryside served as "seeds" for initiating fresh outbreaks. The land they left lay fallow. Despite this, the Roman Empire survived the devastation wrought by the Cyprian plague and was even able to overcome subsequent invasions by the Huns. But by 275, the Roman legions were forced to retreat from the Danube and the Rhine to the city of Rome. The situation was so precarious that the emperor decided to fortify Rome itself to protect against this plague. This also proved to be ineffectual.

The plague of Cyprian strengthened Christianity. Cyprian wrote:

> Many of us are dying in this mortality, that is many of us are being free from the world. This mortality is a bane to Jews and pagans and enemies of Christ; to the servants of God it is a salutary departure . . . without any discrimination the just are dying with the unjust . . . the just are called to refreshment, the unjust are carried off to torture; protection is more quickly given to the faithful; the punishment to the faithless . . . this plague and pestilence which seems horrible and deadly, searches out the justice of each and every one. . . .

This ability of Christianity to deal with the horrors and hardships of a plague made church doctrine an attractive alternative to the stoic and pagan philosophies, which were impersonal, uncompassionate, and ineffectual in explaining the randomness of death due to disease. And so, Christianity strengthened its hold on the Roman people. This attraction to Christianity not only altered the current religious and cultural practices of Romans but also influenced future social and political development.

Saint Sebastian, often depicted as a naked youth wearing a crown, tied to a tree with his body pierced by arrows, is the patron saint of archers, athletes, and soldiers as well as the protector from plague. His answers to prayers for protection from the plague, first in Rome and later in Milan (1575) and Lisbon (1599), were the cause for his elevation to sainthood. Documentation and myth along with the fate of Saint Sebastian have been woven together in *The Golden Legend*, a text from the Middle Ages written in 1275 by the archbishop of Genoa, Jacobus de Voragine (1229–1298). It is written that Sebastian was born of a wealthy family in Narbonne in 257 and was educated in Milan. He became an officer in the Imperial Roman Army during the time of Emperor Diocletian (284–305) and secretly converted to Christianity. Diocletian's name is often associated with the last and most terrible persecutions of the early Christian Church. During Diocletian's persecutions, Sebastian visited his fellow Christians in the prisons, giving them food and comfort. However, in 286, when Diocletian learned of his works in converting others to the faith, he ordered Sebastian to be shot to death by his archers. Sebastian was tied to a tree, shot with arrows, and left for dead. However, the arrows did not kill him, and a Christian widow, Irene, tended his wounds. Upon his recovery he continued to preach, and after he confronted the Emperor Diocletian to denounce his cruelty, the enraged emperor ordered that he be beaten to death by the blows of a club. The destructive effects of the plague caused people to compare their being struck down by death to an attack by an army of archers, and so they prayed for salvation from a divine being; it is claimed that their prayers were answered by Saint Sebastian. So, in time, Sebastian the saint of archers became the people's protector from the plague.

Over the next three centuries, Rome slowly collapsed under pressure from the Germanic tribes (Goths, Vandals) as well as recurrent outbreaks of mysterious plagues such as Antonine, Cyprian, and others. Gradually, the Four Horsemen of the Apocalypse—Disease, Famine, War, and Death—led to a disintegration of the Roman Empire. When the Germanic peoples moved into Italy and Gaul, crossed the Pyrenees into Spain, and even reached North Africa, they too became subject to this plague; by 480, the Vandals themselves were so sickly that they were unable to resist invasion by the Moors.

The Justinian Plague

The historian Procopius of Caesarea wrote of the Justinian plague in *The Persian Wars*, "it embraced the entire world and blighted the lives of all men." He described the symptoms of the disease: great swellings, or buboes, in the groin and armpit as the lymph nodes enlarged; delirium and frantic restlessness; and sometimes coma. There was general hysteria and panic. If the buboes were lanced, there was the possibility of survival, but most died within 5 days. The plague raged in the city of Constantinople (present-day Istanbul), the capital of the Roman Empire in the East, for months; so numerous were the corpses that they were cast into hollow towers of some incomplete fortifications. Procopius's eyewitness account claimed that 300,000 died, but this is probably an exaggeration. However, there is no question that all of Europe suffered from this plague and that 5 years later it had reached Ireland. All told, it killed about 1 million people.

This disease arrived in 541 and raged intermittently in Europe, North Africa, and the Middle East until 757. Called the plague of Justinian, it was probably bubonic plague that came to the Mediterranean from an original focus in northeast India or via central Africa. From Lower Egypt, the disease reached the harbor town of Pelusium in 540, from where it spread to Alexandria and then by ship to Constantinople, the capital of Justinian's empire. It is speculated that Justinian's General Belisarius was unable to accomplish the goal of Justinian, that is, the reestablishment of the Roman Empire in all its glory, because of the outbreak of plague.

The plague-induced losses in the population reduced the taxes and services available to the government. Lacking manpower and money, Justinian was unable to send aid to his beleaguered troops. The survivors of the plague were damaged both physically and mentally and had a diminished self-confidence. As disasters continued to befall the empire, the Roman and Persian forces were unable to offer more than token resistance to the Moslem armies that swarmed out of Arabia in 634.

Bubonic plague recurred every 3 or 4 years for decades, and its effects lasted well into the seventh century. It is estimated that by 600 the plague of Justinian—the first pandemic of bubonic plague—had reduced the population by 100 million people, almost 50% of the population of Western Europe. William H. McNeill has observed that the plague of Justinian resulted in "the perceptible shift away from the Mediterranean as the preeminent center of European civilization and the increase in the importance of more northerly lands." The Justinian plague and the subsequent outbreaks of plague over the next two centuries marked the end of the Classical World—the Greek and Roman civilizations—and ushered in the Dark Ages. Plague so diminished trade in the Mediterranean that it left most countries with only a bartering economy. Cities withered, feudalism grew, religion became more fatalistic, and Europe turned inward upon itself.

Figure 4.1   *The Plague*, by Felix Jenewein (1900), shows a mother carrying a coffin with her child. (Courtesy of the Wellcome Library of Medicine.)

Chapter 4

An Ancient Plague, the Black Death

During the last 2,000 years, three great bubonic plague pandemics have resulted in social and economic upheavals that are unmatched by those caused by any armed conflict or any other infectious disease. In Constantinople, the capital of the Roman Empire in the East, it was the first bubonic plague pandemic (542 to 543) that surely contributed to Justinian's failure to restore imperial unity. In 1346, the second pandemic began. By the time it dissipated in 1352, the population of Europe and the Middle East had been reduced from 100 million to 80 million people (Fig. 4.1). This devastating pandemic, known as the Black Death, the Great Dying, or the Great Pestilence, put an end to the rise in human population that had begun in 5000 BC; it took more than 150 years for the population to return to its former size. Some believe this catastrophic crash in population to be Malthus's prophecy come true, but others, such as the historian David Herlihy, consider the Black Death not to be a catastrophe promoted by "positive checks" (i.e., disease, war, and famine) but an exogenous factor that served to break a Malthusian stalemate. That is, despite fluctuations in population size, relatively stable population levels were maintained over prolonged periods of time owing to "preventive checks" (changes in inheritance practices, delay in the age of marriage, and birth control). The Black Death did more than break the Malthusian stalemate: it allowed Europeans to restructure their society along very different paths.

Although those living in the medieval period recognized that plague was a contagious disease spread from person to person, its cause was not identified. Indeed, most believed it to be "a vicious property of the air" itself. Today, more than 600 years later, we know that the source of the second pandemic was microbes left over from the first pandemic (the Justinian plague), which had moved eastward and remained endemic for seven centuries in voles, marmots, and the highly susceptible black rats (*Rattus rattus*) of the arid plateau of central Asia (roughly corresponding to Turkestan today). Plague-infected rats moved westward along the caravan routes between Asia and the Mediterranean known collectively as the Silk Road; plague traveled from central Asia, around the Caspian Sea, to the Crimea. There, the rats boarded ships and moved from port to port and country to country, spreading plague to the human populations living in filthy, rat-infested cities. Indeed, the story of the Pied Piper of Hamelin (or Hameln) may have had its roots in the plague-ridden cities of Germany (Fig. 4.2). Legend has it that on 26 June 1284, the city became infested with rats. A pied piper came to Hamelin and agreed to rid the city of rats for payment of a large sum of money. He was able to enchant the rats by playing his flute, and he led them to the river, where they drowned. The English poet Robert Browning set the story to rhyme:

> Rats! They fought the dogs and killed the cats
> And bit the babies in their cradles,
> And ate the cheeses out of the vats,
> And licked the soup from the cook's own ladles.
> But then the Pied Piper was hired:
> To blow the pipe his lips he wrinkled
> And green and blue his sharp eyes twinkled.
> And ere three shrill notes the pipe uttered
> Out of the houses the rats came tumbling

Figure 4.2 The rat killer (from a 17th century drawing by Jan Georg van Vliet) through the course of time becomes the children's storybook character the Pied Piper. (Courtesy of the National Library of Medicine.)

Great rats, small rats, lean rats, brawny rats
Brown rats, black rats, gray rats, tawny rats
And step for step they followed dancing
Until they came to the river Weiser.

His work done the Pied Piper said:

"First, if you please my thousand guilders!"
A thousand guilders! The Mayor looked blue
So did the Corporation too.
"Beside" quoth the Mayor with a knowing wink
"Our business was done at the river's brink;
"We saw with our own eyes the vermin sink,
"And what's dead can't come to life, I think.
But as for guilders, what we spoke
Of them, as you very well know, was in joke
Beside our losses have made us thrifty
A thousand guilders! Come take fifty!"

The piper's face fell.
Once more he stept into the street,
And to his lips again laid his long pipe of smooth straight cane;
Soft notes as yet musicians cunning. . .
Out came the children running
All the little boys and girls,
With rosy cheeks and flaxen curls,
And sparkling teeth like pearls,
Tripping and skipping, ran merrily after
The wonderful music with shouting laughter. . .
When lo they reached the mountainside,
A wondrous portal opened wide,
As if a cavern was suddenly hollowed,
And the Piper advanced and the children followed
And when all were in to the very last
The door in the mountain-side shut fast. . .

And the Piper and dancers were gone forever.

Versions of the tale were gathered by the Grimm Brothers (1812). This plague-inspired occurrence has over time come to have a moral interpretation: evil will befall those who do not carry out their promises.

The Black Death is most associated with Florence, one of the great cities of Europe at the time, and because it felt the full impact of the epidemic, it is sometimes called the Plague of Florence. Giovanni Boccaccio (1313–1375), who lived in Florence during the plague, described what he witnessed:

towards the beginning of spring . . . the doleful effects of the pestilence began to be horribly apparent by symptoms . . . an issue of blood from the nose was a manifest sign of inevitable death; but in men and women alike it first betrayed itself by the emergence of certain tumors in the groin or armpits, some of which grew as large as a common apple, others as an egg . . . from the two said parts of the body it soon began to propagate and spread itself in all directions; after which the form changed, black spots . . . making their appearance in many cases on the arm or thigh or elsewhere, now few and large, then minute and numerous . . . almost all within three days from the appearance of said symptoms, sooner or later, died . . . such as were left alive inclining almost all of them to shun and abhor all contact with the sick and all that belonged to them, thinking thereby to make their own health secure . . . they banded together and formed communities in houses where there were no sick, and lived a separate and secluded life, which they regulated with the utmost care . . . eating and drinking moderately of the finest wines, holding converse with none but one another . . . diverting their minds with music and, other delights as they could devise.

The contagious nature of plague led to the belief that the only way security could be achieved was total isolation of the sick. Despite the fact that it was clearly recognized that the enemy was the sick, it was the lack of appreciation that microbes were the cause of infectious diseases that led to the institution of crude and generally ineffectual public heath measures. In 1374 the Venetian Republic required that all ships, their crew, passengers, and cargo had to remain on board for 40 days while tied up at the dock; this gave rise to the term "quarantine" (from the Italian word *quaranta* meaning "forty"). Soon all European ports restricted entry by quarantine, but the disease continued unabated because the rats left the ships by using the docking lines. Cordon sanitaires (literally a ring of armed soldiers ordered to guard against the fugitives of disease) restricted the movement of people and may have reduced the spread of plague, but oftentimes the infected individuals were shut up in their homes with the uninfected members of the family and the flea-infested rats, leading to higher mortality. More effective measures included the burning of clothing and bedding and the burying of the dead in shallow unmarked graves sprinkled with lye. The public, unable to identify the real source of the plague, used "outsiders" as scapegoats. The Black Death also led to societal and religious changes: feudal structure began to break down; the laboring class became more mobile; merchants and craftsmen became more powerful; and guild structures were strengthened. There was also a decline in papal authority, and people lost faith in a Christian church that was powerless to stem the tide of death. The horrors of the plague during this time are depicted in Pieter Brughel's 1562 painting *Triumph of Death* (Fig. 4.3) and are graphically described in the introduction to Boccaccio's classic collection of short stories, *The Decameron*, and in Albert Camus' book *La Peste*. It is even depicted in popular movies such as *The Seventh Seal* by Ingmar Bergmann,

Figure 4.3   Detail from *Triumph of Death* by Pieter Brughel (1562), The Prado, Madrid, Spain. (Courtesy of the Wellcome Library of Medicine.)

Monty Python and the Holy Grail, and the movie based on the book *Arrowsmith* by Sinclair Lewis.

Though the Black Death was undoubtedly the most dramatic outbreak of bubonic plague ever visited upon Europe, it did not disappear altogether. Between 1347 and 1722, plague epidemics occurred in Europe at infrequent intervals, without the introduction from caravans coming from Asia. In England, the epidemics occurred at 2- to 5-year intervals between 1361 and 1480. Half of the population of Milan died in 1630, 60% of the population of Genoa died in 1656 to 1657, and 30% of the population of Marseilles died in 1720. In the Great Plague of 1665, which was described in the diary of Samuel Pepys (and fictionalized in Daniel Defoe's *Journal of the Plague Year*), at least 70,000 Londoners died, out of a population of 450,000.

The third pandemic began in the 1860s in the war-torn Yunnan region of China. Troop movements from the war in that area allowed it to spread to the southern coast of China. Plague-infected rodents, now assisted by modern steamships and railways, quickly spread the disease to the rest of the world. In the three pandemics, it is estimated that rat-borne bubonic plague killed more than 200 million people.

A Look Back

At the beginning of the 12th century, the European population grew quickly after achieving relatively stable numbers during the Dark Ages. One of the reasons for the unchanging numbers of people was poor harvests and famine, but with the introduction of new crops, windmills, waterwheels, horse collars, and the mould board plow, agricultural production increased. Money instead of barter began to be used in trade, and with this prosperity new towns grew and so did the population. But by the 13th century, there was ushered in a sustained period of cold winters and rainy summers. Agriculture could not keep up with the population rise, and over the next century famines occurred every few years. The result was poverty and misery, especially in the crowded and filthy cities.

The tens of thousands of people now living in the dirty, garbage-laden cities had contact with a resident black rat population that lived in the thatch of the roofs. Rats and humans lived in close quarters not only in the cities, but also in the ships that traveled between them. As the human population soared, so did the rat population: it was the humans who provided food and shelter for these rodents. Black rats, highly susceptible to plague, became infected, and with a dying-off of rodents, humans became victims. By 1347 there were outbreaks in the port cities of Kaffa (present-day Feodosia), Constantinople (present-day Istanbul), and Genoa, and, continuing to move by ship, plague spread via North Africa to Spain by 1348. By this time, it had already reached central Europe, including France, Germany, Switzerland, and Austria, and even got as far as Great Britain. From there it traveled to Scandinavia aboard a ship from London that docked in Bergen in 1349 carrying crew, wool cargo, and plague. By 1351, it was in Poland, and when in 1352 it reached Russia, plague had completed its circuit, and a deadly noose had been secured around Europe.

In 1358, Agnolo di Tura, a citizen of Siena, described the situation:

> Father abandoned child; wife, husband; one brother, another; for this illness seemed to strike through the breath and sight. And so they died. And no one could be found to bury the dead for money or for friendship . . . and in many places in Siena great pits were dug and piled deep with huge heaps of dead. . . . And I, Agnolo di Tura, called the Fat, buried my five children with my own hands and so did many others likewise. And there were also so many dead throughout the city who were so sparsely covered with earth that the dogs dragged them forth and devoured their bodies.

The short-term effects of plague were shock and fear. Panic and fright broke the continuity of the existing economic structure and disrupted the routines of work and service. People were terrified, and they deserted the cities and towns. In consequence, communities were thrown into chaos. But the Black Death also had its positive aspects: it contributed to technological advances through the invention of labor-saving devices, as well as a reevaluation of Galenic medicine, and it also led to a more diversified economy with a redistribution of wealth.

Public health

Deaths from plague radically reduced the average life expectancy from 35 or 40 years to 20. The contagious nature of plague led to the belief that security could be found only by total isolation of the sick. To protect the populace, boards of health were established and given sweeping powers that abrogated civil liberties. Since the sick were regarded as the enemy, they had to be segregated from the healthy. Infected individuals were removed from their homes and families and isolated in special hospitals called pest houses. Often, people were sealed up in their houses with the infected rats and left to starve to death or to die from plague. Few were ever cured or survived their time in the pest house; yet their isolation may have helped curb the spread of disease. The dead were removed and quickly buried. The burning of the clothing and bedding of the infected may also have helped. Other public health measures of a crude sort, such as quarantine, were put into play. However, even with quarantine, and with 90% of the sailors dying aboard ship, 50% of the population of Venice died.

Plague undermined confidence in local church leaders, and many people left on pilgrimages to seek salvation elsewhere. Pilgrimages increased the number of travelers on the highways, and these strangers provoked hostility; at the same time, they were the "seeds" for spreading the infection. To restrict the movements of such strangers, cordon sanitaires were established. Towns closed their gates to travelers. These measures circumscribed plague to certain towns while protecting others. Restricting the long-distance movement of people and baggage from infected cities may also have helped to reduce the spread of the black rat.

Discrimination

Restrictions such as cordon sanitaires and quarantine frequently were expanded to limit personal freedom and to identify the culprits. The medieval public wanted to know who was to blame. They found in their community likely sources—strangers, lepers, beggars, the poor, prostitutes, and the Jews. Plague aided in the spread of anti-Semitism, a practice already begun when Jews were slaughtered beginning with the First Crusade in 1096. In some Muslim countries it was the Christians who were blamed; however, in both Christian and Muslim countries the Jews were blamed. Although Jews as the largest cultural minority had been free to practice their religion by both Roman and canon law, once plague claimed the lives of the political leaders of a town, such freedom was soon abrogated. Early in 1348 the rumor arose that Jews were spreading the disease by poisoning the Christian wells. Though this appeared absurd to the kings and popes, and Pope Clement VI of Avignon (France) tried to discredit the charge by issuing a bull calling the accusations "unthinkable," Jews continued to be blamed and persecuted. By the fall of 1348, formal accusations were brought against Jews, and when court officials extracted confessions, the rumors began to be believed. Accounts of "well poisoning" by Jews spread from Avignon to Strasbourg and then to all of Germany and Poland. Riots occurred in Strasbourg, and when the members of the local government tried to protect the Jews, they were thrown out of office. With a new government in place, more than 900 Jews, about half the entire community, were rounded up, and on 14 February 1349 they were burned on the grounds of the Jewish cemetery. At Freiborg, all Jews were placed in a large wooden building and burned to death. In northern Italy, legislation was enacted to identify the suspect groups—they chose the color yellow—and Jews were required to wear a yellow Star of David on their clothing; prostitutes were also forced to wear yellow labels. (It is noteworthy that yellow was traditionally used to identify lepers and criminals.) And because Jews were prohibited from owning land, they had become merchants, bankers, financiers, moneylenders, and pawnbrokers. Although they had the knowledge for mercantile trade, they were still discriminated against, especially in Germany and France. Killing of the Jewish moneylenders and bankers had another benefit—no loans to repay. The king of Poland, Casimir, had a Jewish mistress and needed the expertise of the Jews, so they were invited to settle in Poland; as a consequence, in western Russia and Poland, Jews became a significant part of the population after the Black Death. (Their numbers would once again be reduced three centuries later by pogroms and gas chambers—latter day instruments of anti-Semitism.)

Church

In cities of 50,000 people, more than 500 died each day. Priests who gave last rites had a very high mortality, and there was a loss of faith in the clergy because they seemed so powerless to prevent death or the spread of death from disease. There was also a decline in papal authority. The Roman Catholic Church passed the responsibility for plague on to God, suggesting that this was Judgment Day, that people had sinned and so nothing could be done to reduce the suffering. However, even God's servants were not spared: all of the friars of a monastery near Avignon, and another near Marseilles, succumbed to plague.

Fear of plague also led to a greater consciousness of religion, especially the "magical religion" embodied in the cults of healer saints. These patron saints, who not only knew suffering but also had the power to heal the sick and provide comfort, usurped the veneration of God alone and set the stage for long and divisive debates over the nature of religion. One of these healer saints was St. Sebastian, and another was St. Roch (Fig. 4.4). Although it is not known whether the life of St. Roch is truth or fiction, his story comes to us in written form in 1414 from the Venetian humanist and scholar Francisco Diedo. Roch was born in Montpellier, France, in 1295, and he is said to have had a birthmark in the form of a red cross on his breast. As a young man he gave to the poor and embarked on pilgrimages to Rome. Wherever he traveled there was plague, and he was able to cure the afflicted by placing his hands on the buboes. He eventually contracted the disease, was expelled from Montpellier, and died in 1327. St. Roch is often depicted as a pilgrim carrying a purse and a staff and pointing to a bubo on his inner thigh.

Another threat to the Church was a pilgrim movement of another sort. The Brethren of the Flagellants began in eastern Europe, but its strongest bases were in Germany and France. The flagellants, sometimes as many as a thousand, marched in procession through the town; they were led by a master

Figure 4.4   St. Roch, the patron saint of those suffering from plague. The original hangs in the Galleria Dell' Academia, Venice, Italy.

carrying a banner of purple velvet and were dressed in cloth of gold. Masked and dressed in dark clothing emblazoned with a red cross, each flagellant carried a whip made of leather thongs and tipped with metal studs with which they beat their backs and chests. (This is graphically shown in Goya's painting of the flagellants as well as in Ingmar Bergmann's movie *The Seventh Seal*.) The flagellants were a counterculture to the Church, and they claimed divine authorization for their mission—to alleviate the plague. Pope Clement VI initially encouraged the flagellant movement; until 1349 the flagellants had their way in recruiting other pilgrims, but when the pope saw he could not control them, he issued a bull denouncing the movement and its practices. Eventually, the movement ceased for reasons not fully understood. However, in its time, the flagellant movement did some good: it brought about a spiritual revival, sinners confessed, robbers returned stolen property, hope was raised (albeit temporarily), and it provided a theatrical diversion. In the final analysis, however, the movement did more harm than good. Jews became the special victims of the flagellants, and their persecutions were the forerunners of the pogroms. In Frankfurt in 1349, the flagellants rushed into the Jewish quarter and incited the people to engage in wholesale slaughter. In Brussels, the mere announcement of their arrival triggered a massacre of 600 Jews. Death from the Black Death itself, coupled with that due to virulent anti-Semitism, virtually wiped out Jewish communities in many parts of Europe and also led to permanent shifts in their populations to Poland and Lithuania.

Medicine

Medicine was also affected. Medieval society had four kinds of medical practitioners: academic physicians, who knew theory but did not care for the sick; surgeons, who learned their trade as apprentices and who were the principal caregivers of the sick; barbers, who did bloodletting and minor surgery; and those who practiced folk medicine, mostly women. Academic physicians were generally older men who relied on the teachings of Galen, who believed, as did Hippocrates before him, that disease was a result of an imbalance in the four humors of the body (blood, phlegm, and black and yellow bile). However, since plague appeared to have little to do with humoral changes, confidence in the academic physicians diminished. When these holders of the chairs of medicine at the great European universities died, the newer and younger appointees moved into other clinical areas, such as anatomy. Surgeons, who wore a costume with a beak containing perfume or spices, a cloak of waxy leather, eye lenses, and a wand with incense (Fig. 4.5), died at higher rates than did the other medical practitioners. Because of this, their role in curing disease was little valued. Indeed, the stench of death was so great during the plague years that to "purify" the air, the perfume *eau de cologne* was invented in Germany and named after the city of Cologne. Today the perfume is known as "4711," the address of the household where it was first

Figure 4.5 Doctor Pestis, the plague doctor, in costume. (Courtesy of the Wellcome Library of Medicine.)

made. New prestige fell to the barbers, and bloodletting and surgery (Fig. 4.6) became an integral part of their practice rather than barbering alone. This also led to an emphasis on studies of human anatomy in health and disease, and the Galenic system, which had no clear theory of contagion, declined in importance. Slowly—very slowly—change occurred both in the ideas and in the practice of medicine, and it was the Black Death that instigated that change.

One of the first physicians to advance a theory of contagion was Girolamo Fracastoro (1478–1553) in his book *On Contagion and Contagious Diseases*. Infectious disease, Fracastoro wrote, could be transmitted by semanaria ("germs") in three ways—by direct contact, through carriers such as dirty linen, and through airborne transmission. His theory was put into practice when as physician to Pope Paul III he recommended the transfer of the Council of Trent from Trent to Bologna as a response to plague. Other physicians, however, did not subscribe to Fracastoro's theory, and soon the practices were displaced by misguided suggestions until they were revived in the 19th century by Louis Pasteur and Robert Koch and their associates.

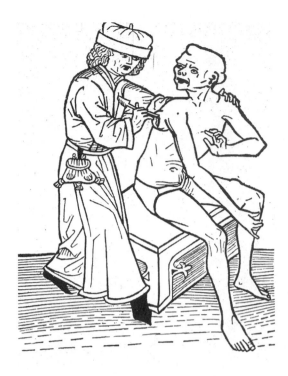

Figure 4.6 A woodcut showing a barber-surgeon as he lances a bubo.

Education

As the death toll from the plague increased, the number of learned individuals decreased. This affected the universities where lawyers, physicians, and clerics were trained. As the number of university students declined, so too did the number of universities. Before 1348, all of Europe had 30 universities, but by the time the plague ended, 5 of these had been wiped out completely. The institution of cordon sanitaires as well as other restrictions on travel prevented students from enrolling at distant universities, and so local universities were established. Cambridge University in England acquired four new colleges, each founded by a bequest of a rich and pious patron and all benefiting poor local scholars and clerics. Similarly, new universities were established throughout Europe in cities such as Vienna, Prague, and Heidelberg, so that it no longer became necessary to travel to Bologna or Paris for an education. This not only diminished the dominance of certain centers of learning, but it also led to curriculum reform, and instruction began to be carried out in the vernacular tongue.

Economy and social order

The immediate effect of the Black Death was paralysis. Trade ceased. The economic effect of the Black Death was inflation and a sharp rise in the cost of food. The Hundred Years War between England and France was halted by a truce in 1349 because of the lack of able-bodied men. In depopulated Europe, even soldiers received higher wages. Another innovation to compensate for depopulated armies was the development of improved firearms; fewer soldiers but with better and more destructive weapons could then successfully engage the enemy in battle.

The nature of farming was altered, too. In medieval times, farming involved the land of lords worked by their serfs. The serfs and their families occupied a small village consisting of several small white-washed two-room huts made of reeds and claylike sod with a thatched roof. In one room lived the family, and in the neighboring room lived their animals. The floor was dirt or covered with leaves or rushes, and in the center were flat stones on which the fire was placed. There was no chimney and no windows, so the smoke of the fire escaped through a hole in the roof or through the open door. Furniture, such as it was, consisted of a table, a few stools, a storage chest, and wooden pallets that served as beds. Adjacent to each hut was a half acre of land used by the serf and his family for the garden; there was a chicken coop and a pigpen, and the nearby stream had ducks and geese. Manure or "night soil" was used as fertilizer, and plow animals such as horses and oxen were shared among several villages.

Medieval villages were organized around large fields that were used for growing grains such as wheat, barley, or oats. The land was cultivated by the serf and his family but owned by the lord of the land (landlord), who usually lived in a large, fortified manor house. The serfs labored for the landlord; their duties were mostly unspecified and subject to the whims of the landlord. The landlord also had almost complete legal power over these poor and illiterate peasants. This feudal system of serfs and landlords declined coincident with the substitution of money for manual services. In effect, salary began to replace labor duties, and rents were paid to the landlord. Since so many peasants had died during the Black Death, if the lord wanted to farm his land he either had to obtain more laborers (usually at a dearer cost) or rent out his land to the survivors from neighboring villages or from the now deserted towns. The government, in league with the landlords, tried to enforce service without payment of the laborers; over time, this engendered great hostility among the peasants as their numbers continued to decline owing to disease. In England, when the crown instituted a series of poll taxes, there was an uprising, called the Peasants Revolt, in 1381. The revolt failed and repressive measures were instituted, but in the end, the landlords had to capitulate because they recognized that without a labor force they would receive no income from the land. The landlords devised another scheme: they would continue to own the land, but now estate managers or stewards were hired to manage the fields and collect the rents. The stewards now lived in a grand house, and the rent-paying peasants were tenant farmers on the lord's land. However, as disease and death continued to deplete the available labor pool, the tenant farmers had to recruit and pay other peasants or landless city-folk. In this way, as the landlord's fields came to be worked by non-land-owning and rent-paying tenant farmers, the feudal system slowly began to change in character. To compensate for their loss in income (due to higher wages), the landlords continued to acquire more and more land, and the tenant farmers in turn began to use less labor-intensive farming practices. This they could accomplish by the invention of the mould board plow and through the conversion of farmland into pastureland. In England, in particular, pasturelands were devoted to sheep farming, which became so profitable that, in some regions, the growing of crops was completely replaced by the raising of sheep for wool.

Mills that once were used for grinding wheat and barley could now be diverted to spinning cloth, operating the bellows of furnaces, and sawing wood. In England, raising of sheep exceeded all other crops, and wool became the basis of prosperity. But the tenant farmers were continually being stressed by ever-increasing rents, as well as taxes imposed by the crown on the land-owning lords. This is immortalized in the nursery rhyme "Baa Baa Black Sheep": "Baa baa black sheep, have you any wool? Yes, marry, have I, three bags full: One for my master, one for my dame and one for the little boy that lives in the lane." (The hardworking peasant had to give one-third of his income to the king, "his master," and one third to the landlord, "my dame," leaving only one third for himself, "the little boy.")

The tenant farmers who had raised sheep for wool became richer and more powerful than their predecessors. By taking advantage of the anarchy caused by the land-grabbing and warring aristocracy (culminating in the War of the Roses, 1455 to 1485), the sheep raisers were able to buy up larger and larger parcels of land from the estates of the ruined lords, becoming wool barons in the process. The smaller labor force in the cities led to their having a stronger negotiating position, so there was a need to pay higher wages; as a consequence, the standard of living was improved. Many individuals, however, fled the cities and moved to the country, where the survivors were induced by landowners to harvest the crops for higher wages. The standard of living among the urban and rural populations improved, and the laboring classes became more mobile. Thus, the Black Death was a contributing factor in altering the social and economic structure in England, where, by the 16th century, landlord and serf ceased to exist. However, feudalism did linger for several more centuries in Europe.

Governments that sought to control wages sowed the seeds of discontent both in the city and on the farm. This led to what is called factor substitution—cheap land and capital were substituted for the more expensive labor. Peasants would not accept a lease unless the landlord provided additional capital in the form of oxen and seeds. In the city, tools and machines (capital) were substituted for human labor. Sometimes the higher labor costs led to technological innovation in the form of labor-saving devices. One such innovation was the invention of movable type and the printing press. As plague continued to deplete the numbers of guild members as well as the pool from which new members could be drawn, the guilds had to take on new apprentices without family connections to a particular trade. Shorter years of apprenticeship, rapid turnover in guild members, and increased recruitment of new members led to a decline in the quality of the product. Scribes who had been employed to copy sheets of manuscript and then to assemble them into a bound volume could not keep up with the demand for their services from an emerging and increasingly more literate population. A way out of this labor-intensive method of bookmaking was to find a cheaper solution. Johann Gutenberg's invention of printing using movable metal type in 1453 was one way that this was accomplished.

Another was to employ an economy of scale in sea- and ocean-going transport vessels. Bigger ships with smaller crews could remain at sea for longer periods of time and would be able to sail directly from port to port; however, this would require better ship construction, improvements in navigational instruments, and new business enterprises, such as maritime insurance to protect the investment in cargo and the ship. As a consequence, merchants such as bankers and craftsmen became more powerful. The new economy became more diversified. There was a more intensive use of capital, technological innovations became more and more important, and there was a greater redistribution of wealth. In time, the aristocracy found it had to yield power to the masses. The social and economic fabric of Europe began to be altered, and it was the Black Death that promoted the change.

Finding the Killer

As plague raged through medieval Europe, it became increasingly apparent that the disease was contagious. However, even if Fracastoro's idea of "seeds of contagion" was accepted, at this time there would still be no means to precisely identify the causative agent. Identifying the "seed" would require not only a technological innovation, but also a change in the concept of infectious (contagious) diseases. The novel concept was that disease could result from the invasion of the body by microbes or germs—the germ theory of disease—and the technological innovation, the microscope, allowed humans to actually see germs. Two schools of thought, one in France under the leadership of Louis Pasteur and the other in Germany, led by Robert Koch, were responsible for firmly grounding the germ theory, and for all of their lives these two microbe hunters remained fierce competitors.

As the plague ravaged China during the third pandemic, Pasteur dispatched Alexandre Yersin (1863–1943), a Swiss-born member of the French medical colonial corps, to Hong Kong to study and attempt to isolate the germ of plague. Yersin arrived when the epidemic was in full force and was given a small table in a dark corridor next to a patients' room where he could leave his microscope, notebook, stains, pipettes, and a few cages holding guinea pigs, mice, and rats. At first, Yersin was not permitted access to the morgue, but through connivance and bribery he was finally able to visit the morgue for a few minutes to examine a sailor who had just died from plague. Yersin punctured the bubo (Fig. 4.7A) on the dead soldier's thigh with a sterile needle and removed some fluid; he then examined the fluid under the microscope, inoculated a few guinea pigs, and sent the remainder to the Pasteur Institute in Paris. On 24 June 1894, he wrote Pasteur that the fluid was full of bacilli (Fig. 4.7B) that stained poorly with Gram's stain, i.e., they were gram negative. (Gram stain consists of a tincture of crystal violet. After bacteria are killed and washed, some lose the dye whereas others retain the purple color; the former are called gram-negative bacilli, and the latter are called gram-positive bacilli.) Yersin also wrote, "without question this is the microbe of plague." A few days later, Yersin found that the guinea pigs he had injected with the fluid that he had aspirated from the bubo had died, and their bodies swarmed with the same bacilli. Yersin was intrigued by the large number of dead rats in the streets of Hong Kong, as well as in the hospital corridors and the morgue. When he examined some of these dead rats, he found the same bacilli to be present. He correctly concluded that plague infects both rats and humans. At about the same time, Koch, convinced that plague was caused by a microbe, sent his associate, Shibasaburo Kitasato, and a large number of assistants as well as abundant equipment to find the plague germ. Kitasato did culture a bacterium from the finger of a sailor who had died from plague, but it was gram positive. Further, Kitasato was never able to prove that this bacillus could produce plague in humans or other animals. Bubonic plague, caused by a rod-shaped, gram-negative bacterium,

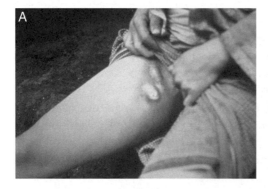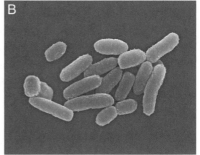

Figure 4.7   (A) The buboes of bubonic plague (courtesy of CDC) and (B) its causative agent, *Yersinia pestis* (courtesy of Dennis Kunkel Microscopy, Inc.).

was named *Yersinia pestis* after its discoverer, Alexandre Yersin (although *Pasteurella pestis* was the name used before 1970).

Yersin discovered the plague microbe but did not find the means whereby the disease could be transmitted from rat to rat or rat to human. He asked himself, was it airborne, or did it have to do with rat feces or dried rat urine? The answer to the question came from Paul-Louis Simond (1858–1947), a French army physician who was sent by Pasteur to Vietnam and India to follow up on Yersin's observations. Simond noted that not only were there large numbers of dead and dying rats in the streets and buildings, but that 20 laborers in a wool factory who had been cleaning the floor of dead rats had died of plague, but none of the other factory workers who had no contact with rats had became ill. Simond began to suspect that there must be an intermediary between a dead rat and a human, and that intermediary might be the flea. He found that healthy rats groomed themselves and had few fleas, but sick rats unable to groom their fur had many; when the rats died, the fleas moved off onto other healthy rats or humans. To prove this, he did an experiment: he placed a sick rat at the bottom of a jar and suspended a healthy rat in a wire mesh cage above it. Although the healthy rat had no direct contact with the plague-infected one, it did become infected by exposure to its fleas (which he determined could jump as high as 4 inches without any difficulty). As a control, Simond placed a sick rat without fleas together with healthy rats in a jar; none of the healthy rats became sick, but when he introduced fleas into the jar, they developed plague and died. On 2 June 1898, he wrote Pasteur that the problem of plague transmission had been solved.

The Disease of Plague

At present, most human cases of the plague are of the bubonic form. Infection results from the bite of a flea, usually the common rat flea, *Xenopsylla cheopis* (see Fig. 1.3C), that previously fed on an infected rodent. The bacteria spread to the lymph nodes (armpits and neck but frequently the area of the groin), which drain the site of the bite, and these swollen and tender lymph nodes give the classic sign of bubonic plague, the bubo. Three days after the buboes appear, there is a high fever, the individual becomes delirious, and hemorrhages in the skin result in black splotches. Some contend that these dark spots on the skin gave the disease the name Black Death, whereas others believe "black" is simply a mistranslation of "pestis atra" meaning, not black, but a "terrible" or "deadly" disease.

The buboes continue to enlarge, sometimes reaching the size of a hen's egg; when these buboes burst, there is agonizing pain. Death can come 2 to 4 days after the onset of symptoms. However, in some cases, the bacteria enter the bloodstream. This second form of the disease, which may occur without the development of buboes, is called septicemic plague. Septicemic plague is characterized by fever, chills, headache, malaise, massive hemorrhaging, and death. Septicemic plague has a higher mortality than does bubonic plague. In still other instances, the bacteria move via the bloodstream to the alveolar spaces of the lungs, leading to a suppurating pneumonia or pneumonic plague. Pneumonic plague, the only form of the disease that allows for human-to-human transmission, is characterized by watery and sometimes bloody sputum containing live bacteria. Coughing and spitting produce airborne droplets laden with the highly infectious bacteria, and by inhalation others may become infected. Pneumonic plague is the rapidly fatal form of the disease, and death can occur within 24 hours of exposure. It is likely that this form of transmission produced the devastating Black Death.

Some have suggested that the nursery rhyme "Ring around the rosies, a pocket full of posies, Achoo! Achoo! We all fall down" refers to bubonic plague in 17th century England: "rosies" refers to the pink body rash, "posies" are the perfumed bunches of flowers used to ward off the stench of death, "achoo" is the coughing and sneezing, and death is signified by "we all fall down." However, since the first time this rhyme was suggested to be plague related was in 1961, and because there is no evidence of a version prior to 1881, it is unlikely that the rhyme relates to the plague.

Y. pestis is one of the most pathogenic bacteria: the lethal dose that will kill 50% of exposed mice is a single bacterium that is injected intravenously. Typically, *Y. pestis* is spread from rodent to rodent by flea bites, but it can also survive for a few days in a decaying corpse and persist for years in a frozen body.

The reasons for the high degree of pathogenicity of *Y. pestis* remain unclear; however, some virulence factors have been identified. The genes of *Y. pestis*, located on one chromosome and three plasmids (circular DNA molecules), direct the synthesis of several proteins involved in disease production. One family of surface proteins, named Yops (*Yersinia* outer proteins), prevents the bacterium from being ingested by the host's white blood cells (macrophages), thereby avoiding the immune mechanisms of the host. Another protein, a toxin, kills other host cells. Another protein is able to degrade fibrin, a material found in the blood clot, allowing the *Yersinia* to move throughout the body and escape from the clot at the site of the flea bite.

While the above description refers to the disease in humans, the disease in fleas also has a distinctive pattern. Small mammals such as urban and sylvatic (wood) rats as well as squirrels, prairie dogs, rabbits, voles, coyotes, and domestic cats are the principal hosts for *Y. pestis*. More than 80 different species of fleas are involved as plague vectors. Fleas are bloodsucking insects; when a flea bites a plague-infected host (at the bacteremic/septicemic stage), it ingests the rod-shaped bacteria. These bacteria multiply in the blood clot in the proventriculus (foregut) of the flea. This bacteria-laden clot obstructs the flea's bloodsucking apparatus; as a consequence, the flea is unable to pump blood into the midgut where it would be digested. As a result, the flea becomes hungrier, and in this ravenous state the flea bites the host repeatedly; with each bite, it regurgitates plague bacteria into the wound. In this way, infection is initiated. *Y. pestis* can also be pathogenic for the flea, and fleas with their foregut blocked rapidly starve to death. If the mammalian host dies, its body cools down, and the fleas respond by moving off the host to seek another live warm-blooded host. However, if there is an extensive die-off of rodents, then the fleas move on to less preferred hosts such as humans, and so an epidemic may begin.

The Plague-Causing Bacterium, *Y. pestis*

Y. pestis has been subdivided into three varieties, or biovars—Antiqua, Medievalis, and Orientalis. Based on epidemiological and historical records, it has been hypothesized that *Y. pestis* biovar Antiqua, presently resident in Africa, is descended from bacteria that caused the first pandemic, whereas Medievalis, resident in central Asia, is descended from bacilli that caused the second pandemic; those of the third pandemic, and currently widespread, are all Orientalis. It is believed that *Y. pestis* probably evolved during the last 1,500 to 20,000 years because of changes in social and economic factors that were themselves the result of a dramatic increase in the size of the human population, which was coincident with the development of agriculture. This increase in food supply for humans also benefited certain rodents and allowed their populations to expand as well. Increased numbers of rodents coupled with changes in their behavior triggered the evolution of virulent *Y. pestis* from the enteric, food-borne, avirulent pathogen *Yersinia pseudotuberculosis*. This required several genetic changes: a gene whose product is involved in the storage of hemin resulted in blockage of the flea's sucking apparatus and enhanced flea-mediated transmission; other gene products (phospholipase D and plasminogen activator) facilitated blood dissemination in the mammalian body and allowed for the infection of a variety of hosts by fleas.

Plague Today

The mortality and morbidity from bubonic plague have been significantly reduced. However, the disease has not been eradicated. Plague remains endemic in regions of Africa, Asia, and North and South America. From 1983 to 1997, there were 28,570 cases, with 2,331 deaths in 24 countries reported to the World Health Organization. In 1997 the total number of cases reported by 14 countries was 5,419, of which 274 were fatal. Epidemics occurred in Madagascar in 1991 and 1997; in Malawi, Zimbabwe, and India in 1994; and in Zambia and China in 1996. In contrast, there were 4 cases and 1 death in the United States in 1997.

The increase in plague in Vietnam during the Vietnam War has been attributed to deforestation that resulted in the movement of animal hosts into areas where humans lived and fought. In the United States, plague has been encountered where houses have encroached on rural areas that harbor plague-infected rodents (sylvatic plague) such as prairie dogs, ground squirrels, mice, and rats. Infection may also be acquired from domestic pets (cats and dogs) that have come in contact with prairie dogs; pets can transmit disease to humans via fleas or in the pneumonic form. In the United States, there has been an increase in the number of human cases of pneumonic plague, especially among veterinarians who have been exposed to infected cats.

Control of plague can be effected by surveying wild populations for infections, monitoring die-offs in the rodent population, making plague-infested areas known to the public, reducing the appeal of residential areas to rodents, and treating rodent burrows with cabaryl dust or bait stations to kill fleas. In some cases, rodenticides have been used.

Because of sylvatic plague and enzootic infections (i.e., an outbreak of a disease in animals other than humans), eradication cannot be achieved.

Though human disease is rare, a feverish patient who has been exposed to rodents or flea bites in plague-endemic areas should be considered a possible plague victim. The potential for the spread of pneumonic plague has been increased by air travel. Passengers who have fever, cough, or chills and come from areas where plague is endemic should be placed in isolation and, if necessary, treated. The diagnosis of plague has remained virtually unchanged since the days of Yersin: Gram staining and culture of bubo aspirates or sputum. The bacteria can also be grown in the laboratory on blood and Mac-Conkey agar.

Unless specific treatment is given, the condition of a plague-infected individual deteriorates rapidly: death can occur in 3 to 5 days. Untreated plague has a mortality of more than 50%. Today, a variety of antibiotics, including streptomycin, sulfonamide, and tetracycline, are effective against bubonic plague. Tetracycline can be used prophylactically, and chloramphenicol is used to treat plague meningitis. No antibiotic resistance has been reported. Earlier treatment was effected by using a serum that Yersin had prepared in horses after injecting them with the plague bacilli he had isolated in Hong Kong in 1894.

The first vaccine for plague was developed by Waldemar Haffkine (1860–1930). Haffkine, the son of a Jewish schoolmaster, was born and educated in Odessa, Russia. While studying at Odessa University, Haffkine came under the influence of Elie Metschnikoff (a future Nobel Prize recipient). Haffkine's efforts to combat anti-Semitism by joining the Jewish Self Defense League resulted in his arrest by the Russian authorities, but he was released after Metschnikoff's intervention. Upon completion of his degree, Haffkine attempted to find a post at the university but was denied because of his refusal to be baptized. In 1889 he emigrated to Paris, where he joined Metschnikoff at the Pasteur Institute to work on a vaccine for cholera. When plague broke out in India in 1896, Haffkine moved to Bombay, where in 1897 he prepared an effective vaccine using dead bacilli. However, in 1902, 19 people out of the thousands who had received the vaccine died; this created a distrust of Haffkine's vaccine for plague. It was later discovered that the deaths had been due to another doctor's mistake in contaminating the vaccine with tetanus bacilli. Unfortunately, Haffkine was blamed, and his career never recovered. He worked at other posts in India and retired to France in 1914. Although he visited Odessa in 1927, he found he could not adapt to the anti-Semitism that had been wrought by the Russian Revolution, and so he returned to France, where he lived until his death.

There are now two types of plague vaccines approved for use in humans. One is a formaldehyde-killed whole-cell vaccine first used in 1942, and the other is a live vaccine used in the former Soviet Union since 1939. A new subunit vaccine that uses the bacterial capsular antigens F1 and V for immunization is under development.

The Black Death that devastated Europe eventually retreated, but why it did so remains somewhat of a mystery. It may have been due to improved standards of health and hygiene (i.e., disposal of garbage), a reduced and less susceptible rat population, or the germ of plague becoming less virulent. What did it take for bubonic plague to become the Black Death? An unusual juxtaposition of circumstances: a dense human population cohabiting with infected rats, a large and susceptible rat population, death of rats, and development of pneumonic plague. The lesson the Black Death teaches us is that ignorance of the contagious agent can lead to misunderstanding, discrimination, persecution, and profound social and medical outcomes.

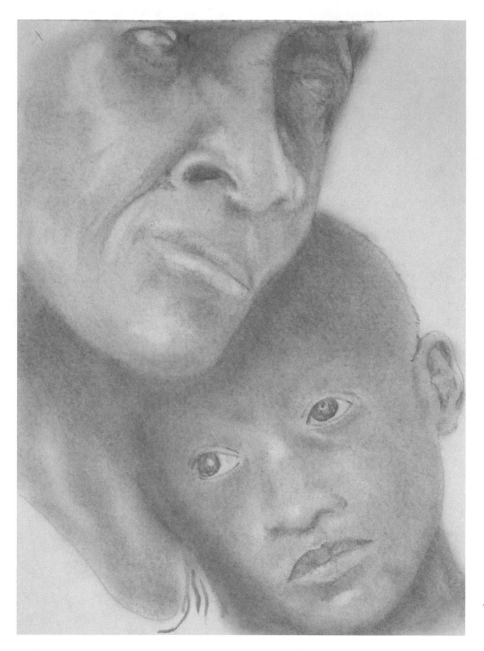

Figure 5.1   A sketch of Lorraine, aged 11, who has AIDS, comforted by her grandmother. (Based on photograph by James Nachtwey—VII Photo Agency and used with permission.)

Chapter 5

A Modern Plague, AIDS

Flesh and muscle melt from the bones of the sick in packed hospital wards and lonely bush kraals. Corpses stack up in morgues until those on top crush the identity from the faces underneath. Raw earth mounds scar the landscape, grave after grave, without name or number. Bereft children grieve for parents lost in their prime, for siblings scattered to the winds. The victims don't cry out. Doctors and obituaries do not give the killer a name. Families recoil in shame. Leaders shirk responsibility. The stubborn silence heralds victory for the disease: denial cannot keep [it] at bay.

So wrote Johanna McGreary as she described the plague that stalks the continent of Africa: AIDS. However, AIDS is not confined to Africa; it is a global disease and the first pandemic of the 21st century (Fig. 5.1). Some in society may look the other away when considering AIDS, believing that neglect will solve the problem. It won't. This terrible depredation threatens us all.

AIDS is a deadly disease for which there is no cure or vaccine. The irony is this: most of us living today, especially those under the age of 25, cannot remember a time when we had to be concerned about an outbreak of typhoid or were endangered by lockjaw (tetanus). The threat of polio and smallpox is virtually nonexistent, thanks to effective vaccines; childhood diseases such as diphtheria and whooping cough are also rare, because as children we were immunized by receiving a DPT (diphtheria, pertussis, tetanus) shot. Vaccination, improvements in sanitation and public health measures, and better nutrition and powerful antibiotic drugs have made disease the exception rather than the rule in the industrialized countries of the world. That is surely not the case with the developing countries, especially those in the tropics, where agriculture is the basis of the economy. AIDS, however, transcends economies. AIDS, according to the World Health Organization, afflicts 15 million people worldwide and occurs both in developed and less developed countries.

The magnitude of the plague of AIDS can best be appreciated by looking at the increase in the number of individuals infected: in 1981 there were

12 cases in the United States, but by 1994 there were 400,000 cases; now there may be 1 million cases. Worldwide, in 1992 it was estimated that 12.9 million were infected. Twelve years later, 30 to 40 million people were infected, and 18 million people had died from AIDS. Since 1999 there have been 15,000 new infections every day!

A Look Back

AIDS began, as most epidemics do, with a small number of cases, a fewer number of deaths, and then an almost geometric increase in the numbers of cases and deaths. In 1981, when AIDS was first described, an estimated 10,000 people worldwide were infected by the virus that causes AIDS—the virus called human immunodeficiency virus (HIV). The discovery came about in an interesting way.

In 1981, the Centers for Disease Control (CDC) began receiving an increased number of requests for pentamidine, a drug used to treat an unusual type of pneumonia. Many of these requests came from Dr. Michael Gottlieb at the University of California at Los Angeles, who began to see an increasing number of patients with this unusual pneumonia caused by a fungus, *Pneumocystis carinii*. (There are also viral and bacterial pneumonias; all cause fever and lung congestion and may lead to respiratory failure.) The pneumonia caused by *P. carinii*, called PCP, is usually found in children with leukemia or in adults with lymphomas or those on immunosuppressive drugs. Gottlieb found, however, that these were not his kind of patients. Instead, they were young, male, and homosexual. In New York City and in San Francisco, there were similar infections as well as the appearance of a rare cancer of the skin and blood vessels that was signaled by a red-purple blotching of the skin, called Kaposi's sarcoma. This sarcoma had been described in Italian and Jewish men from around the Mediterranean and was most commonly found in older individuals. However, in New York City, San Francisco, and Los Angeles, the sarcoma was common and was present in young, homosexual males. At that time the syndrome—a collection of characteristic disease symptoms—was called GRID, for *g*ay-*r*elated *i*mmune *d*eficiency.

Two years later, in 1983, two laboratories—one in France headed by Luc Montagnier and one at the National Institutes of Health in the United States headed by Robert Gallo, identified the causative agent of GRID—a virus that Gallo named HLTV III (*h*uman *l*ymphotropic *T*-cell *v*irus) and Montagnier named LAV (*l*ymph*a*denopathy *v*irus). Today, these are recognized to be the same virus. The virus was renamed *h*uman *i*mmunodeficiency *v*irus (HIV), and the disease complex it produced was called *a*cquired *i*mmunodeficiency *d*isease *s*yndrome (AIDS).

How are HIV and AIDS related to one another? HIV is a virus—an entity that one biologist defined as "a bit of bad news wrapped in protein." A virus such as HIV (Fig. 5.2) is spherical in shape, resembling a 20-sided soccer ball, but a very small soccer ball, with a diameter of only 1,000 Å (an angstrom is one-tenth of a micrometer). Put another way, if you lined up the HIV particles in single file, one inch would contain 240,000 particles. Viruses cannot be seen with the naked eye or the light microscope, but they can be seen with the electron microscope, which magnifies objects 100,000 times. Viruses are not cells, but they are able to reproduce themselves by taking over the machinery of a living cell, and in this sense, they are the ultimate parasite. The material containing the viral instructions for reproduction, that is, its genes, may be composed of one of two kinds of nucleic acid, DNA (deoxyribonucleic acid) or RNA (ribonucleic acid). The viral nucleic acid is packaged within a protein wrapper called the core, which in turn is encased in an outer virus coat or capsid; the outermost layer, called the envelope, is partly of host origin (Fig. 5.3).

The genes of most organisms are made up of DNA. The DNA, localized in the nucleus, can be thought of as a blueprint, and it is copied or transcribed into a message using a molecule of RNA, called messenger RNA; this message is then translated into a product, a protein. The transfer of genetic information can be written: DNA→RNA→protein. In the case of HIV, the genetic material is in the form of RNA, not DNA, so in order for this virus to use the

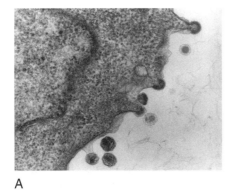

A

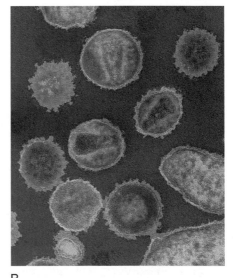

B

Figure 5.2   The appearance of the human immunodeficiency virus (HIV). (A) The transmission electron microscope reveals HIV attached to and being released from an infected lymphocyte. (Courtesy of Dennis Kunkel Microscopy, Inc.) (B) HIV magnified 140,000 times. (Photo by Oliver Meckes/Nicole Ottawa/Photo Researchers, Inc. © 2005 Photo Researchers, Inc.)

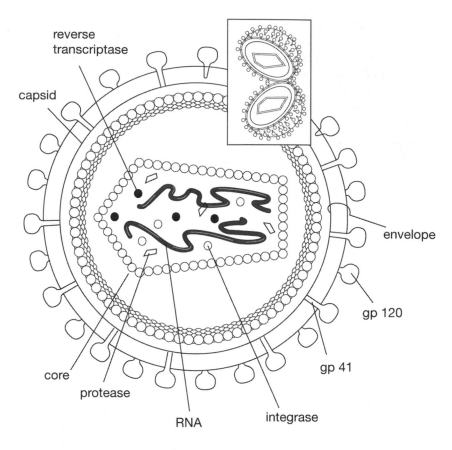

Figure 5.3   A diagrammatic view of the human immunodeficiency virus (HIV) when sliced in half.

machinery of the eukaryotic host cell (which can only copy from DNA), it must subvert the cell's machinery to copy viral RNA into DNA. As a consequence, in HIV the information is written: RNA→DNA→RNA→protein. Because the flow of information appears to be the reverse of what is typically found in cells, these viruses are called retroviruses (Fig. 5.4).

HIV discovered

As problem solvers, scientists constantly ask questions of nature. George Gaylord Simpson, a paleontologist, put it another way: "Cats may be killed by curiosity but science would die without it." More often than not, scientific questions are framed in very specific terms so that the query results in a precise answer. Biologists, as they explore the nature of living things, may ask three kinds of questions: structural, functional, and evolutionary. A structural

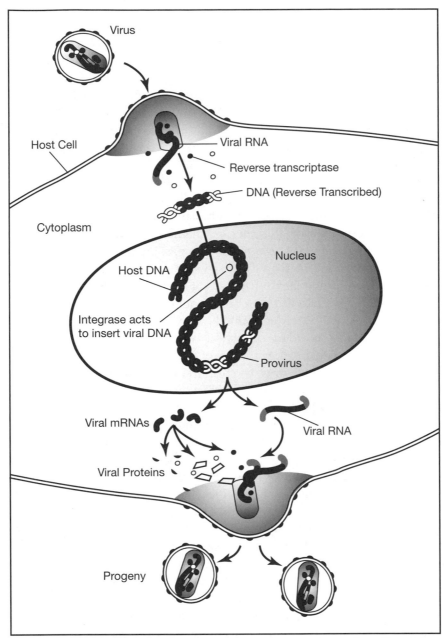

Figure 5.4   The life cycle of a retrovirus. The virus attaches to the cell, and the viral RNA is reverse-transcribed into DNA by the viral transcriptase and integrates itself (as a provirus) in the host DNA (using the integrase). There it gets transcribed into mRNAs coding for viral proteins (integrase, transcriptase, protease) and viral RNA genes. The proteins and the genomic RNA are assembled into virus progeny.

question would be, what is this "thing" made of and how is it put together? A functional question would be, how does this "thing" work? And an evolutionary question would be, how did this "thing" come about? Few biologists start out attempting to answer all three types of questions with a single exploration of the "thing"; rather, it is hoped that, through the efforts of several scientists asking many questions, doing lots of experiments and obtaining lots of information (data), the answers to all these kinds of questions will be found. In short, scientists in general, and biologists in particular, expect that discoveries from different exploratory paths will converge so that a more complete understanding of the "thing" emerges. That is indeed what happened with the "thing" we call HIV/AIDS.

The discovery of HIV began more than a century ago (1884) with the development of a porcelain filter by Charles Chamberland, who was working in the laboratory of Louis Pasteur. Chamberland's filter had very small pores, and it was possible by using this filter, as Chamberland wrote, "to have one's own pure spring water at home, simply by passing the water through the filter and removing the microbes." A few years later (1898) Dmitri Ivanowski took the mottled, dying leaves from a tobacco plant with mosaic disease, ground them up, and filtered the extract through a Chamberland filter. When the recovered sap was introduced into healthy plants, it produced the same symptoms of disease as did the unfiltered sap. Although Ivanowski believed this result was due either to the presence of a toxin elaborated by bacteria dissolved in the filtered sap, or to the bacteria of the tobacco plant passing through the pores of the filter, he had in fact discovered the virus that causes tobacco mosaic disease. Indeed, since bacteria could not possibly pass through the pores of the porcelain filter, the infectious agent had to be smaller than a bacterium; today, we know that these agents of disease, smaller than bacteria, are viruses.

In 1911, a sick chicken was brought to a young physician, Peyton Rous (1879–1970), who was working at the Rockefeller Institute in New York City. The chicken had a large and rather disgusting tumor in its breast muscle. Rous wondered what could cause such a tumor (called a sarcoma), so he took the tumor tissue, ground it with sterile sand, suspended it in a salt solution, shook it, centrifuged it to remove the sand and large particles, and filtered it through a Chamberland filter. The sap (or filtrate) was used to inoculate chickens, and several developed sarcomas within weeks. Rous examined the filtrate and the sarcoma under the light microscope and found that neither contained bacteria; he concluded that he had discovered an infectious agent capable of causing tumors and that the infective agent was smaller than a bacterium. Clearly, he had found a virus that caused a cancer. (Viruses such as Ivanowski's tobacco mosaic virus and the virus that produces "spontaneous" chicken tumors, named the Rous sarcoma virus [RSV], would only be seen after the development of the electron microscope in the 1940s.) Rous had shown that a chicken cancer could result from a virus infection, but failed to find similar virus-causing cancers in mice or humans. He received no support for his beliefs from other scientists, so the sarcoma story was regarded as a biological curiosity. As a result, Rous turned his attention to other aspects of pathology and did not return to the theme of cancer until 1934. (Rous' 1911 work was ultimately recognized by his receipt of the Nobel Prize in 1966, just 4 years before his death at age 91.)

In the 1950s it was discovered that under certain conditions viruses could cause cancers such as leukemia and other tumors in mice. At about this time it also became possible to grow a variety of eukaryotic cells in laboratory dishes (in vitro), and with these tissue cultures it became possible to study the effects of viruses on isolated cells rather than in mice or chickens. Indeed, tissue cultured cells could be transformed into cancerous ones simply by adding virus to the laboratory-grown cells. At that time it was also found that viruses were of two kinds, those that contain DNA and those that contain RNA.

Renato Dulbecco, who began to study DNA viruses in tissue culture cells in the late 1950s, found that, although some viruses caused tumors, on occasion virus could not be detected in the infected cell because its genetic material (DNA) had become inserted or integrated into the cell's chromosomes. In other words, the virus had been incorporated into the host cell genes and behaved as if it were a part of the cell's genetic apparatus. The virus DNA was now a part of the cell's heredity!

Dulbecco worked with DNA viruses, and this made it easier to visualize how the DNA of a virus could be integrated into the DNA of the host cell. However, RSV was an RNA-containing virus, so it was not obvious how the genetic information of RSV could become a part of the tumor cell's heredity. Howard Temin (1934–1994) attended Swarthmore College, majoring in biology, and then went on to Cal Tech (California Institute of Technology) to do graduate work with Renato Dulbecco; his doctoral thesis was on RSV. Temin's experiments, carried out from 1960 to 1964 at the University of Wisconsin in a laboratory located "in the basement, with a sump in . . . the tissue culture lab and with steam pipes for the entire building in . . . the biochemistry lab," convinced him that when the RSV nucleic acid was incorporated into the host cell, it acted as a "provirus." Under appropriate conditions, the provirus would trigger the cell to become cancerous.

A fellow graduate of Swarthmore College, David Baltimore (1938–), did his doctoral work at MIT (Massachusetts Institute of Technology) and at the Rockefeller Institute, where he studied the virus-specific enzymes of RSV. His first independent position was at the Salk Institute in La Jolla with Renato Dulbecco, studying RSV in tissue culture. When he returned to MIT as a faculty member, he continued to study the RSV enzymes. In 1970, Temin and Baltimore simultaneously showed that a specific enzyme, reverse transcriptase, in RSV was able to make a DNA copy from the virus's RNA. They independently went on to show that the replication of RNA viruses involves the transfer of information from the viral DNA copy, and that the viral DNA was integrated into the chromosomes of the transformed cancer cells. Later, other investigators were able to show that purified DNA from the transformed cancer cell, when introduced into normal cells, caused the production of new RNA tumor viruses. Clearly, RSV was a cancer-causing retrovirus. For their discoveries, Dulbecco, Temin, and Baltimore shared the 1975 Nobel Prize.

This was the setting when AIDS was described in 1981. Two years later, Robert Gallo and Luc Montagnier identified the retrovirus responsible for the destruction of lymphocytes (HIV). Upon virus replication, the host immune system was crippled, leading to the clinical syndrome of AIDS. Treatments for HIV infections were not available for several years after HIV was identified, but through the efforts of George Hitchings and Gertrude Elion, who directed an antiviral program at Burroughs-Wellcome, all that would change. Hitchings and Elion were chemists-pharmacologists who had produced active compounds against the smallpox virus. Some of these were sent to the Sloan-Kettering Institute in New York to be screened for anticancer activity as well. In 1947 they had an active compound, 2,6-diamino purine that resulted in remission of acute leukemia. This encouraged the search for other antitumor agents, and by 1948 they had several compounds—analogs of nucleic acid bases—that could interrupt the rampant replication of cancer cells.

When the pandemic of AIDS began, Burroughs-Wellcome already had on hand a wide array of antitumor compounds. One was acyclovir, or AZT, developed in 1964. It wasn't long before it became the first-line treatment for AIDS (in the early 1980s). Hitchings and Elion received the Nobel Prize in 1988. Thus, the paths of discovery from pathology, virology, and pharmacology converged to allow an understanding of the causes of AIDS, the way in which HIV replicates, and how retrovirus multiplication could be blocked.

HIV's Target: the Immune System

To understand how HIV is able to cause AIDS, it is necessary to digress briefly to discuss our immune system. Blood consists of cells suspended in a yellowish, salt- and protein-containing fluid called plasma (after the blood has clotted, the fluid is called serum). There are two kinds of cells in our blood: red blood cells and white blood cells. The red cells contain hemoglobin (responsible for the red color of blood) and are involved in the transport of oxygen to and removal of carbon dioxide from the tissues. The white blood cells have a different function; they are a part of the body's defense system and play a key role in immunity. In our bodies there are about a trillion white cells, of five sorts. There are three kinds of granule-bearing white cells, called eosinophils, basophils, and neutrophils, and there are two kinds that lack granules in the cytoplasm, called lymphocytes and monocytes (macrophages). The lymphocytes and macrophages are produced in the bone marrow but are found in regional centers such as the spleen and lymph nodes as well as in the blood. (Recall that in bubonic plague the bacteria accumulate in the lymph nodes of the inguinal or groin area, and when these become inflamed and enlarged they are called "buboes," from the Latin meaning "inguinal.")

Lymphocytes are divided into two different types, called T and B. The B lymphocytes make antibody either on their own or by being activated by a T-helper cell, called T4. The T4 lymphocytes are so named because they have on their surface a receptor molecule, CD4. Macrophages also have CD4 on their surface. Another kind of T lymphocyte, called a killer cell (also referred to as a T8 lymphocyte), is also activated by the T4-helper cell. T cells, unlike B cells, do not make antibody, but they are involved in what is referred to as cell-mediated immunity. Cell-mediated immunity is the immunity that is responsible for transplant rejection and for delayed hypersensitivity reactions such as those that occur after being stung by a bee. Although T cells do not make antibody, they can communicate with one another using soluble chemicals called chemokines. Chemokines are soluble chemical messengers that attract or activate other white cells, especially T and B lymphocytes and macrophages. To be activated, these white cells must have on their surface a receptor (analogous to a docking station) for the chemokine. One chemokine receptor, called alpha (α), is an entry cofactor for the invasion of T cells by HIV, and it also activates neutrophils. Another chemokine receptor, called beta (β), is an entry factor for the invasion of macrophages by HIV, and it activates monocytes, lymphocytes, basophils, and eosinophils.

Now let us look more closely at the T lymphocytes—the cells that can be infected with HIV—and how the virus and the host cell interact with one another. The glycoproteins of the HIV capsid resemble lollipops: the "stick" is called gp41, and the "candy ball" is gp120 (Fig. 5.3). The functions of these two viral proteins are to bind and anchor the HIV to the surface of the cell. CD4 on the surface of the T cell allows for the docking of gp120; once docked, the gp120 changes its shape so that it can bind to the chemokine receptor, and after binding there is fusion and entry of HIV. So, if a person has a mutation in the chemokine receptor, that individual would not be susceptible to HIV because the virus cannot enter the T cell. Only 1% of Caucasians of western European descent have this trait; the trait is absent from western and central Africans and Asians.

How and when did this mutation arise? One explanation begins with a study carried out in the small English village of Eyam. Each year on the last Sunday in August, the villagers of Eyam celebrate Plague Sunday. The service commemorates the 90 survivors (out of a population of 350) who lived through the bubonic plague epidemic of 1665. Their descendents, who lived to tell the tale of the horrors of this plague, provide the basis for a most-interesting hypothesis about those who are able to resist the modern plague of AIDS.

In August 1665, bubonic plague arrived in Eyam, not by silk caravan traders, but through a parcel of cloth that the tailor George Vicars received from London, where plague was already raging. The cloth was damp, so Vicars hung it in front of the fire to dry out; in the process, plague-infected fleas were released. On 7 September, Vicars died from bubonic plague, and from him the disease spread to the entire village. To minimize further spread of the disease, the villagers of Eyam established a cordon sanitaire and no one was permitted to leave or enter the confines of the village. To minimize the spread of infection, food, medicine, and other supplies were left outside the cordon by people from a plague-free neighboring village, and the residents of Eyam paid for all goods with money washed in the clean well water of the village or left in vinegar-soaked holes. Elizabeth Hancock's husband and six children died a week after Vicars, and the other villagers' fear of contracting the disease forced her to bury the bodies by herself; it is believed that the Hancock family contracted the disease from Elizabeth when she helped others suffering from plague. The plague raged for 14 months, but by 1 November 1666 it had run its course and its last victim had died. But the strangest part of the story was that several of the survivors, including Elizabeth Hancock, had been exposed to the plague bacilli, but they suffered no illness.

In 1996, Stephen O'Brien of the National Institutes of Health examined the parish register of Eyam to determine who had lived and who had died during the plague of 1665. More important, he was able to locate the descendents of the survivors. When the DNA of the descendents was looked at, it was found that they had inherited from their relatives a gene mutation in the chemokine receptor called delta-32. The mutant gene was approximately 10 times more common in Eyam than in other European populations, and was not found in native Africans, Indians, or East Asians. In fact, the frequency of the mutation (14%) was matched only in certain regions of Europe that had been directly affected by the plague or in places in the United States and Canada that had been settled, for the most part, by European plague survivors or their descendents.

O'Brien hypothesized that the delta-32 mutation had protected the people of Eyam from the bubonic plague; however, a more recent study questions this and suggests an alternative hypothesis. According to this hypothesis, the delta-32 mutation arose some time in the distant past among those living in northeastern Europe. Latvia, Lithuania, Estonia, White Russia, and the Ukraine were known as the Pale of the Settlement because they formed the area to which many countries attempted to confine the influx of Jews. This region was, so to speak, in the middle of nowhere, and even today, when something is said to be "beyond the pale," it means "far out." The delta-32 mutation, which alters the β chemokine receptor (CCR5), may have protected individuals not against bubonic plague, but against another plague, smallpox. Indeed, we know that the smallpox virus requires CCR5 to attach to cells so it can invade them. So, the hypothesis states, during the 17th and 18th centuries, those who lacked the delta-32 mutation died from smallpox, and the survivors (who were protected) reproduced and passed the mutated gene on to their descendents.

In addition to acting as a docking site for the smallpox virus, CCR5 also serves as a receptor for HIV. Because of this, the mutant form of CCR5—delta-32—can be considered the "AIDS resistance gene." People with two normal CCR5 genes can easily contract HIV; those with one copy of the mutation can be infected but will develop AIDS more slowly; HIV will never infect individuals with two copies of the delta-32 gene.

Let us now consider what happens after HIV docks on to the T4 lymphocyte (Fig. 5.4). Precisely how virus-host cell membrane fusion occurs is not well understood, but once HIV enters the cell, the capsid breaks down, its coat is lost, and virus RNA and reverse transcriptase are released. The latter begins to synthesize viral DNA from the viral RNA, and the viral DNA is transported to the nucleus of the T lymphocyte, where it is copied (transcribed) into viral RNA; this RNA leaves the nucleus, and in the cytoplasm, viral proteins are made, cut into fragments, and protein and RNA are assembled into a virus particle. Following virus assembly, the viruses leave the cell by budding, and in the process the host lymphocyte is destroyed. The released viruses then travel via the blood to various lymphoid organs. Virus replication can be quite rapid: 10^{10} viruses are produced and cleared from the bloodstream each day, and every two to three days a host cell is destroyed as a result of virus multiplication.

It is because the HIV glycoproteins have a high affinity for CD4 that this virus specifically attacks T4 lymphocytes and macrophages. Although during infection antibody to HIV is produced (because B cells react to the envelope and capsid proteins), the antibody, found in the serum, cannot eliminate the virus because it is unable to act on viruses hidden away inside the T lymphocyte. The antibody, however, can be useful in another way, as a diagnostic test for the presence of HIV. This diagnostic test for HIV became available in 1985. Therefore, if the serum contains antibody to HIV, then the person must have been infected with HIV; such an individual is said to be seropositive (meaning serum positive). Usually, an individual becomes seropositive 6 to 8 weeks after an HIV infection, but a negative test does not mean the individual has not been infected; subsequently, that individual may become seropositive.

HIV and AIDS

In a population of HIV-infected individuals receiving no therapy, an infected individual will have 10 million to 50 million HIV particles, and that amount will cause AIDS in 5 to 10 years. Approximately 90% of people infected with

HIV die within 15 years if untreated. It is the quantity of virus carried in the blood that determines how quickly death will occur. The more virus particles, the more likely AIDS will develop. The reason for this is that the more HIV there is in the blood the faster will be the destruction of the immune system and the quicker will be the development of opportunistic infections.

One of the characteristics of an HIV infection is depletion of T4 cells. Other than direct lysis to release virus, it is not known exactly how the virus depletes T4 cells. In a healthy individual there are ~1,000 T4 cells/mm^3 of blood, and in an HIV-infected individual the number declines by about 40 to 80 T4 cells/year (Fig. 5.5). When the count reaches 400 to 800/mm^3, the first opportunistic infections appear. Opportunistic infections are those that under ordinary circumstances cause no disease; however, in individuals with a weakened immune system, some parasites are able to take the opportunity afforded them by the body's crippled defenses, and the result is clinical disease. The first opportunistic diseases are annoying infections of the skin and mucous membranes, such as thrush (painful sores that usually occur in the mouth, though they can also occur on the vulva and prepuce, due to the fungus *Candida albicans*) and shingles (an infection of the peripheral nerves due to varicella-zoster virus, the causative agent of chicken pox), and usually

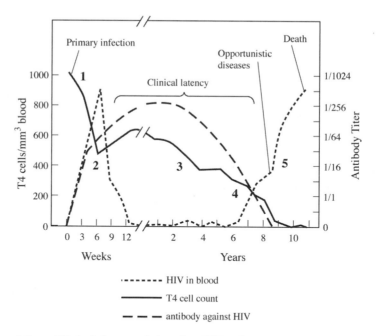

Figure 5.5   Clinical characteristics of an HIV infection. At 1, virus production occurs, and this results in fever, fatigue, diarrhea, and body aches. At 2, antibody is produced, and the number of viruses in the blood begins to decline. At 3, T4 cells decline. At 4, antibody levels decline dramatically, and T cell function is lost. At 5, virus production is re-established.

appear 1 to 3 months after infection. Other infections are severe athlete's foot and white patches on the tongue caused by the Epstein-Barr virus. Once these symptoms appear, the person is said to have ARC—AIDS-related complex. Such individuals may also have lymph-adenopathy, excessive weight loss (10–15% body weight), profuse night sweats, fevers, persistent cough, and diarrhea. Some individuals may be infected but not show ARC for years, yet such individuals are infectious and usually are seropositive.

Once the T4 count drops below 200, the seropositive individual is said to have AIDS. At this T-cell level, the AIDS-defining opportunistic infections are PCP, cryptococcal meningitis, and toxoplasmosis; these infections result in 50 to 75% of deaths from AIDS. When the T4 cell count drops below 100 T4 cells/mm^3, several other opportunistic infections result: histoplasmosis, cytomegalovirus (a herpesvirus leading to blindness and damage to lungs and digestive tract), *Mycobacterium avium* (bird tuberculosis), and chronic cold sores due to herpes simplex virus. In the immunodeficient individual—which is what an AIDS patient is—these diseases are debilitating and deadly; together they cause 90% of the deaths. In addition, some other conditions may occur: encephalopathy leading to hallucinations and dementia, motor loss, and lymphomas.

Another consequence of HIV is the development of Kaposi's sarcoma (KS). KS is due to a herpesvirus (Epstein-Barr) related to those that cause cold sores and infectious mononucleosis. The KS virus causes a cancerous growth in the cells of the blood vessels and gives rise to red-pink blotches on or in the skin. Ordinarily, KS is a minor problem except in the elderly and those living around the Mediterranean, where it can be aggressive. KS does accompany HIV infections and is seen almost exclusively in homosexual men. The reason for this distribution is that it is spread by oral-genital contact. The KS virus lies dormant in the cells of the throat, and the virus particles are most abundant in the saliva during an acute infection; this is how the virus is transmitted. But why are men a favored host? It is speculated that it might be due to a mutation on the male Y chromosome that increases susceptibility, or it could be that testosterone enhances the infection whereas estrogen reduces it.

"Catching" HIV

HIV is an infectious disease, but not a very contagious one. It is not spread very easily by fomites (inanimate objects), since the virus is killed by high temperature, detergents, and 10% chlorine bleach. It is not a vector-borne disease; indeed, if it were transmitted by bloodsucking mosquitoes or flies, it would not be a young person's disease. HIV is principally found in the secretions and body fluids of infected individuals. However, the most infectious sources are blood, breast milk, vaginal secretions, and semen, where there are sufficient concentrations of virus to cause infection. Viruses are found both free and inside T lymphocytes. A tear or lesion in the skin or mucous membranes must occur for infected fluid to initiate an infection. Infection may also result from injection of infected blood into the body, organ transplants, or ingestion of infected breast milk; virus can also cross the placenta.

There are three means of HIV transmission. The first is unprotected sexual intercourse. The most dangerous kind is anal intercourse: the rectal area is susceptible to rips and tears, thereby permitting entry of virus into the bloodstream of the recipient. Lower rates of transmission occur between an infected female and a male, because the amount of virus in vaginal secretions is much less than that in semen. HIV, however, can also be transmitted via oral sex (with cunnilingus being less efficient than fellatio) and through the sharing of HIV-contaminated blood and blood products. It can also be passed from infected mother to child owing to small ruptures in the placenta and/or via breast milk. (The predictability of HIV transmission in the population is described on p. 17; the rapidity of disease spread in a population is given by the multiplier value, R_0). The explosive spread of AIDS in some parts of the world is, in part, a reflection of the increased number of partners involved in the exchange of HIV-laden fluids.

HIV control today

Two possible approaches to controlling an infectious agent are drug treatment and vaccination. At the present time there is no drug that will cure an HIV infection. A few drugs have been shown to be effective in slowing the spread of symptoms and prolonging life. The most commonly used drug, AZT (azidothymidine or zidovudine [trade name, Retrovir]), delays the onset of AIDS by inhibiting viral multiplication. AZT was originally developed for cancer chemotherapy but was found to be too toxic; however, the U.S. Food and Drug Administration (FDA) did approve it for use in AIDS patients. In the case of a rapidly multiplying virus that copies RNA into DNA by reverse transcriptase, the incorporation of the abnormal nucleic acid base (thymidine) in AZT jams the cell's copier and, in so doing, blocks the synthesis of new virus particles. Although AZT is also incorporated into host cell DNA, it is incorporated at a much lower level than in the virus and therefore is not very toxic. However, because AZT inhibits red blood cell formation in the bone marrow, individuals on AZT do suffer from nausea, weight loss, and anemia. Other drugs that block the synthesis of viral nucleic acids, such as ddC (dideoxycytidine) and ddI (dideoxyinosine), work in a similar fashion. AZT and these other drugs, called nucleoside analogs or reverse transcriptase inhibitors, are not cheap and must be taken frequently, usually daily.

A viral enzyme that cuts up viral proteins, called a protease, is encoded by an HIV gene. Viral proteins are synthesized much like a run-on sentence: the protease cuts the long viral protein into shorter fragments, producing, in effect, separate words, so that these can be assembled into a complete virus particle. Some of the newer drugs being used for treatment of HIV target this viral protease and act to inhibit the cutting process; if this inhibition works, then short viral proteins cannot be produced, and a complete virus cannot be assembled. Protease inhibitors (indanivir, saquinavir, and ritonavir) have been combined with reverse transcriptase inhibitors to block viral replication, and the combination can reduce virus production 90 to 99%; However, this treatment can be costly, and even with the combination, drug-resistant viruses emerge. Then there is triple therapy, also called HAART (*h*ighly *a*ctive *a*nti-*r*etroviral *t*herapy), which includes, in addition to a protease inhibitor and a nucleoside reverse transcriptase inhibitor, a nonnucleoside reverse transcription inhibitor. Because AIDS takes a long time to manifest, it can be difficult to assess the effectiveness of a new compound after only a short time. As with most viruses, HIV is hard to kill without doing any damage to the host. A further problem with HIV and AIDS is that an assortment of opportunistic infections must also be treated, because it is the opportunistic infections that actually kill the patient.

Though there have been attempts to develop vaccines by using either a weakened (attenuated) virus or portions of the virus itself, neither has proved to be effective. This is because of the high mutation rate of the virus and the difficulty in developing an effective cell-mediated immune response (see chapter 10). Were such a vaccine to be developed, it would have to be administered to individuals already infected with HIV. Treatment by drug or vaccine is a two-edged sword: it may benefit the individual, but if it does not reduce that person's infectiousness, it might not significantly benefit the community.

In the absence of a vaccine and an effective cure by drug treatment, what can be done to avoid infection and reduce the spread of HIV? There are three tactics: (i) sexual abstinence, (ii) maintaining a monogamous relationship with a previously celibate or uninfected individual, and (iii) changing risky behavior, especially by those in high-risk groups such as those with many sex partners, those engaging in anal intercourse, those using contaminated needles, and those who shun the use of condoms.

The power of risk behavioral changes to control the spread of AIDS has been clearly shown in a theoretical transmission model using data from the San Francisco Young Men's Health Study. The model addressed three questions: (i) If an HIV vaccine were available, what proportion of young gay men in the community would have to be vaccinated in order to eradicate HIV? (ii) How effective would such a vaccine have to be in order to ensure eradication? (iii) What effect would changes in sexual risk behavior have if a mass vaccination program were instituted? In this model the value for R_0 (the number of secondary cases produced when one infectious case is introduced into an uninfected population; see p. 13) was estimated to be between 2 and 5, based on sexual risk behavior data such as the number of receptive anal sex partners and the use of condoms. (At the time of the study, the infection rate in young homosexual men in San Francisco was 18%.) The model predicted that, with $R_0 = 2$, 80% of those vaccinated would have to become immune to ensure eradication; that is, susceptibility to infection would have to be reduced by 95%, and immunity would have to persist with a half-life of 35 years. However, if $R_0 = 5$, then the minimum efficacy of the vaccine would have to be 80%, and coverage would have to be 100% in order to achieve HIV eradication. Since participants in the study indicated they would not participate in a vaccine trial if it were only 50% effective, it is possible that a mass vaccination program could actually increase the severity of the epidemic. In the absence of a change in risk behavior, and with $R_0 = 5$, a vaccine that was 60% efficacious could not eradicate the epidemic, and even with a vaccine of 80% efficacy, 100% coverage would be needed. However, if risk behavior were halved, eradication would be possible with a vaccine that was 60% efficacious. The theoretical model predicted that without a mass vaccination campaign and solely a reduction in risk behavior, HIV could be eradicated. For example, if $R_0 = 2$, the risk behavior levels would have to be decreased by 50%, and if $R_0 = 5$, the decrease would have to be 80%. This model calculates that, even when a highly effective vaccine does become available (and to date there is none), coverage will have to be very high in order to achieve eradication; in San Francisco it is extremely unlikely that HIV eradication could occur without significant reductions in risk behavior combined with an effective vaccine. This model predicts that eradication of HIV will require simultaneous deployment of efficacious prophylactic vaccines and behavioral interventions. Today, behavioral prevention remains the only effective way to stop the spread of HIV.

Although early in the AIDS epidemic the virus was spread via blood products (i.e., preparation of Factor VIII for hemophiliacs) and whole blood used during transfusion, since 1985 it has been possible to test for the virus. Consequently, by screening the blood supply, blood and blood products are unlikely sources of transmission. Health care workers may be at higher risk, but only if contact is made between the body fluids of an infected individual and the bloodstream of the recipient.

"Safer sex" describes those practices that minimize the transfer of those bodily secretions that contain the virus. Using a condom during oral, anal, or vaginal sex can reduce the transmission of the virus, but the condom must be used correctly. It needs to be placed on the penis after it is erect, leaving room at the tip for the ejaculate. Withdrawal should occur before the penis becomes flaccid, and the condom should never be reused. Latex condoms are perishable and can deteriorate with time and heat. Water-based lubricants such as K-Y jelly, not a petroleum-based one such as Vaseline, should be used.

A popular misconception is that condoms have a high failure rate and so there is little benefit in using them to protect against AIDS. The truth is that condoms are very effective in blocking HIV transmission, but they must be used consistently and correctly. Most of the statistics showing high failure rates relate to the use of condoms for birth control, and the failure rate is determined by the number of births when using condoms. However, these failures may be due to inconsistent or improper use of condoms and not an inherent flaw in the condom itself. At least one study shows that most failures come from a small minority of people who did not, on closer examination, follow the proper steps for condom use as outlined above.

Those who avoid the use of condoms sometimes justify this reckless behavior by claiming that HIV is so small that it can pass through the holes in a condom. These individuals know little about HIV or the molecular characteristics of the latex rubber in a condom. First of all, most virus is inside T4 cells, and these cells are larger than sperm. If sperm could pass through a condom, condoms would be useless for contraception. Even the free virus does not pass through the condom wall, because it contains millions of atoms, much larger than even the molecules in water, which do not pass through the intact condom wall either. Finally, some people are concerned about condoms breaking during use. Condom failure is very rare; condoms are rigorously tested during manufacture, and where ripping does occur, it is commonly the result of improper storage or the use of petroleum-based lubricants. The best evidence for protection by condoms comes from a study of couples in Europe in which one member was HIV infected and the other was not. Among 123 couples who reported consistent condom use, none of the uninfected individuals became infected, whereas among 122 couples that used condoms inconsistently, 12 of the uninfected partners became infected. "Safer sex," including condom use, minimizes risk but does not eliminate it entirely. In a similar fashion, while driving a car we may manage to reduce the risk of accidental injury or death by wearing seat belts and driving carefully, but we cannot eliminate all threats to our existence.

The resource needs for HIV/AIDS were estimated in 2005: $9.2 billion annually would be needed to expand support for HIV control in 135 low- and middle-income countries. Of that, $4.8 billion would be used for prevention, such as education, condom use and promotion, treatment, counseling, transfusion screening, and testing; $4.4 billion would be used to provide for care and support, including HAART and orphan assistance.

HIV's Origins

Where did AIDS come from? Where was HIV before the 1980s? Most assume the epidemic began in New York City and San Francisco among the homosexual community and then moved to Europe, but newer evidence points to the first appearance of AIDS in Europe among individuals who had never been to the Americas. One case was a Norwegian sailor who appears to have become infected sometime before 1966. He developed lymphadenopathy with recurrent colds. He died in 1976 at age 29 with dementia and pneumonia. His wife and daughter also died of opportunistic infections. Blood serum samples taken from all three between 1971 and 1973 tested positive for HIV, i.e., they were seropositive. The sailor had traveled to the Cameroon in 1961 and 1962 and had been treated for gonorrhea during the trip; it is likely that he became infected at that time and subsequently passed the infection on to his wife and child. In another case, a sailor in Great Britain in 1959 had all the telltale signs of HIV; his only trip outside of Britain was to Morocco.

Even before the 1950s, there were AIDS-like cases in Europe: in 1939 in Danzig, Germany (today Gdansk, Poland); in 1941 in Switzerland; in Austria in 1942; and in Italy in 1946. By the end of the 1940s, AIDS had reached Scandinavia. In many of these cases, the infections were seen in young children who had pneumonia and died with an acute infection; i.e., PCP had developed in 2 months, rather than being the chronic disease that AIDS is today.

A well-described mini-epidemic occurred from 1955 to 1958 in a Dutch hospital, the so-called "Swedish barracks" in Limburg. This was a midwife-training school devoted to caring for premature infants whose mothers were often unwed local women. Analysis showed that 81 children had PCP, with the first signs developing in 50 days. The data are shown in Table 5.1.

The mothers showed no signs of disease (PCP), but they obviously transmitted the infection to their offspring. It is clear that by 1958 the infection had become more aggressive. It is suspected that infection was transmitted from mother to infant by blood transfusions and injections of nikethamide to improve breathing, as well as antibiotics. At the time, reusable syringes and needles were not sterilized between injections, and single-use needles and syringes were so expensive they were not used. It appears that HIV was spread at the Swedish barracks because the blood of an infected child remained in a syringe used to treat an uninfected child.

A similar scenario occurred in 1981 in the Hospital of Leiden University in The Netherlands, where it was standard for premature babies to receive intravenous nutrition and transfusions of plasma. Eleven children were subsequently diagnosed with HIV infection, and 10 years later eight of them had died. Retrospective analysis showed that all 11 had received plasma from a

Table 5.1 Cases of PCP in Limburg, 1955–1958

Year	Admissions	Cases%	Mortality
1955	230	14	2
1956	0	6	0
1957	194	28	32
1958	56	33	33
Total	480	81	

single source: a homosexual male who had donated blood when he was seropositive for HIV, but without any other symptoms. Although we do not know the precise source of the Limburg "epidemic," it is clear that in the 1960s and 1970s, with increased opportunities for transmission, the virus changed from one that caused a relatively mild infection, and from which there was a high degree of probability for recovery, to the more lethal form— and this process of enhanced virulence did not take a very long time.

HIV and the African connection

There are three types of HIV: HIV-1, discovered in 1983, has spread rapidly from many foci; HIV-2, discovered in 1985, has spread very slowly and is confined mostly to West Africa; HIV-0, discovered in 1990, is found only in the Cameroon and Gabon, and it has barely spread. HIV-1A and HIV-2 are both spread heterosexually, but HIV-1A is more virulent than HIV-2. HIV-1B dominates in Europe and the United States; it has adapted to being spread by anal intercourse. HIV-1B may have started out in Europe as a low-lethality strain (similar to HIV-2), but once it entered the emancipated gay community, it became an efficient killer. Indeed, in the first 10 years of the U.S. epidemic, 50% of all AIDS cases occurred in gay men. It is suspected that gay men introduced HIV into Haiti from the United States.

So where did HIV-1B come from? It probably came to Germany from Africa, and the most suspect region is the Cameroon, which was at one time German East Africa. In the 1880s, the Europeans partitioned Africa into many colonies. The French had West Africa; the British settled in Sierra Leone, Ghana, the Cape, and Egypt; and Germany was active in Togo and Cameroon. European colonialism caused considerable turmoil in Africa, and it was not uncommon for the Europeans to have sexual relations with the native population. We suspect that the soldiers, sailors, and travelers from West Africa to East Africa took the virus with them. By the end of World War I, when the German colonies were redistributed among the victors—the French took 80% of the Cameroon, and Britain took 20%—HIV-1 had already been seeded around Lake Victoria; in time, it would emerge as HIV-1A. HIV-1B, which caused the mini-epidemics in The Netherlands, was probably introduced into Europe in 1939 by the 300 Germans who returned to Germany from the Cameroon via Danzig. These men were part of a group of Germans who had remained in the Cameroon after World War I, but by 1939 when World War II began, they lost their land and were repatriated to Germany by ship. This virus ultimately became HIV-1B.

During the 1980s in the Uganda-Tanzania region around Lake Victoria, a new disease was described—"slim disease"—characterized by diarrhea, weight loss, and fever (Fig. 5.6). The disease appears to have come to Uganda from Tanzania by traders and soldiers. For the first 50 years (1900 to 1950), African HIV-1 appears to have been transmitted only fast enough to maintain it as an

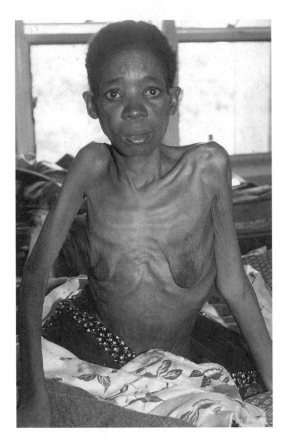

Figure 5.6   Slim disease in a 25-year-old African woman. (Courtesy of Sebastian Lucas.)

endemic disease but not fast enough to increase its virulence. However, in the late 1960s and 1970s, when new opportunities arose through heterosexual activity of traders and soldiers and prostitution, HIV-1A emerged. With its morbidity and mortality increased—typified by "slim disease"—HIV-1A moved from its center around Lake Victoria, spreading beyond Uganda-Tanzania along the truck routes. AIDS then became more and more widespread in Africa.

From **Pan** *to pandemic: HIV-1*

HIV-like viruses occur in cows, lions, horses, sheep, goats, and simians (monkeys), but in most of these natural hosts they cause little or no immunodeficiency; once they infect domestic equines and house cats, however, they do cause disease. The simian immunodeficiency viruses are called SIV. Discoveries in the 1980s in chimpanzees (*Pan troglodytes*) in the Gabon rainforest showed a similarity between the virus of the chimpanzee and HIV: it was called SIVcpz. Serum taken from infected chimpanzees or serum produced against the SIVcpz by injection reacted with HIV-1, and vice versa. SIVcpz is transmitted among chimps sexually and appears to be harmless.

The transmission of disease from animals to humans is called a zoonosis. In the case of HIV, an extremely rare zoonosis established itself as a human-to-human infection. SIVcpz from *Pan* entered the human population, or so we believe, through forest people who engage in the bush meat trade, that is, hunters who use primitive butchery methods to dismember chimpanzees and other apes, sell and eat the meat, either raw or improperly cooked, and in the process of butchery come in contact with chimpanzee body fluids such as saliva and blood. In this way, the hunters exposed themselves to the risk of a zoonotically transmitted disease. This species jump of virus from chimpanzee to human probably occurred in the eastern part of the west equatorial rainforest, that is, in the eastern part of the Cameroon, northern Congo, and the West African Republic. Once the virus had entered the human population (that is, in the hunting people), it was transmitted sexually. It has been speculated that over a period of 200 to 400 years, the virus evolved into HIV-0 and HIV-1—with the latter becoming much more virulent with time.

If the hunters of the African forest had HIV, and it caused an AIDS-like disease, why did it go unnoticed for so many years? Probably because it remained isolated in a small population and because European contact with these forest people did not take place until the 1890s. Further, within Africa a disease of this sort might not be noticed as being different from the wasting that occurs with malnutrition or infections such as malaria, yellow fever, and hookworm. Precisely when the virus spread to other human populations and became diversified remains uncertain; however, it appears to coincide with a time when socioeconomic and political changes were occurring in Africa. During the 1950s and 1960s, colonial rule in Africa ended; there were civil wars; vaccination programs involved reuse of needles and syringes; there was the growth of large cities; there was a sexual revolution; and travel within Africa and between Africa and the rest of the world increased.

The genes of HIV

HIV has in its genetic blueprint ~10,000 nucleic acid bases (adenine, uridine, cytosine, and guanine, or A, U, C, and G); that is about 1/100,000 the number of nucleic acid bases we have in each of our cells. The genes of HIV consist of a long string or strand of nucleic acid bases; built into this strand, in a coded form, is all the information needed for the manufacture of its proteins. Each gene has a particular sequence of bases in the RNA. The genetic code is a triplet code; that is, three nucleic acid bases (called a triplet) code for one amino acid. The triplets are read in a nonoverlapping sequence much as you are reading the words in this sentence. If the sequence of bases in the viral gene is AGCUUCAGA, and if A is the beginning of that gene, then the sequence would be read AGC, UUC, AGA, and each triplet would code for an amino acid. A protein results from linking the amino acids together. The sequence of bases in the RNA of HIV code for about nine genes. Several genes code for the virus core, reverse transcriptase, protease, and envelope; others control viral gene expression. One is an infectivity factor, and there is a long terminal repeat that acts like an "on" switch to allow the expression of other genes that make proteins or copy viral nucleic acids.

The linear array of nucleic acid bases that form the genes of HIV can be used to show how closely related the immunodeficiency viruses are. Let us consider the base sequences as if they were letters forming words in a sentence; if the letters of words in the sentence are similar, then the meaning is also. When the sequences of bases in the viral genes are similar, then there is greater relatedness (homology) between the viruses; conversely, when they are different, then they are less closely related.

For example, if we have a sentence that reads, "The cat ate the rat and met its . . . ," another sentence with a substitution might read, "The rat ate the rat and met its. . . . " One change in 24 letters in the second sentence gives 96% similarity. If the sentence reads, "The bat ate the cat and set his . . . , " then the similarity drops to 80%. If the sentence reads, "The bat bet the cat and got his . . . , " the similarity drops to 67%.

With a change in the sequence of letters, the meaning is changed. Similarly, by the substitution of nucleic acid bases in the gene sequence, a different protein product or function may result. Some of these changes in genetic information may have contributed to the enhanced virulence and specificity of HIV.

The sooty mangabey connection: HIV-2

HIV-1 and HIV-2 are the two major forms of HIV. There are also several SIVs. An SIV was first isolated from Japanese macaques, but there are also SIVs found in African green monkeys (gm), sooty mangabeys (sm), and African chimpanzees (cpz). The closest relative of HIV-2 is the SIV from the sooty mangabey monkey called SIVsm—indeed, there is 80 to 90% genetic homology. HIV-2 probably arose in West Africa in the rainforests of Benin. HIV-1 and HIV-2, and SIVgm and HIV-1, show less genetic homology (~40%), whereas SIVgm and HIV-2 show ~70% genetic homology, and SIVcpz and HIV-1 show ~75% genetic homology.

Further, antibodies to HIV-2 and SIVsm show these two viruses to be indistinguishable. Among West African prostitutes, 10% test positive for both SIVsm and HIV-2, but in the United States and Central Africa there is little reaction with SIVsm. HIV-2 is endemic in West Africa, where there appears to be no clinical epidemic; HIV-2 is less pathogenic, and individuals with HIV-2 are at lower risk for developing AIDS than are those infected with HIV-1.

As mentioned above, HIV-2 spreads poorly by heterosexual contact and remains pretty much confined to West Africa. If HIV-2 originally came from the sooty mangabey SIV, where did the sooty mangabey get its SIV? Perhaps from large cats harboring feline immunodeficiency virus where it was a benign disease. (In the domesticated cat, where it is transmitted in the saliva, it causes an AIDS-like disease if the cats also harbor feline leukemia virus.)

What caused the sudden and increased appearance of AIDS in the 1980s in both Africa and the United States? The emergence of AIDS is probably the result of a change in the modes and frequency of transmission of HIV. This, in turn, is related to changes in human behavior—intravenous drug use, sex with many partners, population movements, prostitution, and anal intercourse. In the 1960s, conditions in Africa also changed: there was war in Tanzania/Uganda; there were waves of prostitution; and East Lake Victoria/Northwest Zaire, the center of HIV, is on the primary truck route from West to East Africa. With greater efficiency of transmission coincident with the rise of large cities, increased road transport, increased prostitution, and the use and reuse of hypodermic syringes, the virulent form of HIV, HIV-1A, emerged. It moved to the United States via the Caribbean, where it was transmitted between gay men with multiple partners (as HIV-1A and HIV-1B), among intravenous drug users, and in the global blood market. Because other sexually transmitted diseases diminished immunity, the appearance of HIV/AIDS was enhanced.

The origins of HIV discussed above are based on the best scientific evidence available; however, other speculations on its origin do exist. Some examples follow. (i) It was genetically engineered by combining human T-cell leukemia virus and Maedi-Visna virus. (The gene sequences show this to be unlikely.) (ii) It was part of a germ warfare program. (The gene sequences of HIV resemble those of viruses in African monkeys, so it is unlikely to be an engineered virus.) (iii) It resulted from the use of monkey kidneys for growing poliovirus. (Most of the kidneys came from Asian macaques; the SIVs closest to HIV are restricted to African species. Further, the first field trials of the polio vaccine in Africa took place in 1957 to 1958, yet the first confirmed AIDS case was in a British sailor in 1959, and he had never traveled to Africa. If the sailor contracted the virus from a source who was vaccinated in 1957 or 1958, then he would hold the record for rapid progression from infection to AIDS. In addition, the monkeys used for polio culture were not from sooty mangabey monkeys but from Asian macaques or African green monkeys. African green monkeys are restricted to West Africa, which is 3,000 miles from the site of the polio vaccine trials.) (iv) During malaria research studies in Africa, blood from HIV-infected monkeys was injected into human volunteers. (This seems most unlikely, since monkeys do not have HIV-1.) (v) The Duesberg hypothesis: Peter Duesberg, a respected retrovirologist at the University of California, Berkeley, has claimed that HIV is benign and AIDS is due to the use of drugs such as poppers, cocaine and heroin, and AZT. The most telling argument against this hypothesis is a study of hemophiliacs before and after HIV infection. In one British study of 2,448 persons with severe hemophilia who were given blood products that were not screened for

HIV (i.e., before 1986), 1,227 became infected with HIV, with a death rate of 8%; in those who received HIV-negative blood (after 1985 when screening for HIV first became available), the death rate was 0.8%—the same as in the general population of hemophiliacs. Further, the death rate for those who received unscreened blood but remained seronegative was also 0.8%. It was concluded that 85% of the deaths in the seropositive hemophiliacs was due to HIV.

Duesberg also has claimed that there appear to be too few viruses in the blood to cause disease; however, he has ignored the fact that HIV resides and multiplies within regional lymph nodes, not in the blood. Further, the taking of AZT and the use of illicit drugs, on their own, do not result in AIDS. Infections with HIV fulfill Koch's postulates: in several laboratory accidents, individuals became HIV positive and subsequently developed AIDS (68 months later), but in these cases there was no administration of AZT. In Thailand, the developing AIDS epidemic showed an increase in HIV-positive individuals prior to the epidemic that appeared to be related to increased levels of prostitution and intravenous drug use. In a study of intravenous drug users in Amsterdam, CD4 counts declined only in those who were HIV positive. Clearly, there is abundant scientific evidence against Duesberg's hypothesis for the development of AIDS, and there is much evidence in favor of AIDS' being a result of infection with HIV.

The Social Context of AIDS

In 1347, infected rats and their fleas were stirred up by traders cutting caravan routes through central Asia, and they brought bubonic plague to Europe. In the space of 4 years, plague in its most virulent form, the Black Death, killed up to 30 million people.

AIDS, like the Black Death, engenders powerful conflicts. What role should the government play in promoting and protecting the public health? How do we weigh individual rights against the good of the community? What limits should be set on identifying and separating the sick from society? What is the social responsibility of the individual? Despite our greater knowledge of how parasites can cause disease and death, and our sophistication in understanding modes of transmission of infectious diseases, the social responses to epidemics, AIDS included, follow a time-honored pattern. Epidemics or plagues are the result of a complex interplay of biological and social factors; despite the uniqueness of each disease and a very different social structure from that of the past, there is a tendency for history to repeat itself. In brief, people do not react kindly to an epidemic disease, even if they understand its etiology. Indeed, it is possible to see parallels between medieval reactions to bubonic plague and some of the contemporary fears regarding AIDS. Some of this stems from a crisis in confidence in physicians, the medical establishment, and the government, and from xenophobia.

The pattern of response seen with an epidemic disease, whether its origins are known or not, is similar. At first, there is first fear and anxiety, followed by flight and denial. The next step may be a justification of the changes to be instituted: "the public must be protected at all costs" becomes the clarion call. For example, there may be requests to establish screening programs to identify carriers as well as the sick; there may be a call for public health measures to remove and isolate (quarantine) infected individuals. A politics of disease may also develop: bureaucratic agencies are established, and they begin to control resources for public health measures, diagnosis, treatment, research, and vaccine development. Emergency measures may be instituted: civil rights may be abrogated, as might the rights to privacy and freedom of movement; the right to confidentiality may be rescinded. Disease may act to buttress social divisions and focus the religious, political, and cultural biases of a society. As a consequence, even in a sophisticated society a cry goes out: find the source of infection. Because the groups with which AIDS is frequently associated within the United States—typically, those held in low esteem and discriminated against in housing, jobs, and everyday social contacts, i.e., those of different sexual persuasion, promiscuous individuals, drug addicts, and prostitutes—we may hear: "Let's get rid of the poor, the homosexuals, the drug users, the unwed mothers, the prostitutes," and so on, and so on. Then there may be a call from the public for voluntary or enforced quarantine for those infected but without clinical signs, for they could be dangerous carriers. Experience, however, shows that sustained quarantine for large numbers of people is not very successful, and in the case of AIDS, since the spread is not through casual contact or droplet infection, quarantine would be of little value in protecting the community. Once we become aware that plagues may produce specific patterns of social and political responses, then we may be better able to deal realistically and effectively with outbreaks of infectious disease. As George Santayana cautioned, "Those who do not remember the past are forced to repeat it."

A Global View of AIDS

AIDS in Africa bears little resemblance to the epidemic in the Americas and Europe. In Africa the disease has a Darwinian perversion: the fittest of society, not the weak, die. Adults in their prime perish, leaving behind the old and the young. Everyone who is sexually active is at risk. Babies are infected by their HIV-infected mothers. Most are unsure as to how they acquired the disease, but with their immune system ruined it is a certainty all will die from opportunistic infections.

The AIDS epidemic has been the worst in sub-Saharan Africa, where the prevalence is 8% in the adult population (ages 15 to 49), and there are 3.8 million new infections each year. In seven countries, all in the southern cone, the adult prevalence rate is 20%; Botswana has the highest rate (36%), followed by Swaziland, Zimbabwe, and Lesotho (25%). The driving force in these countries is the ease of mobility that comes with an extensive transport system and work migration, especially of miners. Other parts of Africa also have high rates of HIV infections: Cote d'Ivoire is among the worst-affected countries in the world, and Nigeria has a prevalence of 15%. Since the average annual income in Africa is $350, the cost of triple therapy is prohibitive. It has been estimated that to treat HIV infections in Africa would require $2 billion per year. Without treatment, those with HIV will sicken and die from AIDS, and without prevention the infection rates will not be checked. The health care systems that are needed are not in place, effective leadership is absent, and the continent continues to suffer from political turmoil.

The epidemic sweeps relentlessly across Africa; its path has shifted from those at highest risk to a generalized one and from one concentrated mainly in urban areas to rural ones as well. Transmission is mostly heterosexual. There are, however, more than 1 million HIV infections in children under the age of 15; these have been acquired from their mothers. The number of children orphaned by AIDS is now over 13 million. At the end of the 1990s, the World Bank calculated that Africa's share of world exports had dropped by nearly 60% to 1.5% (from 3.5% in 1970). This loss is equal to a drop of 21% of the region's total economic output and more than 5 times the $13 billion Africa receives in annual aid. In more personal terms, where subsistence agriculture predominates, livestock is sold to pay for funeral expenses, or orphaned and unskilled children are left to look after the livestock and cannot attend school. Communities are saturated with orphans, many of whom must fend for themselves. They sink into penury, suffer from malnutrition and psychic distress, drift into lives of crime and prostitution, and contribute to social unrest. The human losses wreck already fragile economies, break down civil societies, and lead to political instability. Because AIDS in Africa is generally met with apathy, the infection rates soar, stigma hardens, denial hastens death, and the gulf between knowledge and behavior grows larger.

In Eastern Europe, the epidemic is confined mostly to injecting drug users (IDUs), and it continues to spread rapidly as these countries experience economic and social upheaval; from the IDUs the disease is spread to their partners and to sex workers. Vietnam has a lower HIV prevalence than the United States but, as in the Eastern European countries, the problem is with IDUs. It is estimated that 130,000 HIV infections occur among the 80 million Vietnamese, and there has been little success in slowing the spread among the IDUs. These infections are now spilling over into China. In Myanmar (formerly Burma) and Thailand, the plague of AIDS is fueled by a mix of drug users and sex workers; yet, in Thailand it has been possible to stem the tide by changes in behavior and condom campaigns. Adopting a "consistent condom use" policy among brothel-based sex workers has reduced the incidence of AIDS in this group from 28% in 1996 to 13% in 2002. Myanmar has one of the worst HIV problems in Asia, and it is spread by prostitution and IDUs. Migrant workers—gem miners and loggers—provide the conduit to the general population. A political dictatorship, a lack of clinics, severe poverty, and reluctance on the part of wealthy countries to invest or offer assistance makes the disease problem worse.

These few examples show how HIV contributes to social vulnerability, reduces life expectancy, limits productivity, and stifles economic growth. AIDS exacerbates and prolongs poverty and increases malnutrition. HIV infections demand a greater proportion of the already meager annual income of those living in Asia, Africa, and Eastern Europe, thereby reducing access to food and health care. AIDS has diminished the number of available teachers, and in this way it impacts the educational system. Globally, HIV is one of the five leading causes of death. The modern plague of AIDS will continue to rise in the coming years as a result of infections that have already occurred, and it will decimate the ranks of the young men and women who are in their most productive years. The power a plague could wield was clearly demonstrated 700 years ago by the Black Death. Today, the modern plague of AIDS is a forceful reminder that the global impact of infectious disease is yet to be blunted.

Figure 6.1   Napoleon's troops in Vilna after the Russian Campaign in 1812. Engraving by Eugene Le Roux after A. Raffet. (Courtesy of the Wellcome Library of Medicine.)

Chapter 6

Typhus, a Fever Plague

Those attending a New York Philharmonic concert and listening to the cannons, the ringing of church bells, and the melody of "Les Marseilles" in Tchaikovsky's *1812 Overture* probably knew that it was composed in 1880 to mark the opening of the Moscow Exhibition of 1882 and the consecration of the Cathedral of Christ the Savior built to give thanks for the Russian victory over Napoleon in 1812. They may have also known from the program notes that the first performance was given on 20 August 1882 to celebrate the 70th anniversary of Russia's victory over Napoleon and was a great success. Although, even today, audiences continue to be impressed by the power of the work, Tchaikovsky felt little enthusiasm for it, and had it not been for a very lucrative commission, he probably would never have composed the overture. Everyone at the performance of the Philharmonic conducted by Leonard Bernstein was aware that the defeat of Napoleon, set to music, dealt with the impact of war on the people, the bravery of the Russian soldiers, and the role of the Russian winter and that it celebrated the glory of the people of Russia. Few in the audience, however, would have known that the Russians had another ally—typhus fever—and that it was this disease that contributed in a significant way to Napoleon's downfall (Fig. 6.1).

Typhus fever is sometimes called "war fever" because frequently it is a companion to hostilities. War involves overcrowding and intermixing of populations. Resources are diverted, and often famine and malnutrition increase. These so-called enabling factors in turn lead to decreases in personal hygiene and medical care and frequently a breakdown of the social structure. During wartime, individuals are subjected to increased stress, they become more susceptible to new diseases, and endemic diseases may become more severe. An invading army may introduce a disease into the defending army as well as the surrounding communities, or the endemic diseases of the community may become a source of infection for the invaders. For example, in 1741 the Austrian army surrendered Prague to the French because 30,000 soldiers died of typhus and the Austrians were unable to defend the city. Following the end of

World War I, the returning armies carried typhus with them. Between 1917 and 1923, there were 20 to 30 million cases and 3 million deaths in Eastern Europe. In 1915, an epidemic of typhus occurred in Serbia, and nearly all of the country's 400 doctors contracted the disease; more than a quarter died. In the wake of World War II, a few million people died of typhus.

Today, there is usually little in the way of epidemic disease in the industrialized countries of the world because of improvements in hygiene, but elsewhere typhus can and does surface, especially in places where there is war, famine, and poverty. In 1935, Hans Zinsser wrote in his classic work *Rats, Lice and History*, "Typhus had come to be the inevitable and expected companion of war and revolution; no encampment, no campaigning army, and no besieged city escaped it."

A Look Back

The origin of this plague, called "prison fever" or "ship fever" as well as "war fever," remains obscure, but it is generally believed that typhus was a disease of rats, transmitted by lice. Subsequently, it became a human infection. In 429 BC during the Peloponnesian War (see p. 55), Pericles, the 65-year-old leader of the Athenians, came down with a mysterious illness. His condition included bouts of high fever, chest pains, vomiting, diarrhea, fetid breath, and a bumpy red rash. His illness occurred at the same time as the Plague of Athens, and although some believe the Plague of Athens was due to smallpox, it is entirely possible that Pericles suffered with and died from typhus. Indeed, the word "typhus" comes from the Greek word *typhos*, meaning "hazy" or "smoky," and was used by the Greek physician Hippocrates to describe the confused state and stupor of the victim.

The earliest severe epidemic of typhus was in Spain between 1489 and 1490, when the forces of Queen Isabella and King Ferdinand were at war with the Moors over the possession of Granada. The disease was described as "a malignant spotted fever," and it was assumed to have been introduced by soldiers who came to Granada from the island of Cyprus. In 1490, of the 20,000 soldiers who were unaccounted for, 17,000 had died of typhus and 3,000 were killed by the Moors. A second outbreak of typhus occurred during the time of the civil war in Granada in 1557; it raged unchecked for 13 years.

Perhaps the most dramatic event involving typhus, recounted in Tolstoy's *War and Peace* and memorialized in the *1812 Overture*, was the defeat of Napoleon in Russia. In the spring of 1812, Napoleon was at the height of his power and glory. His Grand Army had experienced almost 20 years of unbroken successes. The Napoleonic Empire spread westward to the Russian frontier and Austria, and in the north, west, and south it was bounded by the North Sea, the Atlantic Ocean, and the Mediterranean Sea. Great Britain was Napoleon's most formidable enemy, and the bulk of her wealth stemmed from trade, especially with India. Napoleon wanted to control India, which was at that time a colony of Great Britain, but the naval superiority of Britain prevented the French from intercepting the trading ships from India. Further, Napoleon could not invade and conquer Britain, because the English Channel formed a barrier his armies could not cross. Therefore, Napoleon's military strategy was to take an indirect approach: conquer India and block the trade routes between India and Britain; then Britain would capitulate. But cutting off the trade routes to India had failed once before, when Lord Horatio Nelson defeated the French navy in 1798 at Abouhir Bay. Thus, it appeared that the only way for Napoleon to conquer India would be through the formation of an alliance. He decided that Russia was strategically located for a land invasion of India, and he sought to make her an ally through a treaty of cooperation. In 1807, Napoleon had defeated the Russians at Tilsit, and the two countries had entered into a treaty; but cooperation soon failed when friction arose over Poland, which the Russians wanted to retain. There were other rifts, so Napoleon decided to go for a dramatic success—the complete and utter defeat of Russia and the saving of Poland. As a consequence of Russia's defeat, Napoleon believed she would become a submissive partner and would permit France to have easy entry to India by land. From 1811 onward, Napoleon planned the invasion of Russia, and in March 1812 he signed agreements with Prussia and Austria to provide troops. Before long, the force of his Grand Army numbered 500,000 men and 1,100 cannons, whereas Russia had less than 250,000 men.

During the summer of 1812, all went well, and the Grand Army moved quickly to the border between Poland and Prussia. In 4 days, without any opposition from Russia, the troops were camped in Vilna, Poland. Poland was filthy and dirty, and the roads were so rudimentary that the column straggled, and supplies of food were slow in arriving. The French troops foraged for food in the countryside, pillaging village after village. The Polish people were infested with fleas, bedbugs, and lice, and their homes were so filthy that one could both see and hear the cockroaches as they swarmed during the day and night. The exceptionally dry summer that year also restricted the amount of water available for bathing, and because of the hostility of the Poles, the men slept together in close groups. Lice in the infested hovels crept everywhere, clinging to the seams of the soldiers' clothing; nits and head lice infested their hair. Within a few days in Vilna, several soldiers developed high fevers, and a pink rash broke out on their bodies. By the time Napoleon's army left Vilna, the troops had contracted typhus, a disease endemic in Poland and Russia. Preventive measures could not be instituted, however, since the cause of the disease and its manner of spread were not known at the time.

The French army marched eastward toward Moscow, unopposed. Indeed, the Russian strategy was to give up land to the invading army, thus drawing the French deeper and deeper into Russia. By the third week of July, at Vitebsk, 80,000 men had died, yet the French were still 150 miles from the Russian frontier and 300 miles from Moscow. As the march to Moscow continued, more and more of Napoleon's men became sicker; not only did they suffer from typhus, but they lacked provisions. By mid-August, at Smolensk, the 174,000 men in the army of Napoleon were still 200 miles from Moscow. Smolensk fell on 17 August, and the battle at Valutino followed shortly thereafter, with 6,000 French casualties. But the Russians continued to retreat. Napoleon could not face withdrawal and humiliation, so he pressed on rather than letting his army rest. On 25 August, the march resumed, and the typhus raged on; Napoleon had now lost more than 100,000 troops. With nearly half of his central column fallen, the strike force was now reduced to 160,000 men. The two armies finally met at Borodino on 5 September 1812. Napoleon had 130,000 seasoned troops, and General Kutusov, the Russian commander-in-chief, had 120,000 troops. Napoleon did not attack, possibly because he was in great pain due to a bladder infection—and a feverish cold. For 2 critical days he delayed, but after the combatants met on 7 September, there were 30,000 French and 50,000 Russian soldiers dead. It was a standoff. The Russians then withdrew beyond Moscow, and the French moved onward to Moscow. The French army suffered another 10,000 casualties due to sickness, and by the time they arrived in Moscow, only 90,000 men were in the central force. Eight out of 10 soldiers of Napoleon's army had fallen on the way to Moscow.

By the time Napoleon's troops entered Moscow, most of its citizens had taken their belongings and evacuated the city. The streets were empty. Fires, deliberately set to deprive the French of food and shelter, had destroyed three-quarters of the city, and typhus continued to ravage the army. A month later, Napoleon commanded his soldiers to leave Moscow, since there was little they could use and it would be impossible to remain there without shelter as winter approached. By the time the army left, another 10,000 men had been lost to disease and wounds. Still the Russians refused to surrender, and so Napoleon and his troops began their march back to Poland. The first snows fell on 3 November, and then the temperature dropped and it became bitterly cold. Many of the troops froze to death. The cold and snow became more intense, and by the time the Grand Army reached Smolensk to meet with the reserves, Napoleon realized that his troops had been wasted with typhus; discipline was breaking down, and food was lacking. When the French army departed the city of Smolensk, another 20,000 were sick in hospitals and living in ruined houses. Upon reaching Vilna on 8 December, another 15,000 had died, and the effective force consisted of only 20,000 men. Vilna was already crowded with the sick, and typhus was rampant. One observer described it, "Men lay on rotten straw soaked with their own excrement, without medical attention, and so hungry that they gnawed on leather or even human flesh." The French army moved on by 10 December, leaving the sick and wounded in Vilna. Typhus remained rampant through the countryside for the entire winter. Dung was burned in the streets because it was believed that smoke would drive off the disease. It did not. At the end of December, about 20,000 men had arrived in Vilna, but by the following June fewer than 3,000 were alive. Word was sent to Napoleon on 12 December that the army no longer existed. By the time the French army had crossed into Germany, there were fewer than 40,000 men, and of those, only 1,000 were ever healthy enough to be fit for duty.

It is estimated that during the Russian Campaign, 400,000 soldiers died from illness, exposure, or wounds; as many as 220,000 may have died from disease alone. The Russians also suffered from dysentery, malnutrition, and typhus, and it is estimated that at least 100,000 soldiers died from either disease or battle wounds. So ended Napoleon's fantasy; the conquest of Russia and India was defeated by cold, hunger, illness, the Russians, and his own physical deterioration. At the Battle of Beresina, on 29 November 1812, Marshal Ney of Napoleon's army wrote to his wife, "General Famine and General Winter, rather than Russian bullets have conquered the Grand Army." He left out General Typhus, the unseen ally of the Russians.

Why did typhus become such an explosive epidemic during the 1812 campaign? There were many factors. The infected peasants served as a source of infection, and their lice were effective vectors. The soldiers were stressed by lack of food and water; the scorched-earth policy of the Russians stressed the soldiers further, and this created movement of lice-infested and typhus-infected refugees. The soldiers were forced to live under crowded conditions for safety and warmth, and they had to sleep on the cold ground; clothing could not be washed or changed. It was a prescription for a military disaster.

Typhus in People

What is typhus? Before the discovery of the causative agent, the presence of typhus in prisons gave the disease the name "jail fever." Because bad smells came from the filthy and unwashed prisoners who were crowded together in jails, it was thought that "bad air" was the source of infection. English judges to this day usually wear a small bunch of sweet-smelling flowers (called nosegays) to ward off the evil odors, thereby preventing them from contracting "jail fever." Typhus is often characterized by the sudden appearance of headaches, chills, a high fever, coughing, severe muscular pain, and a blotchy rash that appears on the fifth to sixth day after infection. The rash initially appears on the upper trunk and then the entire body except, usually, on the face, palms, and soles of the feet. Delirium, confusion, and ultimately death occur. In the human body, the rupture of the cells that line the small blood vessels (capillaries) cause hemorrhages and blood clots, and red cells break down. Untreated, the disease may last up to 3 weeks, with a mortality rate of 30 to 70% under epidemic conditions. Those who recover are immune.

Finding the Killer

In 1909 Charles Nicolle (1866–1936), director of the Pasteur Institute in Tunis, was assigned the task of doing something about typhus, which was decimating the local population. He first transmitted the disease to guinea pigs by injection of blood from an infected human, proving that typhus persists in the blood of an infected individual and remains infectious. He learned how the disease was transmitted quite by accident, however. Tunis had many patients with typhus both within the hospital and on the streets; also, Nicolle observed that the clothing of typhus patients was infectious. On occasion, he found that people who worked in the hospital laundry also became infected, but once the patients were admitted to the hospital, and after they had a hot bath and were dressed in hospital clothing, they were no longer infectious. Nicolle deduced that there must be a vector in the clothes and underwear of the patients, and he suspected that the vector was a louse. He injected a chimpanzee with blood from a typhus patient and allowed the chimpanzee to come down with typhus; lice were then obtained from this infected chimpanzee and transferred to a healthy monkey. The healthy monkey came down with typhus. Nicolle recognized that the typhus agent was highly infectious and could be absorbed through the eye; so if sufficient amounts of dry fecal material from a louse were accidentally placed on the tip of the finger, it could be rubbed into the eye to cause infection. In a similar way, he observed, a person with louse droppings on the skin or clothes can become infectious.

Nicolle also solved another mystery: he noticed that typhus tended to occur in individuals who were unable to clean themselves of their body lice and louse-infested clothing. In particular, criminals lodged in the filthy jails could spread typhus from felon to judge, not by foul-smelling air but through infected lice. In September 1909, Nicolle wrote, "it is obvious that typhus is transmitted by lice." He received the 1928 Nobel Prize for his work on the transmission of typhus; however, the discovery of the causative agent of typhus was left for others.

Typhus is not the same as typhoid fever, which is a water-borne disease caused by a bacillus, *Salmonella*. Today we know the causative agent of typhus to be *Rickettsia prowazekii*, related to *Rickettsia rickettsii*, the organism that produces Rocky Mountain spotted fever. Rickettsias range in size from 0.3 to 0.6 μm and therefore are of intermediate size between viruses and bacteria. Oblong in shape and gram-negative, they are obligate intracellular parasites.

Howard Ricketts (1871–1910) described the infectious agents responsible for both Rocky Mountain spotted fever and typhus. Ricketts graduated from Northwestern University Medical School, and before taking up a position as an Assistant Professor of Pathology at the University of Chicago, he traveled to the Pasteur Institute in Paris. There he became a bona fide microbe hunter. In the spring of 1906, Ricketts went to Montana to study "spotted fever," and while examining stained specimens of blood from infected monkeys and guinea pigs, he found "a bipolar bacillus." He saw the same bacillus in humans suffering from the disease; however, despite his training in standard techniques of bacteriology, i.e., nutrient broth and blood agar, all attempts on his part to grow these bacilli failed. Ricketts did note that "spotted fever" was associated with people who had tick bites. Upon examining the salivary glands and gut of ticks that had fed on infected guinea pigs, he found the telltale microbe. But when ticks were fed on uninfected guinea pigs, no such microbes were seen. He concluded that this "spotted fever" of the Rocky Mountains was transmitted by the bite of ticks that carried the "bipolar bacilli."

In 1906, an outbreak of typhus in Mexico caught Ricketts' attention, and while there (unaware of the work of Nicolle in Tunis) he showed that lice could transmit the disease, that typhus could be transmitted to monkeys, and that after recovery the individual was immune. However, Ricketts was unable to follow up on these leads; he contracted typhus and died. (Not until 1938 did H. R. Cox discover that rickettsias were intracellular parasites that required a living cell; he found they could be grown in embryonated chicken eggs, making it possible to produce large quantities of rickettsias to be used as a phenol-killed vaccine.) A few years after Ricketts' death (1913), a Czech born in Bohemia, Stanislaus Von Prowazek (1875–1915), carried on the work of Ricketts at the Bernhard Nocht Institute in Hamburg, Germany. In collaboration with Henrique Da Rocha-Lima, a Brazilian, he studied typhus in Serbia and Turkey and found rickettsias in lice that had fed on typhus-infected patients. In 1915, while working in a German hospital facing an outbreak of typhus among Russian prisoners of war, Von Prowazek contracted typhus and died. Da Rocha-Lima described the causative agent of typhus in 1916, and he named the organism *Rickettsia prowazekii* after his fallen colleague and Ricketts.

A Lousy Business

It is human lice—including the body louse *Pediculus humanus corporis* and the head louse *Pediculus humanus capitis* (Fig. 6.2A), as well as the crab louse *Phthirus pubis* (Fig. 6.2B)—that are involved in rickettsial transmission of typhus from human to human by means of fecal contamination of the bite wound.

In days gone by, when people bathed and changed clothes less frequently and there were no insecticides, everyone had lice—from the royal

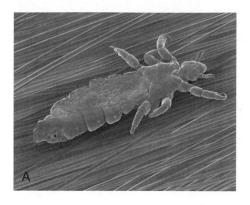

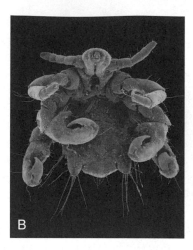

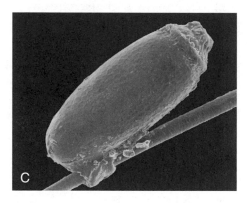

Figure 6.2 Lice. (A) An adult head louse and (B) an adult pubic (crab) louse, as seen with the scanning electron microscope. Notice the differences in the shape of the body and the claws for clinging to the hairs. (C) A nit. (Courtesy of Dennis Kunkel Microscopy, Inc.)

family to the lowly peasant. Archbishop Thomas Becket (b. 1118) was murdered in Canterbury Cathedral on 29 December 1170. The next day, he was dressed for burial in an extraordinary way. He had on a large brown mantle; under it was a white surplice, and under that was a lamb's wool coat, then another woolen coat, and a third woolen coat; under that was a black-cowled robe of the Benedictine order, and under that was a hair-cloth shirt covered with linen. As the body grew cold, the vermin that were living in the many layers of clothing started to crawl out, and it is told, "The vermin boiled over like water in a simmering cauldron, and the onlookers burst into alternate weeping and laughter."

The habit of shaving the head and wearing a wig was probably due in part to an attempt to hold down the vermin population, but even the wigs could harbor head lice and their eggs, nits. Samuel Pepys (1633–1703) wrote in his diary that he had to return to Westminster and complain to his barber that the wig he had newly purchased was so full of nits that he had to have it cleaned because they "vexed me cruelly." George Washington wrote, "Kill no vermin such as fleas and lice in the sight of others, and if you do see any carefully put your foot down on it; and if it is on the clothes of your companion put it off privately, and if it be on your own clothes, return thanks to him who put it off."

Even today, lice are found in jails, mental institutions, schools, and facilities where cleanliness is less than optimal. Lice are more frequently encountered among the crowded poor in the city. For example, in England in 1939 to 1940, preschool and school children in large cities had an incidence of 50%, but in rural areas the incidence was 5%. In Riverside, California, the incidence of head lice today in some of the public schools is 30%. Although it has been claimed that people may be infested with thousands of lice, this appears to be an exaggeration; a more accurate number would be less than a dozen. In a recent survey carried out in Ethiopia among migrants, 39% had body lice (of those, 36% had <10 body lice and 5% had >200 body lice), 65% had head lice, 10% had scabies mites, and 4% carried fleas.

The head louse and the body louse are descendants of lice that were associated with our hairy forefathers and foremothers when we lived in caves. As we became more and more hairless, the lice became restricted to the more hairy parts of the body—the pubic area and the head—and others lived in the clothing next to the skin. A person infested with dozens of body lice may remove his or her clothing and find not a single specimen on the body, but examining the underwear or clothing will reveal them on the threads or the seams.

The life cycle of the body louse begins with an egg laid in the folds of clothing (but nowhere else!). Since the body louse is susceptible to cold, the eggs are usually attached to the inner clothing close to the skin. The egg is held in place by a glue-like substance, and if the clothes are not removed, the egg will hatch in 6 to 9 days. The newly emerged louse moves on to the skin and feeds before returning to the clothing, where it remains until its next meal. Typically, a louse will feed five times a day. The growing louse will shed its outer skin (exoskeleton) three times, at 3, 5, and 10 days after hatching, and after the last shedding (called a molt), it is mature and can live for another 20 days. All stages feed on blood. An egg attached to the hair is called a nit (Fig. 6.2C). A female louse can lay 10 eggs a day and may live 30 to 40 days. Theoretically, a pair of lice can produce 200 offspring during their 1-month life span.

Lice have been recognized as parasites of humans for thousands of years. Nits are about 1 mm in length, so a fine-tooth comb is necessary to remove them; combs were probably invented to remove nits (Fig. 6.3). Nits have been found associated with a 5,000-year-old Egyptian mummy, and preserved lice have been recorded among the remains of the inhabitants of the city of Pompeii (destroyed by the volcanic eruption of Mount Vesuvius in AD 79). The nit has become a part of our language: someone who worries about petty details is called a nitpicker.

Figure 6.3   *The Louse Hunt* by Gerhard Ter Borch (1617–1681). Mauritshuis, The Hague.

The female and male lice puncture the skin and feed on blood, and the red blood cells in the gut are digested rapidly and remain liquid. The louse gut, however, is susceptible to rupture, and the louse may turn red as its gut contents spill out into the rest of the body. This is most frequently seen when the louse is infected, because the rickettsias also destroy its gut cells. Lice that are feeding have the nasty habit of defecating, and louse feces are extremely dry and powdery.

Lice do not survive if starved, and they cannot live at high or very low temperatures, preferring temperatures between 29 and 32°C—approximately the temperature in the seams of clothing close to the body. They tend to avoid temperatures above 40°C. If a host becomes too hot because of fever or exercise, infecting lice will leave. Since body lice die at 50°C, this temperature is critical when washing clothing, as soap and water on their own will not kill lice. Lice are also very susceptible to dehydration and require a humidity of

70 to 90%; death occurs below 40%. The only way the louse can maintain rehydration is by feeding on blood, and lice must feed every day or else they will die. Lice abandon the body of a dead person.

Lice become infected with rickettsias after taking a blood meal from an infected human. Several days later, the rickettsias have multiplied in the louse, and now the powdery feces contain many infective rickettsias. It is the feces of the louse that set up the infection, since feeding is irritating to the human— when the individual scratches his or her skin, the feces carrying the rickettsias are rubbed into the wound. Thus, a typhus infection is due to wound contamination, not inoculation through a louse bite. The destruction of the gut of the lice by the rickettsias results in their death 1 to 3 weeks after infection, but death does not occur before they are able to transmit the rickettsias.

Discrimination and Dissemination

Anne Frank, a German-Jewish teenager, wrote in her diary (*The Diary of Anne Frank*) how she and her family with four others were forced to go into hiding during the Holocaust. They remained hidden from the Nazis for 25 months by living in an annex of rooms above her father's business in Amsterdam, The Netherlands. After being betrayed, all were arrested and deported to Nazi concentration camps. Nine months later, Anne Frank, age 15, died of typhus at Bergen-Belsen. Suffering from malnutrition and living under the crowded and unsanitary conditions of the concentration camp, she was exposed to infected lice, contracted the disease, and succumbed to it.

Millions of other Jews died in Nazi concentration camps, not from typhus but from being deloused. At the end of World War I and before the discovery and deployment of DDT as an insecticide (1939) and a delousing agent (1943), it was a common practice to fumigate entire railroad cars and the holds of ships with steam or cyanide gas. In 1923, German chemists synthesized Zyklon-B (a porous material soaked in hydrocyanic acid with a stabilizer), ostensibly as an improvement on these methods, since it did not require tanks of cyanide gas and clothing, furs, and leather goods could be deloused without damage by steam heat. However, Zyklon-B was put to a more sinister purpose during World War II, when the Germans disguised their extermination facilities at Bergen-Belsen, Buchenwald, Dachau, and Auschwitz as delousing stations with showers, barbers, and laundries to lull the filthy, overcrowded, and lice-ridden Jews, political prisoners, and other "undesirables" into gas chambers filled with Zyklon-B. In this way, the delousing agent, Zyklon B, did not improve the public health by controlling typhus; rather, it became an instrument of death for the "final solution."

R. prowazekii can live in louse feces for up to 3 months, and Russian clothes cleaners have been reported to have acquired the infection by this route. Because body lice live in clothing, the prevalence of typhus is determined by weather, humidity, poverty, and lack of hygiene. Epidemic typhus occurs more frequently during the colder months, especially winter and early spring, since this is the time of the year when people wear multiple layers of clothing; in poor countries, it is also a time when clothing is changed infrequently. Transmission is favored by sexual promiscuity, crowded and cramped conditions, and poor personal hygiene.

Although nits may survive for up to 16 days if kept at low temperatures, leaving clothes unwashed and unworn for a week results in the death of the lice and their eggs. Historically, this may have been the basis for the weekly change of clothes on Sunday after the Saturday night bath. Eradication of the body louse can only be achieved by improving the general level of hygiene. The simplest method for control is delousing by frequent changes of clothing; if this is not possible, then washing or dry cleaning can be used. Boiling is also effective. A more rapid approach is dusting of clothing with DDT (10%), malathion (1%), or permethrin (1%). Since body lice visit the skin only to feed and otherwise spend their time in the clothing, it is unnecessary to delouse the individual. Neither bathing nor shaving is necessary for removal of adult lice, but the adults (not the nits) of head lice can be killed by an oral dose of ivermectin. Nits can be removed by combing or using a chemical remedy such as Nix, Rid, or lindane.

Typhus and the "Immigrant Problem"

On 30 January 1892, the steamship SS Massilia, carrying 750 immigrants and a crew of 50, reached New York City. Shortly after the passengers disembarked, typhus fever broke out in the city's Lower East Side. Almost all the cases occurred among the newly arrived Russian Jewish immigrants. Public health authorities described its victims as foreigners with an odor that resembled the smell of rotting straw. To contain the further spread of typhus, the health authorities established a quarantine.

Avoiding and isolating the sick has been a human response to plagues since antiquity. When Hippocrates and Galen warned that it was "dangerous to associate with those afflicted," there was a clear appreciation of the fact that diseases could travel and that the path they would take followed human migrations. However, before the 20th century, physicians could offer little in the way of treatment or prevention of epidemic diseases. One approach to disease control—even before there was an appreciation of germs as agents of disease—was the quarantine of those discovered to be sick.

The word "quarantine," derived from the Italian *quaranta giorni*, or "40 days," has its origins in response to the Black Death (see chapter 4). But why 40 days? The enforced detention of crew, passengers, and cargo for more than a month before disembarking was permitted may be based on the Hippocratic doctrine distinguishing acute and contagious diseases from those that are chronic, or it may refer to a number frequently mentioned in the Old Testament; others ascribe it to the 40 days Jesus spent alone in the wilderness. However, the most likely explanation is a practical one: during the Renaissance it was noted that after 40 days those stricken with the plague either died or recovered and were no longer contagious. The drive for establishing a quarantine tends to be greatest for those diseases for which the means of transmission is poorly understood. Because of this, any and all precautions against its spread seem reasonable. To wit: bubonic plague, SARS (severe acute respiratory syndrome), AIDS, and in 1892 the New York City typhus epidemic.

Quarantine may have had a medical rationale, but more often than not it has been used to isolate and stigmatize a particular social group. Dr. David Musto wrote, "quarantine is far more than . . . marking off or creation of a boundary to ward off a feared biological contaminant lest it penetrate a healthy population"; it has "a deeper emotional and aggressive character." It blames, shames, and separates, but "when it hits hardest at the lowest social classes or other fringe groups, it provides that grain of sand on which the pearl of moralism can form." Typhus can be that grain of sand.

In September 1891, a group of Jews began their escape from the harsh conditions and anti-Semitism in the Russian province of Volhynia. The exiled Jews boarded a steamer in Odessa that took them across the Black Sea to Constantinople, Turkey, where they remained as fugitives for 3 months. By Christmas, given expulsion orders, they traveled to Smyrna (Izmir), where some went by rail to Marseilles. There they boarded the SS Massilia, bound for New York. The voyage was interrupted by a stop in Naples, where 470 Italians, also immigrating to New York, boarded. Conditions on the ship—especially in steerage—were abominable. The SS Massilia was overcrowded and filthy. The food consisted of decaying herring, rotting potatoes, stale bread, rancid butter, and tea. Open troughs served as toilets that were only occasionally flushed with water or cleaned. Salt water was used for bathing, laundry, and the washing of plates and utensils. After 28 days at sea, the immigrants—malnourished and debilitated—arrived in the New York harbor. At a mandatory quarantine station off Staten Island, the 800 passengers and crew were inspected by two physicians for evidence of typhus, cholera, yellow fever, plague, smallpox, and leprosy. The entire process took less than an hour. Because nothing "medically remarkable" was found, the ship and its passengers were admitted to the port of New York. A medical exam at the Ellis Island immigration station (the port of first landing for more than 75% of immigrants to the United States) was similarly perfunctory, and all were given a clean bill of health. On 1 February, the 268 Russian Jews from the SS Massilia were placed by the United Hebrew Charities in eight boarding houses on New York City's Lower East Side. By 11 February there was an outbreak of typhus in this neighborhood that would spawn panic in New York City and beyond.

The social fabric of New York City in the 1890s could hardly be called the "gay nineties" celebrated in song and story. There were periods of economic depression; social upheavals coincided with the emergence of crowded cities bursting at their seams with factory workers; and there was labor unrest. Many Americans attributed these social evils to the newly arrived immigrants, especially those from Russia, Italy and Austria-Hungary. Further, during this period there were pandemics of cholera and typhus raging in Europe and Asia; with fast-moving steamships, the wide Atlantic Ocean would no longer serve as an impenetrable barrier against the introduction of epidemic diseases by infected foreigners. In addition, the number of immigrants to the United States was accelerating: between 1881 and 1884, there were 3 million immigrants, and between 1885 and 1898 that number had doubled. These new immigrants, "wretched refuse" they were called, were a threat not only to the nation's public health but also to its economic well-being. It was generally perceived that these foreigners would not only drive down wages but would soon become public charges. Another objection to immigration from Eastern Europe was the racist and anti-Semitic sentiments of "native" Americans.

As late as 1892, many New Yorkers were convinced that disease was a result of miasmas—malodorous vapors that could cause epidemics. It was "logical" (though unscientific) to propose that typhus could be spread from the foul-smelling, unkempt, and impoverished immigrants to the healthy and native-born Americans. The chief sanitary inspector of the New York City Health Department, Cyrus Edson, proposed a spontaneous generation theory for the miasma: "typhus cases could have arisen under the conditions such as existed on that ship; where the hatches have been battened down and the people were overcrowded and the conditions filthy, and they have breathed impure air for a certain length of time and had food which was not of a proper character to nourish them; these conditions would tend to breed typhus de novo." To stem this "typhus plague," Edson had to act swiftly. He ordered the sanitary policemen to find "every single Russian Jewish passenger of the Massilia." Rigid quarantine was established in the Lower East Side, but typhus continued to rage.

Edson then resorted to more stringent measures. All healthy contacts were forcibly evacuated to Riverside Hospital on North Brother Island in the East River. (Ironically, this would be the same site where Mary Mallon—"Typhoid Mary"—would be held against her will from 1915 until her death in 1938; see p. 11.) The overwhelming majority (~1,150) of those sent to North Brother Island were healthy people who were simply unfortunate enough to have resided in or near the boarding houses where the sick Massilia passengers had lived. (The Italian passengers on the Massilia who had dispersed beyond New York City were much more difficult to track down, and those who were found were neither infected nor quarantined.) Adding insult to injury, the sanitary conditions on North Brother Island were far from optimal to contain an epidemic. Outhouse privies were rarely cleaned, and because there was limited access to soap and running water, many of the detainees waded into and bathed in the East River. Bed space was at such a premium that "patients" were crowded together in flimsy tents that provided little shelter against the elements.

Hearing of the typhus epidemic in New York City, many American newspapers published articles and cartoons depicting Russian Jews as vectors of the disease. The *Boston Medical and Surgical Journal* stated, "We open our doors to squalor and filth and misery—which means typhus." A *New York Times* editorial of 13 February 1892 read, "Such immigrants are not wanted either in this city or any other part of the United States. They should be excluded. The doors should be shut against them." In late February these complaints led the port of New York to establish a policy that detained and quarantined all East European Jews coming through New York harbor regardless of their port of embarkation. Enforcement, however, was strictly along ethnic lines, since passengers from Scandinavia who traveled on the same ships as European Jews were allowed to land without delay.

Then, by 1 April 1892, the typhus epidemic seemed to be over. Among the Massilia passengers there were 138 cases and 13 deaths, and among the 1,200 people quarantined first on the Lower East Side and then on North Brother Island there were 49 cases and 6 deaths, for an overall death rate of ~12%. No new cases occurred on the Lower East Side. Those who were quarantined and recovered or did not develop typhus after 3 weeks of observation were released. Quarantine was considered to be effective: typhus had been restricted to the Lower East Side, and the disease spread no further. The *New York Times* credited Edson and the health department for "averting a pestilence," and the *New York Tribune* claimed it was "the isolation of all Jewish suspects and their stringent quarantine that was responsible for the success." But quarantine's success had exacted a social cost: civil liberties were violated, there was overt discrimination and ethnic insensitivity, and the inadequate resources provided for medical care contributed to an excessive number of deaths.

There still remains a lingering question: why was it that only the Russian Jews of the SS Massilia were infected with typhus? It appears that the infection was not acquired in Odessa, since the Jews remained healthy for the 3 months they were in Turkey. Most likely, the Jews contracted typhus in Constantinople, where the disease was endemic. Thus, the seeds of infection that were sown in Turkey blossomed in the crowded boarding houses on New York's Lower East Side and set in motion a xenophobia that would restrict the entry of East European Jews and other "undesirable" immigrants. Active anti-immigration sentiments culminated in the passage of the Immigration Restriction Act of 1924 and similar legislation enacted through the

1930s. The message here is clear: epidemic disease can often inspire quarantines, and quarantine can be the tool for social scapegoating.

Typhus Today

Typhus occurs most frequently in the mountainous areas of Ethiopia, Burundi, and Rwanda in Africa; in Peru in South America; and in Nepal and Tibet in Asia, where the climate favors the presence of body lice. A recent study has shown that two-thirds of Ethiopian students are infected with lice. In the United States there is a reservoir for typhus in flying squirrels. Some claim that epidemic typhus occurred in the Americas sometime in the 16th century and that this coincided with the Spanish importing the body louse into the Americas. It may be that after human infections occurred, typhus passed into the flying squirrel population, and then it spread from Mexico, where it was described by the Aztecs.

Typhus is frequently considered a disease of the past, but it is not. Since World War II, the largest outbreaks have occurred in Africa. In Ethiopia, the number of annual cases ranged from 7,000 to 17,000. In 1976, after a 12-year absence and following a civil war, typhus broke out in refugee camps in Burundi, and 100,000 were infected. In 1997, there was a large outbreak in Burundi, sporadic cases in northern Africa, and a smaller outbreak in Russia; in 1998 there was a small outbreak in Peru and a case in Algeria.

When symptoms such as headache, fever, and rash occur in patients with body lice or persons living in cold, crowded, and unhygienic conditions, this should suggest typhus. Though traditional methods used in bacteriology cannot be used to grow the intracellular *R. prowazekii*, the rickettsias can be grown in embryonated eggs or tissue cultures. After being grown in chick embryos or tissue cultures, the rickettsias can be stained, or they can be detected by molecular methods such as the polymerase chain reaction.

At one time, protection against typhus was achieved through vaccination. Professor Rudolf Weigl (1883–1957) and colleagues at the University of King Jan Casmir in Lviv, Poland, developed the vaccine by using material from typhus-infected louse intestines. One hundred lice were needed for a single immunization. The preparation of the vaccine itself was not without risk, and several of Weigl's staff contracted the disease and died from it. The Weigl vaccine was used in the 1920s and early 1930s, and later a phenol-killed vaccine developed by Cox in 1938 was used; however, today most patients are more efficiently treated with an antibiotic such as doxycycline, tetracycline, or chloramphenicol.

So what are the lessons of typhus now that we know its cause, its mode of transmission, how it can be controlled, and how it is treated? The words Hans Zinsser wrote 75 years ago still ring true: "Typhus is not dead. It will live on for centuries, and it will continue to break into the open whenever human stupidity and brutality give it a chance, as most likely they occasionally will. But its freedom of action is being restricted and more and more it will be confined, like other savage creatures, in the zoological gardens of controlled diseases."

Figure 7.1   A Yanomami mother watches over her sick infant child dying from cerebral malaria. (Courtesy WHO/TDR/Mark Edwards.)

Chapter 7

Malaria, Another Fever Plague

I wanted to sit up, but felt that I didn't have the strength to, that I was paralyzed. The first signal of an imminent attack is a feeling of anxiety, which comes on suddenly and for no clear reason. Something has happened to you, something bad. If you believe in spirits, you know what it is: someone has pronounced a curse, and an evil spirit has entered you, disabling you and rooting you to the ground. Hence the dullness, the weakness, the heaviness that comes over you. Everything is irritating. First and foremost, the light; you hate the light. And others are irritating—their loud voices, their revolting smell, their rough touch. But you don't have a lot of time for these repugnances and loathings. For the attack arrives quickly, sometimes quite abruptly, with few preliminaries. It is a sudden, violent onset of cold. A polar, arctic cold. Someone has taken you, naked, toasted in the hellish heat of the Sahel and the Sahara and has thrown you straight into the icy highlands of Greenland or Spitsbergen, amid the snows, winds, and blizzards. What a shock! You feel the cold in a split second, a terrifying, piercing, ghastly cold. You begin to tremble, to quake, to thrash about. You immediately recognize, however, that this is not a trembling you are familiar with from earlier experiences—when you caught cold one winter in a frost; these tremors and convulsions tossing you around are of a kind that any moment now will tear you to shreds. Trying to save yourself, you begin to beg for help. What can bring relief? The only thing that really helps is if someone covers you. But not simply throws a blanket or quilt over you. The thing you are being covered with must crush you with its weight, squeeze you, flatten you. You dream of being pulverized. You desperately long for a steamroller to pass over you. A man right after a strong attack . . . is a human rag. He lies in a puddle of sweat, he is still feverish, and he can move neither hand nor foot. Everything hurts; he is dizzy and nauseous. He is exhausted, weak, and limp. Carried by someone else, he gives the impression of having no bones and muscles. And many days must pass before he can get up on his feet again.

This is malaria as described by the Polish journalist Ryszard Kapuscinsk in his personal story of Africa, *Shadow of the Sun*. His experience is not unusual. Malaria is a fever plague, and it has been said that this disease has killed more than half the people who have ever lived on this planet. It still kills (Fig. 7.1). Today, every 10 seconds a person dies of malaria—mostly children under the age of 5 living in Africa. The total number of cases of malaria is estimated to be 300 million to 500 million, with approximately 10% of those occurring outside Africa. Annually, 2 million to 3 million deaths are caused by malaria. Worldwide malaria infections are on the rise, and as the fever plague spreads, it will continue to affect us in the places we live, work, travel to, and fight in.

A Look Back

The antiquity of human malaria is reflected by the records in the Ebers papyrus (ca. 1570 BC), in clay tablets from the library of King Ashurbanipal (692–627 BC) and in the classic Chinese medical text the *Nei Chang* (2700 BC). These records describe the typically enlarged spleen, periodic fevers, headache, chills, and fever. Malaria probably came to Europe from Africa via the Nile Valley or by means of closer contact between Europeans and the people of Asia Minor. The Greek physician Hippocrates (ca. 470–380 BC) discussed in *Of the Epidemics* the two kinds of malaria: one with recurrent fevers every third day (benign tertian) and another with fevers on the fourth day (quartan). He also noted that those living near marshes had enlarged spleens. Although Hippocrates did not describe malignant tertian malaria in Greece, there is clear evidence of the presence of this malaria in the Roman Republic by 200 BC. Indeed, the disease was so prevalent in the marshland of the Roman Campagna that the condition was called the "Roman fever." Since it was believed that this fever recurred during the sickly summer season due to vapors emanating from the marshes, it was called by the Italian name *mal'aria*, literally, "bad air." Over the centuries, malaria spread across Europe, reaching Spain and Russia by the 12th century; by the 14th century, it was in England. Malaria was brought to the New World by European explorers, conquistadors, colonists, and African slaves. By the 1800s it was found worldwide.

On 20 October 1880, Charles Louis Alphonse Laveran (1845–1922), a physician in the French Foreign Legion in Algeria, examined under the microscope a drop of blood taken from a soldier suffering with malaria fever (Fig. 7.2). He found within the red blood cells transparent globules containing black-brown malaria pigment and, on occasion, mobile filaments emerging from clear spherical bodies (a process he called exflagellation). He also found that some malaria patients had blood cells shaped like crescents. In effect, Laveran had discovered an animal parasite with different developmental stages. Initially, his work was received with skepticism, but by 1884 the Italian workers E. Marchiafava, A. Celli, and C. Tomassi-Crudeli, working in the malaria-infested Roman Campagna, confirmed his findings.

The significance of Laveran's observation of exflagellation went unappreciated until 1896 to 1897, when William MacCallum and Eugene Opie,

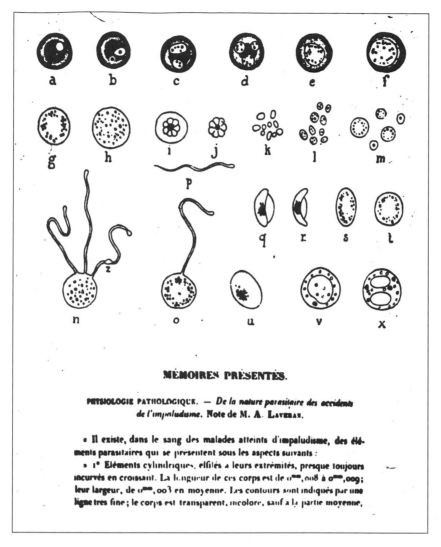

Figure 7.2 Laveran's drawing of what he saw under the light microscope when examining a drop of blood from a soldier with chills and fever.

students at Johns Hopkins University, found that the blood of sparrows and crows infected with *Haemoproteus* (a bird parasite closely related to malaria) contained two kinds of crescent-shaped gametocytes (male and female sex cells). They also found that exflagellation reflected the release of microgametes from the male gametocyte. They correctly interpreted their observations: the gametocytes of *Haemoproteus* in the blood, when ingested by a mosquito, release the gametes in the mosquito's stomach, where fertilization takes place, producing a wormlike zygote, the ookinete.

There remained a mystery: how did "bad air" cause malaria? Ronald Ross, a surgeon-major in the Indian Medical Service, was probably the least likely person to solve the puzzle of how humans "catch" the disease. Ross was born on 13 May 1857, in the foothills of the Himalayas, where his father was an officer in the British Army stationed in India, which at that time was part of the British Empire. As a boy of 8, his parents shipped him to England to receive a proper British education. He was a dreamer, and although he liked mathematics, he preferred wandering around the countryside, observing and collecting plants and animals. At Springhill Boarding School, which he attended from the age of 12, he began to write poetry, painted watercolors, and thought of becoming an artist. However, his father had other ideas and insisted that he study medicine in preparation for entry into the Indian Medical Service. So, at age 17 (following his father's orders), young Ronald began his medical studies. He was not a good student, not because of laziness, but because he had so many other interests and could not concentrate on medicine. He preferred composing music to learning anatomy, and he wrote epic dramas rather than prescriptions. Publishers rejected his "great works," and he had them printed at his own expense. He eventually did pass his medical examination (after failing the first time), worked as a ship's doctor, and then entered the Indian Medical Service. Although India was rife with disease—malaria, plague, and cholera—Ross busied himself writing mathematical equations, taking long walks, writing poetry, playing the violin, and studying languages. Occasionally, he used his microscope to look at the blood of soldiers ill with malaria, but he found nothing. He shouted to all who could hear: "Laveran is wrong. There is no germ of malaria."

In 1894, Ross returned to England on leave. By that time, he had spent 13 years in India and had few scientific accomplishments: he published a few papers on malaria in the *Indian Medical Gazette*, claiming that malaria was primarily an intestinal infection. On 9 April he visited with Dr. Patrick Manson (1844–1922), who, while working as a medical missionary in China, had shown that mosquitoes carry filaria—the roundworms that cause elephantiasis. Dr. Manson, an expert microscopist, took a drop of blood from a sailor ill with malaria and showed Ross the parasite peppered with the black-brown malaria pigment. One day, as the two were walking along Oxford Street, Manson said,

> Do you know Ross, I have formed the theory that mosquitoes carry malaria . . . the mosquitoes suck the blood of people sick with malaria . . . the blood has those crescents in it . . . they get into the mosquito stomach, shoot out those whips . . . the whips shake themselves free and get into the mosquito's carcass . . . where they turn into a tough form like the spore of an anthrax bacillus . . . the mosquitoes die . . . they fall into water . . . people drink a soup of dead mosquitoes and they become infected.

Though this was a romantic story and mostly a guess on the part of Dr. Manson, Ross took it as fact.

Manson became Ross's mentor and encouraged him to study mosquito transmission. Ross left England in March 1895 and reached Bombay a month later. In June, encouraged by Manson's passionate plea, he captured various kinds of mosquitoes, although he hadn't a clue what kind they were. He set up an experiment: he put an infected mosquito in water, where she laid her eggs; when they hatched, all were allowed to die. The water with its dead and decaying mosquitoes was then given to a volunteer to drink. The volunteer came down with a fever, but after a few days, no crescents could be found in his blood. The same experiment was repeated with two other men who were paid for their services. Failure again. So it appeared that drinking water containing mosquito-infected material did not produce the disease. Ross began to think that perhaps the mosquitoes had the disease, but they gave it to human beings by biting them and not by being eaten. He began to work with patients whose blood contained crescent-shaped malaria parasites, and with mosquitoes bred from larvae. The first task was to get the mosquitoes to bite the patients. It was like looking for a needle in a haystack. There are more than 2,500 different kinds of mosquitoes and, at the time, there were no good means for identifying most of them. Initially, Ross worked mostly with the kind with the gray and striped wings. When these mosquitoes were dissected, there were the whiplike extensions in the mosquito stomach, but there was no further development. This result was no more informative than that which Laveran had seen in a drop of blood on a microscope slide nearly 20 years earlier.

Today, we understand (as Ross did not) why this was so: the gray mosquitoes are *Culex*, and those with striped wings are *Aedes*; neither carries human malaria. The one he should have used was the brown, spotted-winged mosquito, *Anopheles*, but Ross did not recognize this for the entire year he dissected mosquitoes. Thousands of mosquito carcasses were all he had to show for his labors, and each mosquito dissection required hours of effort at the microscope.

Then Ross, now 40 years old and having spent 17 years in the Indian Medical Service, turned from the incompetent species of mosquito to the competent, brown, spotted-winged *Anopheles*. On 16 August 1897, his assistant brought him a bottle in which mosquitoes were being hatched from larvae. It contained "about a dozen big brown fellows, with fine tapered bodies hungrily trying to escape through the gauze covering of the flask which the angel of fate had given my humble retainer!" He went on, "My mind was blank with the August heat; the screws of the microscope were rusted with sweat from my forehead and hands, while the last remaining eyepiece was cracked. I fed them on Husein Khan, a patient who had crescents in his blood. There had been some casualties among the mosquitoes, and only three of the *Anopheles* were left on the morning of August 20, 1897. One of these had died and swelled up with decay." At 7 a.m., Ross went to the hospital, examined patients, attended to correspondence, and dissected the dead mosquito, without result. Then—a significant reminder of one of those unknown quantities in the equation to be solved—he examined an *Anopheles*. No result again. He wrote in his notebook:

At about 1 p.m., I determined to sacrifice the last mosquito. Was it worth bothering about the last one, I asked myself? And, I answered myself, better finish off the batch. A job worth doing at all is worth doing well. The dissection was excellent and I went carefully through the tissues, now so familiar to me, searching every micron with the same passion and care as one would have in searching some vast ruined palace for a little hidden treasure. Nothing. No, these new mosquitoes also were going to be a failure: there was something wrong with the theory. But the stomach tissues still remained to be examined—lying there, empty and flaccid, before me on the glass slide, a great white expanse of cells like a large courtyard of flagstones, each one of which must be scrutinized—half an hour's labor at least. I was tired and what was the use? I must have examined the stomachs of a thousand mosquitoes by this time. But the angel of fate fortunately laid his hand on my head, and I had scarcely commenced the search again when I saw a clear and almost perfectly circular outline before me of about 12 microns in diameter. The outline was too sharp, the cell too small to be an ordinary stomach cell of a mosquito. [See Fig. 7.3A.] I looked a little further. Here was another, and another exactly similar cell.

The afternoon was very hot and overcast; and I remember opening the diaphragm of the substage condenser of the microscope to admit more light and then changing the focus. In each of these, there was a cluster of small granules, black as jet, and exactly like the black pigment granules of the . . . crescents. I made little pen-and-ink drawings of the cells with black dots of malaria pigment in them. The next day, I wrote the following verses and sent these to my dear wife:

This day relenting God
Hath placed within my hand
A wondrous thing; and God
be praised. At his command,

Seeking his secret deeds
With tears and toiling breath,
I find thy cunning seeds,
O million-murdering death.

I know this little thing
A myriad men will save.
O death, where is thy sting?
Thy victory, O grave?

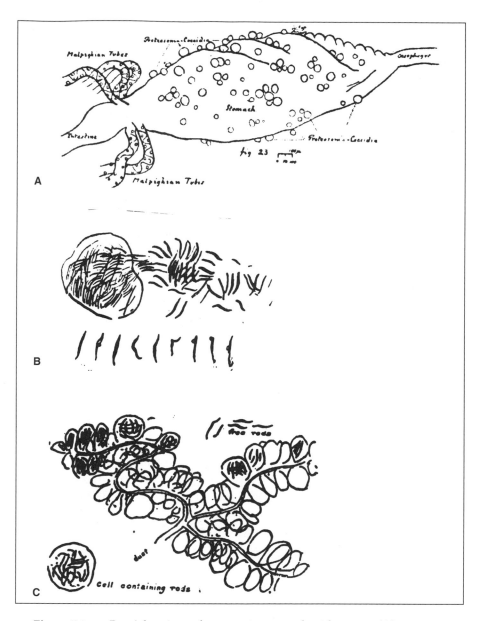

Figure 7.3 Ross' drawings of a mosquito stomach with oocysts (A), an oocyst bursting to release sporozoites (B), and a salivary gland with sporozoites (C).

Here was the clue. I had shown that, four or five days after feeding on infected blood, the mosquito had wartlike oocysts on its stomach. But, did these keep on growing, and how did these mosquitoes become infective? I planned to answer these questions shortly but, before that work could begin, I reported my findings to the *British Medical Journal* in a paper entitled, "On some peculiar pigmented cells found in two mosquitoes fed on malarial blood." It appeared December 18, 1897.

Ross knew he could wrap up the unfinished work in a matter of a few weeks, but then he was struck by a blow from the Indian Medical Service by being ordered to proceed to Calcutta—immediately. As soon as he arrived in Calcutta, he set his hospital assistants the task of hunting for mosquito larvae and pupae. Soon he had a stock of the brown mosquitoes and set about getting them to bite patients who were suffering from malaria. By flooding the ground outside the laboratory, he hoped to imitate rain puddles and see whether something could be learned about mosquito breeding. "If I am not on the pigmented cells again in a week or two," he wrote to Manson, "my language will be dreadful."

In Calcutta, Ross was given a laboratory. There were two Indian assistants who had already been working there when he arrived, but they were old men and not very intelligent, so he engaged two younger men, Purboona and Mahomed Bux. Both of these he paid out of his own monies. However, there was not a large number of malaria cases in the Calcutta hospitals. So Ross turned to something that Manson had suggested earlier: the study of mosquitoes and malaria, as seen in birds. Pigeons, crows, larks, and sparrows were caught and placed in cages on two old hospital beds. Mosquito nets were put over the beds and then, at night, infected mosquitoes were put under the nets. Before much time had passed, the crows and pigeons were found to harbor malaria parasites in their blood; also, he found the pigmented cells on the stomachs of mosquitoes that had been fed on infected larks.

Ross became certain of the whole life history of the malaria parasite, except he had not actually seen the zygotes, those spherical cells into which the flagella had penetrated, turning into oocysts. This was the last stage in the study. He found that the size of the pigmented cells on the stomach depended exactly on the length of time since the mosquitoes had been fed on infected blood. The parasites grew to their maximum size about 6 days after they had fed on infected blood. They left the stomach after this time, but what happened to them then?

One day, while studying some sparrows, he found one that was quite healthy, another that contained a few of the malaria parasites, and a third that had a large number of parasites in its blood. Each bird was put under a separate mosquito net and exposed to a group of mosquitoes from a batch that had been hatched out from grubs in the same bottle. Fifteen mosquitoes were fed on the healthy sparrow; in their stomachs, not one parasite was found. Nineteen mosquitoes were fed on the second sparrow; every one of these contained some parasites though, in some cases, not very many. Twenty insects were fed on the third, badly infected, sparrow; every one of these contained some parasites in their stomachs, and some contained huge numbers of them.

Ross wrote in his *Memoirs*,

This delighted me! I asked the medical service for assistance and a leave, but was denied this. I wanted to provide the final story of the malaria parasite for this meeting; but, I knew that time was very short. I still did not have the full details of . . . the change from the oocysts in the mosquito's stomach into the stages that could infect human beings and birds. Then, I found that some of the oocysts seemed to have stripes or ridges in them; this happened on the 7th or 8th day after the mosquito had been fed on infected blood. I spent hours every day peering into the microscope and wrote to Manson: The constant strain on mind and eye at this temperature is making me thoroughly ill. I thought no doubt these oocysts with the stripes or rods [see Fig. 7.3B] burst— but then what happened to them? When they burst, did they produce the same stages that infected human blood?

Then, on 4 July 1898, Ross got something of value. Near a mosquito's head, he found a large branch-looking gland. It led into the head of the mosquito. Ross said to himself, "It is a thousand to one that it is a salivary gland. Did this gland infect healthy creatures? Did it mean that if an infected mosquito fed off the blood of an uninfected human being or bird, then this gland would pour some of the parasites [called sporozoites (Fig. 7.3C)] . . . into the blood of the healthy creature?"

During 21 and 22 July of that year, Ross took some uninfected sparrows, allowed mosquitoes (which had been fed on malaria-infected sparrows) to bite them, and then, within a few days, was able to show that the healthy sparrows had become infected. This was the proof—this showed that malaria was not conveyed by dust or bad air. After all, men and birds don't go about eating dead mosquitoes. On 25 July, now sure, he sent off a triumphant telegram to Sir (Dr.) Patrick Manson, reporting the complete solution; 3 days later, Manson spoke at the British Medical Association meeting in Edinburgh, describing the long and painstaking piece of research that Ross had been carrying out for years. Ross's findings were communicated on 28 July 1898, and he left India for England in February 1899.

Ross's discovery of infectious stages in the mosquito salivary glands in a bird malaria appeared to be the critical element in understanding transmission of the disease in humans. However, Manson rightly cautioned, "One can object that the fact determined for birds do not hold, necessarily, for humans." Ross and Manson wanted to grab the glory of discovery for themselves and for England, but they were not alone in such a quest. The German government dispatched a team of scientists under the leadership of Robert Koch to work in the Roman Campagna, an area notorious for endemic malaria. They isolated a bacillus from the air and the mud of the marshes and, rejecting the claims of Laveran, named the causative agent for malaria *Bacillus malariae*. However, when the bacillus could not be grown in the laboratory, Koch discarded it as the cause of the disease; undeterred, he continued to look further. Koch then visited the laboratory of Professor Giovanni Battista Grassi at the University of Rome and told him of his failure with the "germ" of malaria and mentioned Ross's communication. At that moment Grassi had what is today called an "ah-ha" moment.

Where Ross was patient and perseverant and willing to carry out a seemingly endless series of trial-and-error experiments, Grassi was methodical and analytical. He was also able to distinguish the different kinds of mosquitoes. Grassi observed, "there was not a single place where there is malaria—where there aren't mosquitoes too, and either malaria is carried by one particular blood sucking mosquito out of the forty different kinds of mosquitoes in Italy—or it isn't carried by mosquitoes at all." He recognized that there were still two tasks left: identify the mosquito that transmitted human malaria and then demonstrate the mosquito cycle for human malaria. Working with Amico Bignami, Giovanni Bastianelli, Angelo Celli, and Antonio Dionisi, he went into the highly malarious Roman Campagna and the area surrounding it, collecting mosquitoes and recording information on the incidence of malaria among the people. (Grassi was, in effect, carrying out an epidemiological study.) It soon became apparent that most of the mosquitoes could be eliminated as carriers of the disease because they occurred where there was no malaria. But there was an exception. Where there were "zanzarone," as the Italians called the large, brown, spotted-winged mosquitoes, there was always malaria. Grassi recognized that the zanzarone were *Anopheles*, and he wrote, "It is the anopheles mosquito that carries malaria. . . ." Grassi and his team were able to infect clean *Anopheles* mosquitoes by having them feed on patients with crescents in their blood, and he was able to trace the development of the parasite from the mosquito stomach to the salivary glands. The life cycle in the human was, as Ross had correctly surmised, similar to that of the bird malaria with which he had worked. Their work showed that the association of the disease with swampy, marshy areas of the world is due to the fact that these are ideal breeding sites for mosquitoes.

With this work, Grassi demolished the theory of Koch and was able to prove that "It is not the mosquito's children, but only the mosquito who herself bites a malaria sufferer—it is only that mosquito who can give malaria to healthy people." Grassi wanted his beloved country Italy to receive recognition for his work, but it did not. Instead, it provoked a bitter and very nasty disagreement with Ross. Ross claimed that it was only after Grassi had read his work on the transmission of malaria using birds that he recognized that only in those areas where there was *Anopheles* was there human malaria— that *Culex* was not involved, but Grassi did not publish this or the development of the parasite in these mosquitoes until late in 1898. Ross wrote in his *Memoirs*, "They . . . had this paper of mine before them when they wrote their note. Their statement was . . . a deliberate and intentional lie, told in order to discredit my work and so to obtain priority. . . . Many of the items . . . are directly pirated from my . . . results . . . stolen straight from me. . . ." Ross did not complete the proof of mosquito transmission with human malaria; the Italians did that. But he flagged the dapple-winged mosquito as the vector. Grassi's contribution was to recognize the vector as *Anopheles*. In effect, Ross was the explorer at the helm of the ship, and the Italians rode the decks and helped make a landing. Ross, not Grassi, received the Nobel Prize in 1902, and Laveran received the prize in 1907.

Although malaria can be induced in a host by the introduction of sporozoites through the bite of an infectious female mosquito, the parasites do not immediately appear in the blood. This was surprising in view of the fact that in 1903, Fritz Schaudinn claimed to have seen sporozoites directly invade erythrocytes. Schaudinn was a dominant figure in parasitology, so few doubted his word, but over time, since no one could repeat his observation, it was called into question. In 1948, H. E. Shortt, P. C. C. Garnham, and their colleagues in England inoculated rhesus monkeys with sporozoites that they had obtained from the salivary glands of mosquitoes infected with *Plasmodium cynomolgi* (a parasite similar to the benign tertian malaria, *Plasmodium vivax* of humans), and in one week, parasites, called pre-erythrocytic stages, were found in the livers of the monkeys. Later, they demonstrated similar stages in biopsy material taken from the livers of human volunteers who had been infected by the bite of mosquitoes carrying *P. vivax*; after infected mosquitoes had fed on other volunteers, this stage was found at the same site in malignant tertian malaria, *Plasmodium falciparum*. It was now clear that, when an infected female anopheline mosquito feeds, it injects sporozoites that go first to the liver (where they live and multiply for several weeks).

The Disease Malaria

Although some 170 species of *Plasmodium* have been described, only four are specific for humans. The human malarias caused by *P. falciparum*, *P. vivax*, *P. ovale*, and *P. malariae* are transmitted through the bite of an infected female anopheline mosquito when, during blood feeding, she injects sporozoites from her salivary glands (Fig. 7.3C). The number of sporozoites inoculated is usually less than 25. They travel via the bloodstream to the liver, where they enter liver cells. The entire process takes less than 1 hour. Within the liver cell, the parasite multiplies asexually to produce 10,000 or more infective offspring. They do not return to their spawning ground, the liver, but instead invade erythrocytes. It is the asexual reproduction of parasites in red blood cells and their ultimate destruction with release of infectious offspring (merozoites) that are responsible for the pathogenesis of this disease. Merozoites released from erythrocytes can invade other red cells and continue the cycle of 10-fold parasite multiplication, with extensive red blood cell destruction. In some cases, the merozoites enter red cells but do not divide. Instead, they differentiate into male or female gametocytes (the crescents of Laveran). When ingested by the female mosquito, the male gametocyte divides into eight flagellated microgametes, which escape from the enclosing red cell (exflagellation). They swim to the macrogamete, and one fertilizes it; the resultant motile zygote, the ookinete, moves either between or through the cells of the stomach wall. This encysted zygote, resembling a wart on the outside of the mosquito stomach, is an oocyst, and through asexual multiplication, threadlike sporozoites are produced in it (Fig. 7.3B). The oocyst bursts, releasing its sporozoites into the body cavity of the mosquito; the sporozoites quickly find their way to the salivary glands. When this female mosquito feeds again, the transmission cycle has been completed.

All of the pathology of malaria is due to parasite multiplication in erythrocytes. The primary attack of malaria begins with headache, fever, anorexia, malaise, and myalgia. This is followed by paroxysms of chills, fever, and profuse sweating. There may be nausea, vomiting, and diarrhea. Such symptoms are not unusual for an infectious disease, and it is for this reason that malaria is frequently called "the great imitator." Then, depending on the species, the paroxysms tend to assume a characteristic periodicity. In *P. vivax*, *P. ovale*, and *P. falciparum*, the period between fever bursts is 48 h, and for *P. malariae*, it is 72 h. The fever spike may reach up to 41°C and corresponds to the rupture of the red cell as merozoites are released from the infected red cell. If the infection is not synchronous and there are several broods of parasites, the periodicity may occur at 24-h intervals. Anemia is the most immediate pathological consequence of parasite multiplication and destruction of erythrocytes, and there can also be suppression of red cell production in the bone marrow. During the first few weeks of infection, the spleen is palpable (see Fig. 3.4) because it is swollen from the accumulation of parasitized red cells and the proliferation of white cells. At this time it is soft and easily ruptured. If the infection is treated, the spleen returns to normal size; however, in chronic infections, the spleen continues to enlarge, becoming hard and blackened in color owing to the accumulation of malaria pigment. The long-term consequences of malaria infections are an enlarged spleen and liver and organ dysfunction.

P. falciparum infections are more severe and, when untreated, can result in a death rate of 25% in adults. Complications of malaria include kidney insufficiency, kidney failure, fluid-filled lungs, neurological disturbances, and severe anemia. In the pregnant female, falciparum malaria may result in stillbirth, lower than normal birth weight, or abortion. Nonimmune individuals and children may develop cerebral malaria, a consequence of the mechanical blockage of microvessels in the brain owing to sequestration of infected red cells. If relapse occurs in falciparum malaria, it is due to the increase in numbers of preexisting erythrocytic forms, previously too low to be detected microscopically; this type of relapse is termed recrudescence. Falciparum malaria accounts for 50% of all clinical malaria cases and is responsible for 95% of deaths.

P. vivax and *P. ovale* malarias also have the capacity to relapse; i.e., parasites can reappear in the blood after a period when none were present. This type of relapse, called recurrence, is due to the delayed liberation of merozoites from preerythrocytic stages in the liver called hypnozoites. *P. vivax* results in severe and debilitating attacks but is rarely fatal. It accounts for about 45% of all clinical malaria cases. The benign quartan malaria, *P. malariae*, may persist in the body for up to 4 decades without signs of pathology. *P. malariae* and *P. ovale* infections contribute to approximately 5% of all clinical cases.

"Catching" Malaria

Malaria is by far the most important of the world's tropical parasitic diseases, but it can and does exist in temperate areas. At present, 90 countries or territories in the world are considered malarious, with almost half in Africa south of the Sahara. As Grassi correctly observed, "Mosquitoes without malaria . . . but never malaria without mosquitoes." Historically, malaria occurred anywhere that the vector and parasite could live, generally between latitudes 64°N and 32°S where the temperatures were between 16 and 33°C and the altitude was below 2,000 meters. Mosquito transmission of malaria is dependent on a complex array of factors, including the incidence of infections in the human population, the suitability of the local anopheline population—density, breeding and biting habits, the availability of susceptible or nonimmune hosts, climatic conditions, and the local geographic and hydrogeographic conditions that contribute to mosquito breeding sites. However, malaria can also be transmitted without mosquitoes. Introduction of infected blood by a blood transfusion or through contaminated needles are two of the less common ways of "catching" malaria.

Malaria Today

When the journalist Ryszard Kapuscinsk described his experiences with malaria, he could not be certain that he wasn't suffering from some other disease, such as a bad case of the flu. How could he be sure it was malaria? A drop of blood from the fingertip could serve as the principal specimen source for diagnosis. In expert hands, a stained blood film examined by light microscopy can distinguish between the various species and can detect 5 to 10 parasites per microliter. However, several new techniques have been developed: fluorescent microscopy, detection of parasite-specific nucleic acid sequences, and parasite antigen detection.

Once Ross and Grassi had flagged the *Anopheles* mosquito as the vector of malaria, methods for controlling it were possible. Of the 450 species of *Anopheles*, only 50 are capable of transmitting the disease, and of these only 30 are considered efficient vectors. In Africa the most efficient vector is *Anopheles gambiae*, which can breed in small temporary pools of water such as those formed by foot or hoof prints or tire tracks. In other areas, *Anopheles stephensi*, which can breed in wells or cisterns, is the vector. In the 1900s, larvicides in the form of oil and Paris green (bright-green powdered copper acetoarsenite that is extremely poisonous and sometimes used as an insecticide or fungicide), as well as drainage, were introduced to limit mosquito-breeding sites in water. This had outstanding success in reducing transmission in some parts of the world. Later, DDT was introduced as a component of an eradication campaign; however, by the early 1960s it became clear that eradication could not be accomplished owing to the emergence of DDT-resistant mosquitoes and the negative ecological side effects of DDT. By 1969, the World Health Organization had formally abandoned its eradication campaign and recommended that countries employ control strategies. Today, attempts at control involve insecticide-impregnated bed nets and spraying with ecologically less disruptive (but more expensive) insecticides.

Since all of the pathology of malaria is due to parasites multiplying in the blood, most antimalarials are directed to these rapidly dividing stages. The earliest of the antimalarials was quinine, derived from the bark of the Cinchona tree and isolated by two French chemists, Pierre Pelletier and Joseph Caventou, in 1820. Quinine continues to be used, but completion of the 5- to 7-day regimen for cure is poor owing to unpleasant side effects such as bitter taste, tinnitus, nausea, and vomiting; it remains limited to parenteral use. Chloroquine and amodiaquine are synthetic antimalarials developed in the 1940s. They were the mainstay of the unsuccessful malaria eradication program of the 1950s. Both act rapidly and can be taken prophylactically, by mouth, once a week. But the problem with chloroquine is drug-resistant *P. falciparum*. Mefloquine (trade name, Lariam) and halofantrine (trade name, Halfan) were synthesized in the 1960s to counter strain resistance to chloroquine. Mefloquine, which acts similarly to quinine, can be taken orally, once a week, before, during, and after exposure. Primaquine is a drug used against the stages in the liver in the vivax malarias; if treatment is successful, it prevents relapse. A newer analog of primaquine called tafenoquine has been developed by a collaboration between the U.S. Army and Smith Kline and Beecham; it will be the first new replacement drug since primaquine was introduced more than 60 years ago.

Regrettably, little is known about the molecular mode of action of many of these drugs or the mechanisms of resistance, though it has been contended that multidrug resistance for halofantrine, mefloquine, chloroquine, and quinine is due to mutations. In the case of some antimalarials, we know quite clearly that they block the synthesis of parasite DNA similarly to the action of zidovudine (AZT) in HIV infections. In the late 1980s, Burroughs Wellcome began a program for the rational design of antimalarials. The result was atovaquone, a mimic of the vitamin coenzyme Q or ubiquinone. It also blocks DNA synthesis by the parasites. Atovaquone is too costly to treat those living in areas where malaria is endemic, but it is used to treat *Pneumocystis carinii* pneumonia in AIDS patients in developed countries.

Qinghaosu, a Chinese herbal medicine, is both the newest and the oldest in the arsenal of antimalarials. It has been used in China for 2,000 years to reduce fever. Qinghaosu, or artemisinin, is derived from the leaves of the wormwood *Artemisia annua*. Artemisinin itself is poorly absorbed, so three derivatives are widely used outside the United States: artemether, artemotil, and artesunate. These act rapidly to clear the blood of parasites, but there is a high rate of relapse.

The major threat of malaria today is not an increasing range of endemicity, but rather a rise in the intensity of antimalarial drug resistance. Drug resistance in malaria has been defined as the ability of a parasite strain to survive and even multiply despite the administration and absorption of a drug in doses equal to or higher than those usually recommended but within the limits of tolerance of the subject. Thus, resistance is a characteristic of the particular parasite strain. First recognized more than 40 years ago with chloroquine in South America and Southeast Asia, drug-resistant malaria poses one of the greatest challenges for controlling morbidity and mortality and is dependent on prompt and accurate diagnosis. In an attempt to overcome or delay the emergence and spread of drug-resistant strains, combination therapy (e.g., Fansidar [sulfadoxine and pyrimethamine], Fansimef [Fansidar and mefloquine], Maloprim [pyrimethamine and dapsone], and Malarone [atovaquone and proguanil]) has been deployed. Since 1991, GlaxoSmithKline Biologicals has been developing LAPDAP, a combination of lapudrine or chlorproguanil and dapsone.

In addition to resistance, there is another treatment constraint: cost. Chloroquine was a very cheap antimalarial, costing about 8 cents per treatment. A course of treatment with a newer drug, such as mefloquine, may cost 80 cents; halofantrine may cost $1.60 to $2.40; and atovaquone may cost up to $30. LAPDAP, released in 2002, costs less than a dollar per tablet and may provide a cheaper alternative. A third constraint for the drug treatment of malaria is the reluctance of the pharmaceutical industry to invest their capital in developing antimalarials that they believe will not yield profits.

Prevention of Malaria

Generally speaking, the measures for prevention of malaria are to keep infected mosquitoes from feeding on humans, to eliminate the breeding sites of mosquitoes, and to kill mosquito larvae and reduce the life span of the blood-feeding adult. Prevention of contact with adult mosquitoes can be accomplished by using insect repellents, wearing protective clothing, and using impregnated mosquito netting and screening of houses. Breeding sites can be controlled by draining water, changing the salinity, flushing, altering water levels, and clearing vegetation. Adult mosquitoes can be killed by using sprays, and larvae can be destroyed by larvicides. Prevention can also be accomplished by education and treatment of the human population. Employing all these strategies has helped to eradicate malaria from many temperate parts of the world, but in the tropics and in developing countries, especially those with limited budgets for mounting public health campaigns in the face of parasite and insecticide resistance, malaria is on the rise.

Genetic resistance to malaria: sickle cell trait and Duffy factor

In 1949, J. B. S. Haldane hypothesized that β-thalassemia, which is caused by mutations in the β globin genes and which results in a decrease or loss of hemoglobin production, may offer protection against malaria. Haldane's hypothesis is supported by the fact that the geographical distribution of the thalassemias overlaps that of endemic falciparum malaria, and populations with particular thalassemias have less malaria (Fig. 7.4). It has been proposed that the protection afforded by thalassemia may be due to a modification in surface receptors for *P. falciparum*. In the case of *P. vivax*, α-thalassemia may be associated with higher levels early in life, thereby heightening immunological defenses against subsequent attacks with the more dangerous *P. falciparum*.

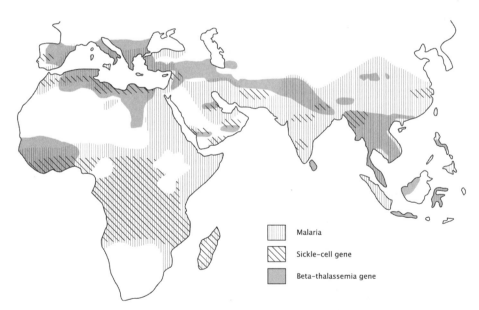

Figure 7.4   The world distribution of malaria (prior to the eradication campaigns of the 1950s) and that of the sickle-cell and beta thalassemia genes. The geographic coincidence provided the suggestion that resistance to falciparum malaria might be of evolutionary significance and that the presence of malaria tended to maintain genes responsible for some deleterious blood diseases at higher frequencies in some populations.

In 1910, James Herrick, a physician in Chicago, examined a drop of blood from a medical student who came from the West Indies and who had pains in the joints and anemia. When he placed a drop of the student's blood on a slide, covered it with a cover glass, and ringed the glass with Vaseline to prevent drying out, he also excluded oxygen; when observed under the microscope, the normally biconcave red cells were sickle shaped. It was later shown that this is due to the presence of sickle hemoglobin. Patients with sickle cell anemia may show clogging of blood vessels, pneumonia, rheumatism, heart disease, painful episodes, inflammation of the hands and feet, and anemia.

One of the curiosities of malaria in Africa was that the native populations seemed to suffer less from the disease than Europeans did. Why? And if malaria is so lethal, why hasn't the human species been eliminated by this fever plague? Malaria as a disease can be controlled not only by drugs but also by the development of immunity—usually requiring repeated exposure to infection (acquired immunity). But there is also resistance to malaria that does not require exposure, infection, and recovery—this is natural or innate immunity—and is a result of genetics. Some of us are born with protective traits conferring resistance to malaria that does not require previous infection.

A glance at the map of the world showing the distribution of malaria prior to the eradication campaigns of the 1940s and 50s shows that malaria is a tropical and semitropical disease (Fig. 7.5). In the 1950s, Anthony Allison observed that the distribution of the sickle hemoglobin gene overlapped the distribu-

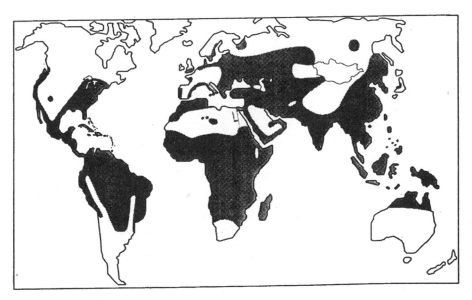

Figure 7.5   The worldwide distribution of malaria in the 1930s prior to the eradication campaigns of the 1940s and 1950s. From I. W. Sherman, *Malaria: Parasite Biology, Pathogenesis, and Protection* (Washington, D.C.: ASM Press, 1998).

tion of falciparum malaria in Africa (Fig. 7.4). Indeed, in some parts of Africa, the prevalence of the sickle cell gene was as high as 20%, and in some areas could be 40%. But what is sickle cell trait and what is sickle hemoglobin?

When sickle cell disease was studied further, it was shown to be inherited. Persons who had the sickle gene produced a slightly different hemoglobin from those with normal hemoglobin. The Nobel laureate Linus Pauling called it a molecular disease. The hemoglobin molecule is a heme-containing protein that gives our blood its red color and is made up of 4 chains of amino acids: 2α, 2β. The α chain has 141 amino acids, and the β chain has 146 amino acids. In people with sickle cell hemoglobin, hemoglobin S, only one amino acid in the β chain is changed from that of hemoglobin A. Hemoglobin A contains the amino acid glutamic acid at the same position that hemoglobin S has the amino acid valine. Since the genetic code word for glutamic acid is CTC (*cytidine thymidine cytidine*) or CTT (*cytidine thymidine thymidine*) and valine is CAC (*cytidine adenine cytidine*) or CAT (*cytidine adenine thymidine*), it is clear that the change from hemoglobin A to hemoglobin S results from a single base change of T to A, a mutation that leads to an alteration in one amino acid out of 287!

How is sickle cell hemoglobin inherited? Humans have 23 pairs of chromosomes; 22 pairs are similar in size and shape in both males and females, and these are called autosomes. The remaining pair is similar in females but differs in size and shape in the male, and these are called the sex chromosomes, X and Y. The gene for hemoglobin is found on the autosomes, that is, it is not carried on the X or Y chromosome and is transmitted without dominance—it is inherited in the same way that ABO blood groups are. An individual with a double dose of the S gene, called a homozygote, has sickle cell anemia; an individual with a double dose of the A gene, also called a homozygote, has normal hemoglobin A. An individual with one A and one S gene is a heterozygote, has sickle cell trait, and has equal amounts of hemoglobin S and hemoglobin A in the red blood cells.

The inheritance pattern when both parents have sickle cell trait is shown in Fig. 7.6. The frequency of AS, or sickle cell trait, is 1 out of 12.5 American blacks; the frequency of SS, or sickle cell anemia, is 1 out of 600 American blacks. In the case of the mating shown, the probability is that 50% of the offspring will have sickle cell trait, 25% will have sickle cell anemia, and 25% will be normal.

Sickle cell anemia, brought on by a double dose of the S gene, can be a debilitating disease, with blockage of blood vessels; severe pain in the extremities, chest, back, and abdomen called crises; and anemia. Clearly, these individuals are sick. Since that is the case, why wouldn't such a deleterious gene be eliminated from the population? The reason is that the heterozygote, of AS genotype, is an individual with sickle cell trait and is protected against falciparum malaria. Because falciparum malaria generally kills children, this allows the most sensitive age group to survive, to develop

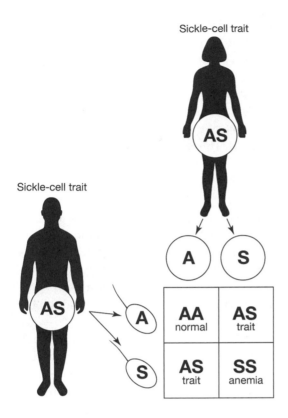

Figure 7.6   The inheritance of sickle-cell hemoglobin. The mating of two individuals with the sickle-cell trait (AS) produces three possible offspring: normal (AA), sickle-cell trait (AS), and sickle-cell anemia (SS).

acquired immunity, and to reproduce. This is an example of natural selection in action, that is, survival of those most fit to reproduce. In this case, natural immunity allows the child to survive and develop acquired immunity to falciparum malaria. Such a survivor can go on to reproduce.

The superior fitness of the AS individual (heterozygote) over either the AA or SS individual (both homozygotes) is called hybrid vigor or heterosis. The S and A genes are both maintained in the population because the loss of the homozygotes due to death either from malaria or anemia is balanced by the gain that results from the enhanced survival and reproduction of the heterozygotes, a phenomenon known as balanced polymorphism.

Of course, without the selective pressure of malaria, the heterozygote has the same advantage as that of the normal homozygote, and under these conditions the frequency of the sickle hemoglobin gene would be expected to decline—and it does.

Today, it is possible to grow malaria in glass petri dishes. The "taming" of *P. falciparum* to a life in the laboratory took place in 1976 in the Rockefeller University laboratory of William Trager. In the 1970s, Trager began a systematic evaluation of the suitability of commercially available tissue culture media for growing the parasites, and one, RPMI 1640, was found to be superior to all others. He also found that an atmosphere of 7% carbon dioxide and 5% oxygen with the balance nitrogen was better than 7% carbon dioxide and 93% air, and that HEPES buffer was effective in maintaining the pH of the serum-containing medium. In 1975 a postdoctoral fellow, James Jensen, joined Trager's lab, and together they set up stationary cultures in petri dishes and held them in a candle jar. (In the candle jar, a white candle is placed in a desiccator equipped with a stopcock; the candle is lit; the desiccator cover is replaced; and, when the candle flame goes out, the stopcock is closed. The atmosphere in the candle jar is 1 to 3% carbon dioxide and 4 to 15% oxygen.) Maintained at 37°C, the parasites grow with a 48-h asexual cycle, as they do in the human body.

The Trager-Jensen method has been used to answer the question, how does sickle cell hemoglobin protect? When *P. falciparum* is grown in petri dishes, and the O_2 levels are lowered to 3%, the sickle hemoglobin in the SS cells forms solid rods, and those rigid "spears" kill the parasite. The AS cells do not sickle, but they lose their intracellular water, and this suppresses parasite growth. In AA cells, the parasite grows and divides normally. When a malaria parasite gets into an AS cell, and that infected cell passes through the capillaries in the deep tissues where the oxygen concentration is low—very much like the drop of blood examined by Dr. Herrick—the infected cell may sickle, or lose water, prematurely. Phagocytes in organs like the spleen and liver recognize such infected cells and remove them. Consequently, the parasite is destroyed. In this way, when a person with sickle cell trait is infected with falciparum malaria, there is selective removal of malaria-infected AS red cells. The protection—innate immunity—is especially important in the young (those under 10 years of age) since they have not had sufficient time to build up immunity as a result of infection; in young people, the infections are also milder.

Let us now turn to another trait that protects against another kind of malaria—vivax malaria. The surface of our red cell bears certain distinguishing molecules called antigens. Since the 1920's, we have known that transfusion success depends on the proper match of donor and recipient red cells. AB and MN are antigens on the red cell surface. If one lacks the A or B antigen, the person is of blood type O. And if one lacks rhesus antigen (called Rh factor), the person is Rh negative. There is another antigen on the surface of our red cells called Duffy factor. Persons with Duffy factor are Duffy positive, and those lacking it are Duffy negative. There is no health hazard associated with being Duffy negative. Where is the Duffy negative antigen most prevalent? The majority of West Africans, and many East Africans and Bedouins, are Duffy negative. When such individuals were studied retrospectively, it was observed that they were never infected with vivax malaria. Why not? The answer has to do with the manner by which a malaria parasite gains entry into the red cell. The invasive stage—the merozoite—must "dock" onto a surface receptor of the red cell before it can get into that cell. For *P. vivax*, that receptor is Duffy factor. If the Duffy factor is missing (Duffy negative), the merozoite cannot attach or dock, and it cannot invade the red cell. A red cell that cannot be invaded does not allow the parasite to reproduce itself, and as a consequence the individual does not suffer from malaria. Clearly, the incidence of Duffy negative antigen in Africa, and in those of African ancestry, suggests that the mutation that led to a loss of Duffy factor protected our ancestors from malaria and promoted their survival.

Persons who are Duffy negative are protected against vivax malaria but not against falciparum malaria. This is because there is a different receptor for the merozoites of *P. falciparum*, called glycophorin. When this receptor is absent, the individual lacking it is resistant to falciparum (but not vivax) malaria. Individuals who are genetically deficient in red cell glycophorin are normal in all ways except they cannot be infected by falciparum merozoites.

The elusive malaria vaccine

What makes scientists believe that a vaccine against malaria is both practical and possible? First, there is abundant evidence from natural infections in humans that immunity is acquired. Although immunity is incomplete (i.e., nonsterilizing) and ineffective in preventing reinfection, it does result in a reduction in mortality. Second, adults in areas where malaria is endemic produce antibody and have low fatality rates. Third, immunization with radiation-attenuated sporozoites induces sterile immunity and protects >90% of human recipients for more than 10 months. Fourth, when passively transferred, immunoglobulins purified from adults who are immune protect the recipient against disease.

A malaria vaccine has yet to be developed; however, when it does become available, it will have to be safe and easy to administer. To be fully protective, the vaccine will have to produce a strong immune response. In those parts of Africa where malaria is endemic and the value for R_0 is 50 to 100, it would require 99% coverage using a lifelong vaccine given at 3 months of age to eliminate malaria. Making the situation even more difficult is that we do not have an in vitro correlate for protective immunity, and the availability of suitable monkeys for testing a vaccine is not great.

Nevertheless, work goes on to develop a malaria vaccine. Some of the potential parasite targets for a vaccine are (i) liver-stage vaccines to reduce the chances of a person's getting sick (this would be suitable for travelers and the military); (ii) blood-stage vaccines to reduce disease severity and the risk of death; and (iii) mosquito-stage vaccines to prevent the spread of malaria through the community. These are also called transmission blocking or altruistic vaccines. The vaccines being contemplated are not based on killed or attenuated stages but are focused on subunit vaccines consisting of selected antigens. Subunit vaccines may consist of synthetic peptides, recombinant proteins, or parasite DNA, packaged in a virus or naked. The World Health Organization claims that over 100 malaria vaccines are under development.

One of the Four Horsemen of the Apocalypse, Disease, in the form of malaria, has long been a companion of War. Alexander the Great, having conquered most of the known world, did not extend his conquests into the subcontinent of India, in large part because in 323 BC he died of malaria at age 32. Malaria repelled foreign invaders from sacking ancient Rome, and Caesar's campaigns were disrupted by malaria. Frederick Barbarossa's army in the 12th century was also prevented from attacking Rome because it was felled by "bad air." Malaria (then called ague) was prevalent in the fens of England and in colonial America, debilitating the farming population. Indeed, in some instances it was the major obstacle, both socially and economically, in the way of the growth and development of the American colonies.

During the Revolutionary War, malaria played a role in several critical battles. During the Civil War, there was a high incidence of malaria among the troops from the North stationed in the South: an estimated one-half of the white troops and four-fifths of the black troops contracted malaria annually. Malaria plagued the French and British engaged in battle during World War I and again in World War II. In 1943, Sir William Slim, the British Field Marshal, said, "For every man evacuated with wounds we had one hundred and twenty evacuated sick." Malaria, the most significant disease problem facing the Allies, prompted one writer to claim that "the triumph over malaria was one of the most important victories of allied forces in the Southwest Pacific." Again, during the Korean and Vietnam wars, malaria was "the great debilitator": *P. vivax* caused the greater morbidity in Korea, and *P. falciparum* caused the greater morbidity in Vietnam. This prompted the military to provide better antimalarial measures and also stimulated the search for new antimalarials. The problems facing medical planners in future conflicts will be even greater, because the next time around, multidrug-resistant strains of *P. falciparum* and insecticide-resistant mosquitoes will have to be overcome. Further, U.S. forces have only infrequently had to contend with malaria in sub-Saharan Africa, where the number of infected bites can be 100 times greater than that in other areas where the United States has fought.

Today, malaria ravages the continent of Africa. Its target is the indigenous peoples, and slowly and inexorably it is killing them (see chapter 15). This is especially true in the rainy and low-lying agricultural areas, where transmission is high. In other areas, the problem is famine and malnourishment. Malnourishment increases the susceptibility to malaria and a variety of other diseases, leads to lower productivity, and puts a further strain on the already fragile health care system. Sadly, in Africa, the riders of the Apocalypse have as their principal target the children; one out of three children die from malaria, and a like number die from AIDS. Thus, it is a certainty that in the 21st century, Disease and Death will continue to ride together and influence the course of history.

Figure 8.1   "Death's dispensary." From the *Illustrated London News*, 1860.

Chapter 8

King Cholera

It was 1901. . . . The revolver was cold, like the grave. . . . The bullets nestled in their chambers. Max von Pettenkofer caressed the metal with liver-spotted hands. The gun was heavy. He put the muzzle to his temple . . . there was no audience. The shot echoed off the walls and down the hall. The last of the miasmatists slumped forward, blood pouring onto the desk.

What provoked Herr Professor Doktor Max von Pettenkofer, the father of public health, head of the Hygienic Institute, favorite of Kings Ludwig and Maximillian II, and recipient of gold medals from the British Institute of Public Health, the German Chemical Society, and the city of Munich, to take his own life? It was cholera morbus, the "cholera of death," but it wasn't the disease itself but his failed theory of its cause. He was now without honor for his "insightful" theory for the causation of cholera—a theory he had staked his life upon to prove. Pettenkofer's theory postulated that there was a factor in the air, called x, but x could not cause disease by itself. What was required, he said, is factor y, which was to be found in the soil. By itself, y too did not cause disease, but when x got into the soil and united with y, it gave rise to z, a "miasma." That is what caused cholera.

Who were these fellow travelers of Pettenkofer, the miasmatists, who believed that disease arises from "bad vapors" and not contagious agents—animalcules and microbes? They were scientists, and their beliefs about disease held sway until 1900—until the microbe that caused cholera was identified. In 1835, cholera swept across Russia and into Munich, Pettenkofer's home city. The epidemic began in the slums, and the poor suffered the most, but it also killed enough of those with money that terror gripped the city, causing panic. Cholera was believed to travel in the air, breathed in by all. Physicians such as Max Pettenkofer thought that its causes were filth, laziness, and neglected hygiene. He was certain it was due to an unhealthy environment. The solution to this disease, he reasoned, was not cure but prevention. In Munich, his city, the public water was so filthy that only the poor drank it, and they died from diarrhea and fever (Fig. 8.1). Those who were better off drank good Munchen beer! Pettenkofer's theory was that stinking miasmas rose from the low-lying marshes and polluted streams to cause disease, and he devoted all of his energies to improving the quality of the drinking water in the belief that the cholera epidemics which physicians could do nothing to control would be stopped. Cleaning up the water, as well as good food and fresh air, would, he was convinced, restore health. In this approach to disease control, he became the first scientific hygienist.

When cholera broke out in 1854, Pettenkofer sought to identify the environmental factors. He made a map of where the victims lived (and died), and found that those neighborhoods in low-lying marshlands had the highest incidence of disease. Clearly, he reasoned, it must be the soil. He proposed that clean air went into the damp soil, where it activated a chemical reaction that resulted in a poisonous vapor, and people who lived nearby came down with cholera. A year later, John Snow, physician to Queen Victoria of England, wrote a small pamphlet, "On the Mode of Communication of Cholera," in which he discussed the 1854 outbreak in London's Soho neighborhood. Snow had made a map of the affected houses and recorded the number of deaths and the number of survivors in each building. In effect, he carried out an epidemiological study. Snow's conclusions on the cause of the London cholera epidemic, however, were contrary to those of Pettenkofer. Snow wrote, " I found that nearly all the deaths had taken place within a short distance of the Broad Street pump . . . and five victims had always drunk water from the pump." Snow took a sample of water from the pump, and on examining it under the microscope found it contained "white flocculent particles." He was convinced that these were the source of the infection. Pettenkofer dismissed Snow's findings, saying he had failed to consider the soil, the water table, and the amount of decaying material. He asked, rhetorically, if the contagious agent was in the water, why didn't everyone get sick? Pettenkofer said in a derisive manner that Snow's explanation was overly simplistic.

In 1857, Louis Pasteur, a French chemist, demonstrated that fermentation (in the production of beer and wine) required a living microbe (yeast) and that wine soured because of the action of certain bacteria. Pettenkofer dismissed these findings. Cholera, he said, had nothing to do with spoilage and fermentation. Ever committed to his miasma theory, Pettenkofer mounted a vigorous public health campaign and tried to bring clean water to Munich, as the British had done in London. He proposed the building of an aqueduct that would carry fresh water from the foothills of the Alps to the faucets of the people of Munich. By 1865, the great aqueduct was finished. Now he wanted the wastewater removed by sewers, as it was in London. Appealing to German national pride, he called on his countrymen to demand clean water; he was determined that Germany would match or exceed the sanitary measures of London, where the death rate from cholera was lower than that of Munich. He was appointed head of the Hygienic Institute in 1879, and within its laboratories, he and his colleagues studied ventilation, poisons, nutrition, and the environment, but not microbes.

Then in 1882, a fellow German, Robert Koch, caused a stir. Koch had previously identified the microbe that causes consumption, the tubercle bacillus. When Koch claimed that the soils of Bombay and Genoa, where cholera occurred, did not meet Pettenkofer's specifications, the two scientific titans were on a collision course. Cholera broke out again in India, and by 1883 it was in the Mediterranean region, brought there by Muslim pilgrims traveling from Mecca. Thousands were dying in Cairo, and Koch (not Pettenkofer) was dispatched to Egypt to isolate the microbe responsible. He did. Koch used his microscope to see bacteria swimming in the feces of 12 cholera patients and 10 corpses of people who had died from the disease. He called it *Vibrio* because of its vibrating wiggles.

Koch was celebrated as a hero when he returned to Berlin on 2 May 1884, and Pettenkofer raged. Pettenkofer insisted that the *Vibrio* did not cause cholera, and he asked Koch to send him a culture prepared from a patient who had died from the disease. Koch complied with a flask, and Pettenkofer, after estimating the number of bacteria to be a billion, put the flask to his lips and drank it in front of an audience composed of his adoring students. Pettenkofer did not become ill. He wrote to Koch, "Herr Doktor Pettenkofer presents his compliments to Herr Doktor Professor Koch and thanks him for the so-called cholera vibrios which he was kind enough to send. Herr Pettenkofer has now drunk the entire contents and is happy to inform Herr Doktor Professor Koch that he remains in good health." Koch did not reply. (Pettenkofer failed to mention that he did have diarrhea and that he had suffered a bout of cholera some years earlier. Perhaps the reason he did not die was his partial immunity by previous exposure to the disease.)

This one demonstration, heroic and dramatic as it was, was insufficient to refute the germ theory of disease, i.e., that diseases are caused by tiny creatures—microbes—that invade the body. The germ theory, announced by Pasteur in 1862, was supported by many years of accumulated evidence of other disease-causing microbes obtained in painstaking fashion by Pasteur and his associates, Koch and his coworkers, and Hansen: leprosy (1873), anthrax (1876), typhoid fever (1880), bacterial pneumonia (1881), tuberculosis (1882), diphtheria (1883), and cholera (1884). By 1901, Pettenkofer was an old man with old ideas that stood against progress in understanding disease transmission. The scientific tide had turned, and Pettenkofer's theory of the cause of cholera was without support. Tragically, he felt there was little honor left, and so he took his own life. The sad story of Pettenkofer, the last of the miasmatists, illustrates how scientists—even very good ones—are sometimes unable to shake themselves free of old (and unsupported) theories because they are blinded by their concept of Nature, a model they create in their own mind and with which they fall in love.

A Look Back

Considering the fact that epidemics have been such a scourge of humankind, it is sometimes difficult to understand why there was such reluctance on the part of scientists like Pettenkofer to believe in microbes as causative agents. In ancient times, outbreaks of disease were considered to be caused by divine intervention, a punishment for the wicked. Later, it was believed (by Pettenkofer and others) that there were vapors or "plague miasmas" that caused disease. In Italy during the bubonic plague (1346 to 1350), this belief led to special physicians' costumes—robes of linen cloth coated with a paste of wax so that the slippery surface would not allow the miasma or "bad vapors" to enter. The plague doctors wore a goggled mask with a beaklike nose that contained strong-smelling perfumes and sweet-smelling herbs to neutralize the disease spirits (see p. 77). It was believed that aromatic materials could ward off disease, so people burned tar and old shoes. One of the practical consequences of miasmatist thinking was that disease might be avoided if one shunned the sick and if the dead were disposed of quickly, with a minimum of human contact. Burning then became the preferred method of corpse disposal, and in some cases it actually worked to stop the spread of some diseases.

But why did people not make the connection between microbes and disease, instead of implicating bad vapors? The reason is that, for many centuries, the microbial agents of disease could not be identified by the senses. We are all inclined to disbelieve that which we cannot detect by sight, hearing, touch, taste, or smell. We are primarily visual animals; we affirm this when we say "seeing is believing." As early as 1500, the Italian Girolamo Fracastoro (1478–1553) wrote about contagion, stating that the "seeds or germs" of disease could be spread by touching contaminated clothing or other inanimate objects, or by contact with the infected individuals themselves. Few believed him, because the intellectual climate during the time he lived did not foster an understanding of the significance of his work: the "seeds or germs" could not be seen. As a result, Fracastoro's theory of contagion was neglected for three centuries.

When did visible proof for the existence of microbes occur? First, for there to be proof there had to be a technological advance—the microscope. Once the microscope was invented, then microbe hunters could begin to find the agents of disease. Microbes were first seen in 1674 by Antonie van Leeuwenhoek (1632–1723), a Dutch linen merchant who had a hobby of grinding magnifying lenses. Using a crude microscope (with a magnification of 300 times), van Leeuwenhoek discovered bacteria and the first human parasite. In a letter he wrote to the Royal Society on 4 November 1681, he stated,

> I weigh about 160 pounds . . . and I ordinarily a-morning have a well-formed stool; but now and then hitherto I have had a looseness, at intervals of 2, 3, or 4 weeks, when I went to stool some 2, 3, or 4 times a day. . . . My excrement

being so thin, I was at diverse times persuaded to examine it. . . . I will only say that I have generally seen, in my excrement, many irregular particles of sundry size. All particles aforementioned lay in a clear and transparent medium wherein I have sometimes also seen animalcules a-moving very prettily. Their bodies were somewhat longer than broad, and their belly, which was flattened was furnished with sundry little paws, wherewith they made such a stir in the medium, and among the globules.

van Leeuwenhoek had discovered the cause of his own diarrhea, though he didn't realize it: it was a result of being infected by *Giardia*, and van Leeuwenhoek's "animalcules" are today called protozoa (see Fig. 1.2D). Giardiasis (a disease caused by being infected with *Giardia*) is a cause of diarrhea, and there are millions of human infections each year. Giardiasis is endemic in the Rocky Mountains, in places such as Aspen and Snowmass. It is also a chronic disorder in St. Petersburg, Russia. An outbreak occurred in the American gymnastic team when it visited Leningrad in 1971, and another occurred in 1990 in workers at a Wisconsin insurance company because a food handler was infected.

Once it was possible to see that there were microscopic organisms—microbes such as bacteria and protozoa—that could cause a culture medium such as chicken soup or meat infusion to spoil and become rancid, the question became: where did these "germs" come from—vapors, inanimate objects, or other bacteria? Up until the 1850s the theory of spontaneous generation, that is, life arising from the lifeless, was popular.

Francesco Redi (1626–1697) was an Italian physician. In 1668 he performed a simple but classic test that shook the foundations of the theory of spontaneous generation. Where others were content to observe nature and suggest imaginative explanations (such as those of Pettenkofer 200 years later), Redi was not content to observe natural phenomena as they occurred but set out to test ideas and to arrange some of the components of nature so that analysis of phenomena could be made; in short, he did an experiment. Redi arranged three jars with decaying meat: one of the jars he covered with paper, one he covered with gauze, and the other he left uncovered. Flies were attracted to the meat samples in the gauze- and paper-covered jars but could land only on the meat in the open jar; maggots developed in this jar but not in any of the others. Decaying meat in itself, said Redi, did not give rise to maggots. It was necessary for flies to land on the meat and deposit their eggs, and the eggs gave rise to maggots (fly larvae). This simple refutation of the generation of life from substances such as rotting meat (lifeless matter) held sway for only a short time, however.

After van Leeuwenhoek discovered microbes in 1675, the proponents of spontaneous generation argued that, although one could not get flies from the nonliving world, one could get living microbes from nonliving broth, rainwater, hay infusions, and other such nutritive sources. The first attack on this micro-level doctrine of spontaneous generation came from Lazzaro Spallanzani (1729–1799), who showed in 1767 that if one boiled the meat broth in a flask and then sealed the neck of the flask, the broth remained sterile. If, however, the neck of the flask was broken, in a short time the broth swarmed with microbes. Boiling the broth (sterilization) prevented the growth of microbes; contamination of it with unheated broth or another substance apparently provided the source of these tiny living creatures. Therefore, he concluded that microbes must arise from other microbes. However, the adherents of spontaneous generation, ever resourceful, stated that Spallanzani's closed flasks eliminated the "vital force," and that was why the broth remained sterile. The issue remained in doubt for more than a century, for it was well known that many microbes required oxygen, a so-called vital principle, and the absence of oxygen in a sealed flask could be expected to inhibit the spontaneous development of living bacteria or protozoans. For a short while, it appeared as if the adherents of spontaneous generation were able to refute the microbial origin of life.

But in 1862, Louis Pasteur (1822–1895), the French chemist-microbiologist, dealt the spontaneous generation theory its death blow by performing a simple and elegant experiment. Pasteur contended that not only do microbes (yeasts, bacteria, and protozoa) cause disease in animals and humans, but they are involved in decay and fermentation (such as occurs in beer and wine) also. Pasteur discovered the microbes that fermented milk, made butter rancid, spoiled wine, and killed off silkworms. He also devised a method for reducing spoilage by heating and then cooling—a technique called, after its discoverer, pasteurization. (Pasteurization is used to preserve milk, cheese, and beer without changing the flavor. This is done by heating to 161°F [72°C] for 15 seconds and then cooling; Pasteur's original method was to heat at 145°F [63°C] for a half hour and then to chill quickly to 50°F [10°C]. Milk must be kept cold after pasteurization, because it does not kill all the bacteria but simply slows down much of their multiplication. To kill the bacteria would require boiling, but that would destroy the flavor of the milk). Without microorganisms, Pasteur said, there should be no change in the meat broth. Pasteur placed some meat broth in a flask and boiled it until it was sterile; then he drew out the neck of the flask so that it was formed into the shape of an S. He did not seal the neck, so air could pass freely into the broth by moving through the twisted neck (Fig. 8.2). The long, curving swan-like neck of the flask, however, trapped airborne microbes and prevented them from reaching the broth. Despite the ease of access of the "vital force"—whatever it was—the flasks remained sterile. However, if Pasteur tipped the flask so that some of the broth ran into the bend of the S and then back into the flask, the broth became contaminated with microbes. Obviously, more than broth plus air was needed to produce life. His swan-necked flask experiment was convincing proof that a vital force did not exist and that microbes come from other microbes. A contemporary of Pasteur the Frenchman (and his archenemy because of fierce nationalistic pride) was the German Robert Koch (1843–

Figure 8.2 One of the swan-necked flasks used by Louis Pasteur in his experiments to disprove the theory of spontaneous generation of life. (Courtesy of Corbis.)

1910). He and his colleagues were able to identify the microbial agents responsible for anthrax, cholera, tuberculosis, diphtheria, and tetanus. He also designed a strategy for unambiguously identifying the causative agent for a microbial-induced disease called Koch's postulates (Table 8.1). Together these great microbe hunters established the firm foundation upon which the germ theory of disease was built.

Thus, by the middle of the 19th century (a time when Pettenkofer was in his scientific prime), it was clear that life gives rise to life and that microbes could cause disease. "Science," one writer wrote, "is a study of errors slowly corrected." And so it was with germs and disease. However, it should be recognized that, at times, even when scientific evidence is presented (and it may be persuasive to some), unless it can be implemented or accepted into the prevailing culture, there will be no error correction. This is how it was with cholera and Pettenkofer. The contagionists (exemplified by Pasteur and Koch) believed that diseases were spread from person to person by an infectious agent, and the miasmatists (exemplified by Pettenkofer) believed that disease was caused by " bad vapors." Until the 1900s, the view of disease causation espoused by the miasmatists (which was the theory of spontaneous generation in a new guise) was the prevailing one.

Table 8.1 Koch's postulates

Koch's postulates for establishing the causal significance of "germs" can be stated as follows:
1. The parasite occurs in every case of the disease.
2. The parasite does not occur in other diseases.
3. After isolation and repeated growth in pure culture, the parasite is able to produce the same disease when introduced into a healthy animal.

The Disease Cholera

The historian William H. McNeill described cholera this way: "The speed with which cholera killed was profoundly alarming, since perfectly healthy people could never feel safe from sudden death when the infection was anywhere near. In addition, the symptoms were particularly horrible: radical dehydration meant that a victim shrank into wizened caricature of his former self within a few hours while ruptured capillaries discolored the skin, turning it black and blue." This was what was called the blue stage of cholera. McNeill continued: "The effect was to make mortality uniquely visible: patterns of bodily decay were exacerbated, as in a time-lapse motion picture, to remind all who saw it of death's ugly horror and utter invincibility."

Cholera is a disease that causes severe diarrhea, producing "rice water" stools. There is also vomiting, convulsions, and muscle cramps but little abdominal pain, and the diarrhea results in a loss of water and electrolytes but not of protein, leading to shock if untreated. In some cases, the individual may die within a day. The origin of the word "cholera" is generally ascribed to the Greek word meaning "gutter of a roof," probably because the symptoms of the disease suggested the heavy flow of water on roof gutters after a thunderstorm. However, another possible origin is from the Hebrew and Arabic words *choleh rah*, meaning "bad illness."

Cholera is a water- or food-borne disease that does not manifest itself until the "germ" enters the human digestive tract. During ancient times, cholera-like diseases were recorded from China, Europe, and Asia, but whether this was true cholera remains in doubt. However, the presence of cholera in India is clearly documented by the writings of early Portuguese settlers in the 16th century, and it is probable that it was there even earlier. Cholera was in Greece by 400 BC. Hippocrates described the symptoms: "An Athens man was seized with cholera. He vomited and was purged . . . and neither vomiting nor purging could be stopped . . . his eyes were dark and hollow . . . he became cold." Aretus of Capodoccia (AD 81–ca. 138), writing in what is present-day Turkey, recorded similar symptoms, adding, "What is first vomited is like water, but what passes by stool is dunglike fluid and of ill odor . . . there are spasms and drawing together of the muscles of the calves of the legs and of the arms. The fingers are twisted . . . there is cold refrigeration of the extremities . . . if he rejects everything by vomiting . . . and becomes cold and ash-colored, and the pulse approaches extinction . . . it is well under such circumstances [for the physician] to make a graceful retreat." In AD 900, Rhazes, an Arab writer from Baghdad, said the disease was incurable. The English physician Thomas Sydenham (1624–1689) first used the term "cholera morbus" to distinguish the disease from "cholera," which meant "state of anger." Other accounts of outbreaks of severe diarrhea were recorded from France, Germany, Brazil, and England in the 17th century, but whether they were actually caused by the same deadly "germ" is unknown.

The history of cholera from 1816 to 1923 is usually described as the history of pandemics. Indeed, there is little to dispute the claim that the greatest epidemic disease of the 19th century was cholera. But that is not the end of it. We are currently in a cholera pandemic that began in 1961 in the Celebes Islands, spread in Asia and the Middle East, and reached Africa in 1970. In 1991, after an absence of almost 100 years, cholera reappeared in the Americas. The first confirmed cases occurred in Peru, and by year's end there were nearly 400,000 cases and 4,000 deaths worldwide.

In 1992, there was a mild outbreak involving 76 people aboard a flight from South America to the United States; the outbreak was due to undercooked or raw fish and vegetables. In the 1990s, more than 200,000 people in Southeast Asia were sickened by cholera. In 1994, a cholera outbreak in Goma, Zaire, killed 50,000 of the half-million Rwandan refugees, mainly Hutus, who were escaping from the Tutsi rebels. This took only 21 days! In 2000, there was another epidemic in Africa, this time in an area of KwaZulu-Natal; interventions such as the provision of chlorinated water curbed the spread of the disease, and mortality was low (<1%) owing to oral rehydration therapy. In 2001, there was a cholera outbreak in southern Africa (South Africa, Swaziland, Mozambique, Zambia, and Zimbabwe) coinciding with the start of the summer rains; thousands were infected, and 63 died. However, this was the exception rather than the rule: mortality from cholera can be high, and during an epidemic the death rate may be 20 to 50%. The El Tor strain of cholera (which began in Indonesia and started the seventh pandemic) is characterized by ~75% of those infected showing no symptoms.

Prior to the first pandemic, cholera was associated with Hindu pilgrimages and holy festivals that drew great crowds to the river Ganges in India, where the disease was endemic and remains so to the present (Fig. 8.3). As a consequence, it is assumed that these practices confined the range of cholera mostly to the Indian subcontinent for centuries. In 1816, when a cholera epidemic began in India, British ships and troops were already in the region. Their presence and troop movements in and out of Calcutta brought the disease to new places with naive populations. British soldiers fighting on the northern frontier of India carried cholera with them from their headquarters in Bengal;

Figure 8.3 A Hindu pilgrimage and bathing in the River Ganges. (Courtesy of Wallace Peters and Anthony Bryceson.)

the troops then transmitted it to their Nepalese and Afghan enemies. Between 1816 and 1823, cholera moved eastward by ship to Sri Lanka, Burma, Thailand, Singapore, and Indonesia. It reached China by land in 1817 and again by sea from Burma and Bangkok, causing a major epidemic in Mainland China from 1822 to 1824. Cholera arrived in Japan by ship in 1822. Muscat, in southern Arabia, encountered the disease in 1821 when British troops were sent there to suppress the slave trade. From there, it moved south along the east coast of Africa, following the slave traders. It entered the Persian Gulf, Mesopotamia (Iraq) Iran, Turkey, Syria, and Egypt and moved along the shores of the Caspian Sea to southern Russia. Then, mysteriously, its spread stopped.

In the next pandemic (1829 to 1851), cholera emerged either from its endemic base in India (Bengal) or from Astrakhan (Russia) as a recrudescent infection which had persisted from the first pandemic. Cholera retraced its previous path beginning in southern Russia, but this time it went further, reaching Europe, where it became rampant. Russian troops fighting in Persia (1826 to 1828) and Turkey (1828 to 1829) and the Polish revolt (1830–31) carried cholera to the Baltic region, and from there it spread by ship to England (1831) and Ireland (1832). Irish immigrants carried the disease to Canada, and then it moved southward to the United States (1832) and Mexico (1833). In North America, beginning in the 1830s (during the second pandemic) and onward, cholera outbreaks occurred with great regularity in port cities such as New Orleans, New York, and Philadelphia. The death rates were so high, they caused panics. In 1849, cholera was called "America's greatest scourge," but still no one understood its cause. In 1831, in Cairo, Egypt, 13% of the population died from cholera. In Europe and especially in Great Britain the disease was known as "King Cholera." Between 1832 and 1833, more than 60,000 died in England of cholera. The number of deaths being so high, and the fear among the people so great, victims of cholera were buried in separate graveyards.

Cholera also established itself in Mecca in 1831 at the time of the Muslim pilgrimages. From there it spread, as it did in India, along the pilgrim routes from Mecca to Medina; these followers of Mohammed then carried the disease back to their homeland. Until 1912, when cholera broke out in Mecca and Medina for the last time, epidemics of this dread disease were a common accompaniment of the Muslim pilgrimages, appearing no fewer than 40 times between 1831 and 1912, or every other year, on average. As cholera added Muslim pilgrimages to its older Hindu pilgrimage dispersal routes, the exposure of peoples beyond India's borders to the new disease became chronic. On top of this, after mid-century, the swifter movement of steamships and railroads became increasingly able to accelerate the global diffusion of cholera from any major world center.

During the third pandemic (1852 to 1859), cholera was found in Africa, the United States, and the Middle East, as well as Europe and India. It was during this pandemic that John Snow published his epidemiological studies and Pettenkofer began his war against the contagionists. The fourth pandemic began in 1863, moved along its old path, and burned itself out 10 years later for unknown reasons. The fifth pandemic lasted 15 years (1881 to 1896) and was widespread in China, Japan (and other countries in Asia), Egypt, Germany, and Russia. Hygienic measures helped to stop its spread in North America, but there were outbreaks in South America and East Africa. The sixth pandemic (1899 to 1923) missed the Western Hemisphere and most of Europe, but there were outbreaks in the Balkans, Hungary, Russia, and the Far East. The seventh pandemic (now ongoing) has a pattern of spread much the same as that of the sixth, but there is evidence that it is slowly creeping into the Western Hemisphere.

Vibrio Discovered

When Louis Pasteur stated that "science favors the prepared mind," he was surely referring to himself, not Robert Koch, But Koch as a preeminent microbe hunter did have a prepared mind. First working alone and then with many capable colleagues in Berlin, he identified the germs of anthrax and tuberculosis. It was therefore natural that he would seek out the "germ" of cholera. In 1883, when cholera broke out in Egypt, Koch had the intuitive insight to suggest that the disease was caused by a microbe that produced a special poison and that this toxin caused the profuse, watery diarrhea. Obtaining fecal samples and examining these microscopically before and after staining, he discovered the gram-negative, comma-shaped bacillus (Fig. 8.4) he called *Vibrio cholerae*. Subsequently, he was able to grow the vibrio in pure culture, thereby satisfying (in part) his postulates for identifying a disease-producing organism. The French had also sent a medical team to Egypt from Pasteur's laboratory in Paris; they arrived at about the same time as Koch and immediately began to take samples of blood from cholera victims; they then injected this blood into rats and guinea pigs. No disease resulted. When one of the members of the team came down with cholera and died, his colleagues were so unnerved that the research was abandoned and they returned to Paris. Thus, German science was victorious over French science, and Koch returned to Berlin a hero.

Cholera and Evolution

The cause of the diarrhea in cholera is a toxin that affects the adenyl cyclase of the gut cells, producing what is called secretory diarrhea—which is of great benefit to an intestinal parasite, since the bacteria are rapidly expelled from the host and move on to infect other hosts. Before a person can come down with cholera, that individual must be infected by a *V. cholerae* that is itself infected with two viruses. One of the viruses has a gene that codes for the cholera toxin, and the other has a gene that codes for the receptor that allows the toxin-coding virus to enter the bacterium. Only when both viruses are present is the disease-causing toxin produced. People with type O blood are more

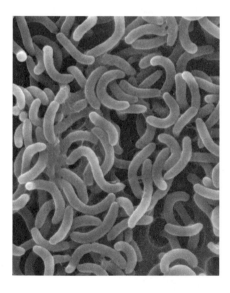

Figure 8.4   *Vibrio cholerae*, as seen with the scanning electron microscope. (Courtesy of Dennis Kunkel Microscopy, Inc.)

susceptible to cholera than other blood groups, but no one knows why. Those living in regions where cholera is endemic show some acquired immunity. For example, adults in Bangladesh have been shown to be 10 times less likely to die from a cholera infection than children younger than 5 years of age.

Cystic fibrosis is a lethal disease caused by a recessive mutation in a single gene and occurs with a frequency of 1 in 2,500 Caucasians. It has been argued that such a high frequency must confer some advantage in the heterozygous (one copy) state. But what is the advantage? At first, it was hypothesized that it may be linked to resistance to cholera. Since the most common mutation appears to be present at significant levels only in Caucasians of northern Europe, it appeared that the hypothesis was supported. The fly in the ointment is this: the cholera bacterium originated in the subcontinent of India and arrived in Europe relatively recently, only after Europeans engaged in trade and military actions in Asia. Another hypothesis suggests that any mechanism that could reduce the threat of excessive dehydration, such as occurs with cholera, would have a selective advantage. There are two tantalizing bits of information: those heterozygous for cystic fibrosis produce, on average, half the volume of sweat per unit time as do their normal counterparts, and mice with the same mutation secrete half the amount of saliva as their normal littermates. But these facts do not explain why the frequency of the mutated gene is higher in Caucasians and why it is not maintained at similarly high levels in other populations. Paul Quinton, a cystic fibrosis researcher, suggests (without any data) that what made the Caucasians different from other populations was their social habits. What the habit is that made the Caucasians, but not other populations, so susceptible to the selective pressure from enterotoxic microbes has not been identified. Another suggestion is that selective pressure against mutations in the gene is greater in other populations than in Caucasians. Thus, heterozygotes lose slightly more sweat on average than their normal counterparts, and this, Quinton suggests, could be a disadvantage, especially if "Caucasians confined to cold climates of northern Europe required much less sweating for thermoregulation than populations in more temperate zones; thus, in colder zones, the combination of the positive effects from protection against secretory diarrhea and deleterious effects of salt loss may favor a higher incidence of mutation, whereas in warmer zones the converse may be true." However, we still do not know whether the selective pressure was the vibrio of cholera or another enterotoxic bacterium, such as *Escherichia coli*.

"Catching" Cholera

How might one get cholera? An infected individual excretes a trillion *V. cholerae* each day, so water and food can easily be contaminated. By eating vibrio-contaminated food and bathing in (and swallowing) contaminated water, one can develop cholera. Notwithstanding current advertising about the woes of a person having high stomach acidity, it has been found that stomach gastric juice is lethal to *V. cholerae*. Therefore, it takes about a billion bacteria to infect a person with normal acidity, whereas in people with low stomach acidity, only 100,000 bacteria would be needed to establish an infection.

Since the days of Robert Koch, there have been two perplexing questions: where do the bacteria go between epidemics, and how could the disease arise spontaneously if people are not infected? The answer, according to Rita Colwell, is that the bacteria go dormant and enter a sporelike state when conditions for their reproduction are not favorable—the water is too cold, or nutrient poor, or it has the wrong salinity. The shrunken vibrios cannot be grown in the laboratory, but they do survive in brackish water, where they are eaten by tiny crustaceans called copepods (Fig. 8.5). Inside the copepod intestine, they cause no damage, but using the world's ocean currents the copepod-enclosed vibrios may be transported to distant shores. Under the right conditions they come out of dormancy, multiply and spread their misery. It may be that the seed for the next epidemic is a bloom of phytoplankton (triggered by global warming or El Niño) that leads to a rise in the cholera-causing bacteria in and around the Bay of Bengal or elsewhere.

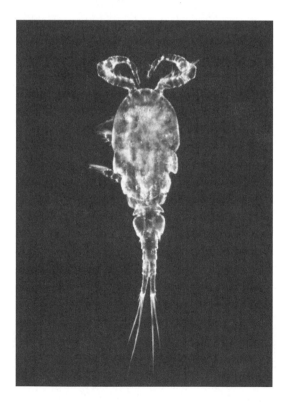

Figure 8.5 The copepod (a small crustacean) serves as a reservoir for cholera and is shown as it appears under the light microscope. (Photo by Laguna Design/ Photo Researchers, Inc. © 2005 Photo Researchers, Inc.)

Cholera Today

The diagnosis of cholera remains much as it was in the time of Koch and Pettenkofer. However, light-microscope examination of stools usually does not show distinctive features. To see the vibrios requires special methods, such as dark-field or phase-contrast microscopy. When viewed with the dark-field microscope, the vibrios move about using their whiplike flagellum and look like shooting stars. However, microscopy and culture from stools or a rectal swab are not necessary in order to treat cholera. Most often, clinical diagnosis is based on the history of the time of appearance of acute symptoms, such as "rice water" diarrhea. Then treatment is begun.

To treat the violent diarrhea of cholera in 1831, health authorities in Hanover, Germany, recommended a variety of nostrums and quack remedies: vinegar, camphor, wine, horseradish, mint, mustard plaster, leeches, bloodletting, laudanum, calomel, steam baths, and hot baths. None was effective, and most patients, treated or not, died. Since cholera was believed to be due to a miasma, the houses where cholera victims lived (and died) were fumigated using a smoke pot. This too did not curb the spread of disease. But, in that same year, a 22-year-old British physician, William O'Shaughnessy, boldly proposed that by the application of chemistry, cholera might be cured. After analyzing the blood of a patient with the blue stage of cholera, he suggested that therapy be "to restore its deficient saline matters . . . by absorption, by inbibition, or by injection of aqueous fluid into the veins." Dr. Thomas Latta carried out the first practical application of intravenous injections of a salt solution on 15 May 1832 in patients suffering with cholera. All improved. He cautioned that, "although by injection of water and salts . . . we may restore the efficient fluids of the body and bring back the blood to its normal state . . . we must still remember that the unknown remote cause, and other agents are still in operation, and require to be remedied before a perfect cure can be performed." Though the cause of cholera was still unknown (and would be for 50 more years), this practical treatment saved lives (8 out of 25) and was regarded as "the working of a miraculous and supernatural agent. . . . " When this pandemic subsided, however, the therapy fell into disuse. Clearly, it was a simple and effective therapy that was far ahead of its time. Latta died in 1833, and O'Shaughnessy joined the East India Company, went to India, and never again concerned himself with cholera. He died in 1899, and his obituary mentions nothing of his pioneering work on rehydration therapy for cholera.

Margaret Kaufman honored the work of O'Shaughnessy and Latta in her poem "Blue Cholera":

> Wanting to acquaint myself
> with the celebrated cholera,
> I traveled down to Sunderland
> from Edinburgh, prepared,

yet unprepared, dear sirs.
I saw a face, a girl
I never can forget, even
were I to live beyond man's natural age.

The girl lay on a pallaise
In a low-ceilinged room.
I bent down to examine her.
The color of her skin—
a silver blue, lead-colored,
ghastly tint; eyes
sunk deep into the sockets
as though driven back
or counter-sunk like nails,
her eyelids black, mouth
squared as if to bracket death;
fingers bent, inky in their hue.
Pulse all but gone at the wrist,
her bosom damp with sweat.

In the corner, a tea kettle
misted the low room, hissing reprimand.
I gazed at the girl,
withdrew my gaze, watched the steam
re-compose itself to drops of water
on the brick lattice
as if the walls in sympathy
were fevered too.
I could not watch the girl for long.
I cast about, only twenty-two myself.

Somehow her face has shadowed
much that I have seen.
Blue cholera, Blue, so unexpected,
as when the evening falls swiftly in winter
before lamplighters arrive, that blue
gray pall cast over everything
that moments before seemed full of light.
I saw that girl, I see her now,
white cap tied about her chin,
her head against the linen rest.
She did not look on me though all
my gaze was full of pity.
What she saw lay beyond my sight,
and if the kettle boiling on the fire
filled the room with thickened steam
to ease her labored breath,
it was as little good to her as I.

Watching the kettle I chanced to think
how it must be re-filled before it burned,

a copper-kettle must be filled,
and from there, as a grasshopper
jumps from one slim blade to another,
my mind fatigued, just—leapt
to form a thought, hung
delicately above the chance
that this blue cholera
could perhaps be stayed by adding fluids.

I wrote the Lancet,
reported to the Central Board of Health,
suggested intravenous application—
to restore the blood
to its normal constituency
its natural specific gravity.
Colleagues undertook this treatment,
using a slender silver tube
to replace deficient saline
in patients desperate to the last degree.

I knew that every treatment
had its season of celebrity;
bloodletting, leeches, calomel,
emesis. What remedy had not been tried?
The girl died, other patients too.
Sadly, it was years before
the treatment gained wide acceptance.
Latta died who might have pled its cause.

A man is proud of such a gain,
even close to death as I am now.
Still, staring at a glass of port,
sunk deep in a chair before the fire,
although I've been to India and back,
knighted for the work I did out there,
I see that girl, the one I could not help,
skin a silver blue, her sunken eyes past pleading.

One hundred and sixty years later, research showed that the cholera toxin acts on the absorptive cells of the intestine to shut down one major route of sodium transport which allows sodium into the cells together with chloride, but the toxin leaves unaffected the transport system that brings sodium and glucose simultaneously into the cells from the lumen of the intestine. This transport system only works when both sodium and glucose are present, and it remains active during diarrhea. Armed with this, it was possible (in 1940) to make the appropriate solution with salts and glucose to rehydrate the patient suffering from severe cholera-induced diarrhea. Today, oral rehydration treatment (ORT) involves oral or intravenous administration of a solution containing glucose, sodium chloride, potassium, and lactate, and the cost is $5 per quart. Another type of ORT, called food-based ORT, substitutes starches and proteins for the glucose. Cereal grains and beans are the source of starches and proteins, and they work effectively to reduce both diarrhea and mortality.

Controlling Cholera

In the early part of the 19th century, few people appreciated that cholera was spread from person to person via contaminated drinking water. Indeed, few (except perhaps for Pettenkofer and his students) appreciated the value of sanitation as a means of controlling disease. Quite remarkably, it was cholera that ultimately provoked the development of sanitary measures to control its spread. Health, as Herr Doktor Pettenkofer, director of the Hygienic Institute, reminded the burghers of Munich and their Austrian kings, depends on an adequate supply of clean water. This, he said, could be provided by an aqueduct such as those built during the time of the Roman Empire. The architecturally magnificent aqueducts, even to this day, remain monuments to the grandeur that was Rome. The first aqueduct that brought clean water to Rome dates from 312 BC. By AD 1, there were six aqueducts, and 100 years later the number had increased to 10. Together, this system of aqueducts could supply the city with 250 million gallons daily and every Roman with 50 gallons—about the same amount Londoners and New Yorkers use today. The water brought by the aqueduct was used for more than drinking. It was also used for bathing. Baths were believed to have curative powers, and the Romans built baths throughout their Empire. The baths of Caracalla (AD 200) could accommodate 1,500 bathers at a time, and a bath built during the reign of Diocletian (AD 245 to 313) had 3,000 rooms. The sanitation system of Ephesus, one of the three largest cities in the Roman Empire, was known to Pettenkofer because its ruins had been excavated in the 19th century by British and Austrian archeologists. (Ephesus, in modern Turkey, is a tourist attraction with over 2 million visitors each year.) The seaport city of Ephesus was first settled in 1400 BC and then colonized by the Ionian Greeks (1000 BC). In a succession of battles, the city fell to the Persians (546 BC) and then to the Romans (133 BC). By AD 27, Emperor Augustus declared Ephesus the capital of the Asian province. During this time there is evidence that St. John and Mary fled Jerusalem to escape persecution and found safe haven in Ephesus. Paul's confrontation with the Ephesians, who worshipped the fertility goddess Artemis, was one of the conflicts that led to his decapitation on the outskirts of Rome in AD 64. By AD 100, Ephesus had 250,000 inhabitants and streetlights! Water came to the Ephesians via four aqueducts, with one source 25 miles away. With flow averaging a quart a second, and with a reservoir, the city never suffered from a lack of water. Ephesus had paved streets, elegant houses and shops, temples, a library, fountains, baths, and a public toilet; in the public toilet there were 40 seats with no separation between them. The toilet walls were faced with marble, the floor was covered with mosaics, and a pool in the center received its water from spouting dolphin statues. Music played to accompany the patrons passing their effluvia.

In the 1800s, the sanitary conditions in London, when compared with the cities of the Roman Empire, were as different as chalk is to cheese. London's water supply was much worse than that which had existed in Rome and Ephesus more than 1,000 years earlier. Why? The explosive growth of the human population led to people being crowded together in cities. But different as they were, these cities had a nexus: none of their citizens was aware that disease could be spread by the water that flowed along the aqueduct or in the river if that source were to become contaminated. Indeed, without such understanding, the most magnificent baths and toilets were useless in preventing a water-borne disease—one of the greatest threats facing the inhabitants of Rome, Ephesus, and London.

Today, we tend to take clean water for granted. A drink of water is obtained by simply turning a faucet. When we want to bathe or shower or cook, we turn on the tap. To get rid of our bodily wastes, we flush the toilet. It wasn't always that way. It is claimed that Queen Elizabeth I, who was born 7 September 1533 and reigned from 1558 to 1603, said, "I take a bath once a year whether I need it or not." In 1832, the local publicity in New York City was that its water was the best—not only could you use it for cooking, bathing, and drinking, but it was also a wonderful laxative! We marvel at the elegance of the palace at Versailles built by Louis XIV, the Sun King (1638–1715), but when Versailles was built in 1661, it did not have a single toilet. The guests at Versailles simply urinated below the staircase, and if they had to defecate they went to the basement or relieved themselves in one of the outdoor privies. The cesspool, which was used to dispose of waste, was covered up once it was filled, and a new one was dug nearby. There is a myth that the flush toilet was invented in England by a plumber named Thomas Crapper (1836–1910), because many of the porcelain toilets and manhole covers bore his name. In fact, it was Albert Giblin who received the patent for the flush toilet in 1819. (In England, the flush toilet is frequently in a room with a "W.C." sign on its door; during World War II, Americans thought it stood for "Winston Churchill," but it is nothing more than an acronym for "water closet.") As much as the flush toilet appeared to provide a more sanitary means for disposing of human waste (and conserving water), it was a public health disaster. All of the effluents were dumped into the rivers, which were not only the exit for waste but also the source of water for drinking and washing water; thus, both entry and exit were in the same locale. No one got the connection between infectious diseases such as cholera, sewage, and drinking water until Dr. John Snow (1813–1858) carried out his epidemiological study of cholera in London. The importance of his findings, however, were neglected for a quarter of a century.

Cholera, Sanitation, and Public Health

How did we get to this sorry state of hygiene? When human populations were much smaller and people lived in villages, water was drawn from the nearby streams and rivers, or shallow wells were dug. Human excrement was disposed of in a privy or by spreading it over a field, transforming it into "night soil." The system worked reasonably well in most of Europe and North America until urbanization and industrialization came about in the 18th century. With the industrial revolution, the human population soared. Villages became towns, and towns became cities; as there were fewer places on the land for the "night soil," its disposal became more difficult. The solution: dump it into the rivers and streams, where it would be carried away by the current. The swifter the flow of water in the stream and river, the quicker the pollutants were removed, but sometimes the rivers (such as the Thames) were tidal, and flow was intermittent. More serious, the same rivers and streams that served for waste removal were also the source of water for drinking, washing, and bathing. The textile and steel mills of the expanding industrial revolution of the 18th and 19th centuries required more and more factory workers, creating an in-migration from the villages to the city. To accommodate these burgeoning numbers, houses were built; these urban dwellings were arranged in back-to-back rows separated by the narrowest of streets and alleys, thereby using the smallest amount of land for the largest number of people. Families, consisting at times of 10 or more people, occupied a single room, and the row houses had a common outside water supply and privy. Under such crowded conditions, there was little privacy, cleanliness was difficult to achieve, and clean water was often limited or unavailable (Fig. 8.6). Unlike the privies at the palace at Versailles, these privies were used by dozens of people and were rarely cleaned out; wastes overflowed into the surrounding back alleys and the shallow wells. Thus, as F. Cartwright stated in *A Social History of Medicine*, "every evil of the industrial town originated in the agricultural village and resulted from the transfer not of people only but of their manner of living. This way of life did not become intolerably dangerous in the countryside because communities were small, cottages widely separated, and work performed in the open air. The disaster of epidemic illness became inevitable when the community was cramped for space, houses lay cheek by jowl and factory hands worked close together for long hours in an enclosed atmosphere." In this way, the urban residents in England and the Americas came to be deprived of the benefits of hygiene.

When cholera arrived in Liverpool, England, in 1832, death came quickly and horribly. Sewage and "night soil" were deposited in the streets and alleys and left either to stagnate or to flow into the open ditches and then to the rivers. Drinking water obtained from these same rivers was untreated. Industrialization had created an overcrowded city with inadequate sanitation, and

Figure 8.6 "A Court for King Cholera" in 1852. From *Punch*. (Courtesy of the Wellcome Library of Medicine.)

some people were forced to live below ground level in cellars. No one knew the cause of the disease. Physicians could do little for those afflicted except dispense opium and brandy, and their bloodletting and purgatives worsened the dehydration of those who were already sick. Three percent of the population (165,000) were ill, and 1,523 died (a mortality of 31%). Those most affected were the cellar-dwelling poor, most of whom were Irish immigrants. Over a period of weeks, the response to the outbreak was a series of riots called by the press "the cholera riots." The seeds of discontent and unrest among the people lay in the inability of those in the medical profession to render any real help. Indeed, there were so many more cholera victims dying in the hospitals than at home that people feared to enter the Liverpool hospital. There was also the suspicion that cholera victims who were removed to the hospital were likely to be killed by the doctors and their bodies sold for use in anatomical dissection. Victims went into hiding to keep from being caught by the body-snatching doctors.

Over a 10-day period, mob violence broke out. Most of the rioting was directed at the physicians, who were called murderers and "bunkers," a term used for those who sold bodies for dissection. At times, there were thousands in the street, mainly women and young boys. They threw stones at the hospital building, chased the physicians, and prevented the removal of cholera patients from their homes to the hospital. Police were needed to restore order, and some rioters were arrested. Social unrest in response to cholera was not confined to England. Civil disturbances also occurred in Russia and elsewhere in Europe. In Hungary, castles were attacked and nobles were murdered by mobs who believed the rich were responsible for the cholera deaths among the poor. The notoriously conservative Liverpool newspapers referred to the rioters as "misguided" and "ignorant" and representing the lower orders of society, particularly the Irish. Though the "cholera riots of 1832" were long ago and were out of touch with what is known today about this disease (and other infectious diseases), they show how fragile the interface can be between physicians and the public. And even now, when there is fear and frustration among the people, a recurrence of social unrest is likely.

The horror of cholera, never experienced before the 19th century in Europe (Fig. 8.7), elicited demands for action from those in authority lest there be more riots. Max von Pettenkofer, the first sanitary scientist, said that sanitation, that is, the disposal of wastes and sewage in such a manner that water and food are incapable of spreading disease, is key to improving health. And forthwith, he began to institute sanitary measures to restore, preserve, and improve the health of the public. Edwin Chadwick (1800–1890), Pettenkofer's equivalent in England, led the movement to improve sanitary conditions. In 1842 he produced a "Report into the Sanitary Conditions of the Labouring Population in Britain" demonstrating that life expectancy in the cities was much lower than that in the countryside. Chadwick and his boards

Figure 8.7 Street scene during a cholera epidemic by Honore Daumier (1808–1879). (Courtesy of the Wellcome Library of Medicine.)

of health produced "sanitary maps" showing the relationship of disease—especially cholera—to overcrowding, lack of drainage, and defective water supply. Chadwick was not entirely altruistic: he believed that a healthier population would be able to work harder and would cost less to support. In the 1840s, piped water was in use, but the companies that supplied the water used bored elm trunks as conduits; since they split under the pressure required to efficiently move water into the city, the water was supplied at very low pressure by ground-level standpipes. Sewage disposal was solved by the use of narrow-bore self-cleaning drains (instead of the brick channels with stone covers), but these drains required a constant flow of water to be cleansed. This led to the development and installation of flush toilets instead of privies. However, the outflow from the sewers was the river, and almost without exception this was also the source of the drinking water.

Over a 20-year period, Edwin Chadwick, a lawyer-journalist but not a scientist, as secretary of the Poor Law Commission became the most powerful member of the general board of health, which had sweeping powers: house-to-house visits in search of infectious diseases, removal of the sick from overcrowded tenements, vaccination of smallpox contacts, investigations into the causes of disease outbreaks in schools and factories, inspection of sanitary improvements, and determining the causes of unexpected deaths. Chadwick was arrogant, loudmouthed, and dictatorial in his approach; his bigoted outlook was more characteristic of a totalitarian state than a democracy. He came to be hated for his board of health policies, the Poor Law Commission guardians who had the sweeping powers of "cleaning stagnant pools and ditches, inspect[ing] lodging houses and prosecute[ing] any person failing to abate nuisances," and his demand from the unions for vital statistics on deaths and numbers of cases of infectious diseases. But, despite his and the board's unpopularity, these and other measures—paving roads, cleaning the streets, providing clean water, and carrying off wastes—did contribute to improving the public health. However, he and other members of the general board of health were dismissed because of their inability to arrest the 1853 to 1854 cholera epidemic. And, like so many in his time, Chadwick believed that cholera was caused by air pollution!

In July 1849 (during the second cholera pandemic) there was an outbreak in London, and John Snow traced the incidence of the disease to two companies supplying water to the city. He wrote, "During the 1849 epidemic, the water of both companies (Lambeth and Southwark & Vauxhall) was drawn from the same contaminated region of the River Thames, and the death rates among their consumers were similar." Then, as mysteriously as cholera had appeared, the epidemic subsided, and between late 1849 and August 1853 Londoners were spared from a cholera epidemic. However, when an outbreak occurred in 1854, Snow had an opportunity to conduct what he called "The Grand Experiment." Of the two companies that had supplied water to London, one of the companies, Lambeth, had moved its intake upstream of the site where sewage was deposited in the Thames, "thus obtaining a supply of water quite free from the sewage of London." Snow found that most of those affected drank water supplied by the Southwark & Vauxhall Company, whereas fewer cholera cases were found among people who drank water supplied by the Lambeth Water Company. Since the Southwark & Vauxhall Company (as Snow correctly observed) drew its water from the River Thames close to the outlet of the sewage system——it was malodorous and contained organic materials—he suspected that this was the source of cholera. At that time, however, since no bacteriologic examination was conducted, it was inductive logic based on epidemiologic evidence that allowed Snow to conclude that contaminated water was the source of the disease cholera.

In July of 1854 (during the third cholera pandemic) Snow's hypothesis that this disease was waterborne was put to the test when there was a severe outbreak of cholera in the Soho neighborhood of London: 500 people died in 10 days. Snow, in detective-like fashion, made a map of the affected area where most of the cholera cases occurred and found that most centered on Broad Street, where there was a pump that provided water to many of the houses in the neighborhood. The farther away from Broad Street, the fewer were the victims. There was a sewer pipe that ran right next to the Broad Street well that had broken and was contaminating the drinking water being drawn by the pump from the well. Snow finally got the local magistrate to "take the handle off the Broad Street pump" and the epidemic ceased. One of the idiosyncrasies of his Broad Street survey was that there were no cases of cholera in the employees of the local brewery, which was located close to the pump. Closer inspection showed that the brewery workers did not drink well water, they drank beer!

Snow's observations and conclusions were published in an 1855 paper, "On the Mode of Communication of Cholera," in which he postulated that cholera was spread by the drinking of contaminated water. Despite Snow's admonitions for sanitation, including boiling of water, cholera outbreaks continued, largely because most physicians were not convinced that drinking water was the cause and because this theory was against the prevailing doctrine of the miasmatists. Indeed, Snow's findings were completely rejected at the time, and it took decades for it to become generally accepted that cholera was a disease carried by water, not air. Acceptance of Snow's theory was eventually aided 35 years later, when the "germ theory" of Pasteur and Koch was established.

Cholera and Nursing

Florence Nightingale (1820–1910) was the founder of modern nursing, as we know it today. At the time she announced she wanted to become a nurse, the nursing profession was equivalent to the "oldest profession"—which was prostitution. In fact, most nurses of that time were women of loose morals, and they drank excessively. They slept on the wards with their patients, and the surgeons had sex available on demand. The hospitals themselves were filthy, and there was virtually no sanitation. Without any real means for alleviating pain, most patients died in anguish on the crowded, stench-filled, and dirty wards.

Florence Nightingale began her nursing training (such as it was) at the Institution for the Care of Sick Gentlewomen in London in 1853. When the Crimean War broke out, she was able to put her nursing skills to the test. The Crimean War raged from October 1853 until February 1856. The direct causes of the war were Russia's demands to exercise protection of Orthodox subjects in Turkey and the privileges of Russian and Roman Catholic monks in the holy places of Palestine. However, the Turks resisted and were able to take a firm stand against the Russians, largely because the Russian threat to take the Dardanelles in Turkey also threatened British sea routes in the Mediterranean Sea. Consequently, the British backed up the Turks. When the Russians moved into Romania, the British fleet was sent to Constantinople. The Turks then declared war against the Russians, but the Turkish fleet was quickly defeated by Russia in the Black Sea. To protect Turkish shipping, British and French troops entered this region of the Black Sea. With the British, French, and Turkish troops allied against Russia, and hoping to prevent Austria from entering the war, the Russians abandoned Romania. The British landed in the Crimea in 1854, beginning a year-long siege of the Russian fortress at Sevastopol. One of the major battles, at Balaklava, is immortalized in Alfred Lord Tennyson's poem "The Charge of the Light Brigade":

Theirs not to make reply
Theirs not to reason why
Theirs but to do and die
Into the Valley of Death rode the 600

Though the poem glorifies the responsibility of the soldier, the British cavalry brigade that attempted to maintain a hold on Balaklava failed because the Russians sent in a large reserve force from Sevastopol. Fewer than 30% of the British troops survived, partly due to British stupidity and rivalry between the two highest-ranking officers. Settlement of the Crimean War took place at the Congress of Paris in 1856.

How did Florence Nightingale get involved in nursing? In 1837, Nightingale believed she had heard the word of God, informing her of her mission. She was not exactly clear about what that mission was, but in 1850 she attended the German School of Nursing, and by 1853 she had risen through the ranks and become the superintendent of London's Institution for the Care of Sick Gentlewomen. She assisted with the cholera epidemic in London in 1854, and she instituted changes that were revolutionary: a bell whereby the patient could call for help, bulk buying of supplies, and an improved diet. When the Crimean War broke out, she was appointed Superintendent of Nursing for

Army Military Hospitals. The Crimea was rat and flea infested. Most of the casualties were due to gunshot wounds and disease—mostly cholera. At one point, more than 70% of the British troops were ill from disease and unfit for duty. Clean clothing was largely unavailable, and the hospital bedding consisted of straw mattresses. In addition, the hospital water allowance was limited to a pint per day. Florence Nightingale attributed the defects in hospital care to four things: (i) the agglomeration of a large number of sick under the same roof; (ii) a deficiency of space; (iii) deficiencies in ventilation; and (iv) deficient lighting. She set about to correct these defects. She recruited 38 nurses, asked for 200 scrubbing brushes, and washed the patients' clothes outside the hospital building. She organized the hospital kitchens, cleaned and painted the wards, disposed of the wastes in a more sanitary way, and spent time visiting the patients in the evening—hence she became known as the "Lady with the Lamp." In 2 weeks, the death rate had dropped by 80%. Pettenkofer would have been proud of her accomplishments.

After the war, she returned to England, and in 1860 she established the Training School and Home for Nurses—the forerunner of all nursing training. The wartime experiences of Florence Nightingale convinced her that the health administration of the British Army was in need of reform. She had studied mathematics and statistics as a young woman, and she put this to good use in assembling data on how administrative inadequacies affected patients' health. Her analysis of mortality for men showed that, even in peacetime, mortality was higher among men in the military than among civilian males of similar age. This higher mortality she ascribed to contagious disease, and she emphasized that this could be prevented by improved sanitary conditions. She published a thousand-page report, "Notes on Matters Affecting the Health, Efficiency and Hospital Administration of the British Army," using graphs to illustrate statistical relationships. Although it is unlikely that she would have considered herself an epidemiologist, she was one of the first to apply statistics to health data and to use these findings to promote health care reform in the interest of preventing disease and death.

Cholera and the "Immigrant Problem"

Grosse-Île, lying 30 miles east of Quebec in the middle of the St. Lawrence River, is the Laurentian gateway to Canada. Today, this picturesque island with its background of majestic peaks is a tranquil national park; its significance, however, lies in its being Canada's most visible link with Ireland's Great Famine. Grosse-Île, where thousands upon thousands of men, women, and children died of "famine fever," was the stage upon which the tragedy "quarantine" was played out.

The quarantine station at Grosse-Île was established after the Napoleonic Wars ended in 1815 and when growing numbers of people, especially the

Irish, left the British Isles, to make new lives for themselves in North America. The second cholera epidemic had taken 5 years to travel from India to Moscow to the British Isles, where in late 1831 it developed into a severe public health problem. Reports to the colonial government in Canada that people from the Old World with the dreaded disease were about to arrive via the St. Lawrence River prompted the Assembly of Lower Canada (as Quebec was then called) to pass a resolution on 23 February 1832 that made Grosse-Île a place of detention for cholera. Cholera was the most-feared epidemic disease of the 19th century and was commonly believed to originate "in the homes of the human riff-raff." In early June of 1832, the number of immigrants arriving in Canada had doubled, and with each ship's arrival so did the risk of a cholera epidemic beginning in North America. The Irish immigrants were profitable for the ships' owners, who had lost the African slave trade and now were able to fill the cargo bays on the outward-bound journey with other "wretched humans" in exchange for carrying Canadian timber to England. Because the numbers of arriving passengers was so great and the space on Grosse-Île so limited, many ships passed the island and entered Quebec harbor directly. By mid June, over 25,000 people had landed, and by summer's end the numbers exceeded 50,000. Soon cholera was in Quebec City itself and was spread beyond its borders by the newly arrived immigrants, who, believing they were healthy, made their way to Montreal and the cities of Ontario.

Grosse-Île lacked a proper jetty, so some of the debilitated passengers who had endured a miserable passage of more than 60 days were forced to wade ashore; many drowned, and the sandy bottom became their grave. For those who reached the island, death might have been a better fate. The writer Susannah Moodie, who emigrated to Canada from England in 1832, described her own prejudices:

> I looked up and down the glorious river; never had I beheld so many striking objects blended into one mighty whole! Nature had lavished all her noblest features in producing that enchanted scene. The rocky isle in front, with its farmhouses at the eastern point, and its high bluff at the western extremity, crowned with the telegraph . . . the middle space occupied by tents and sheds for cholera patients and its wooded shores dotted over with motley groups added to the picturesque scene. . . . Never shall I forget the extraordinary spectacle that met my sight the moment we passed the low range bushes which formed a screen in front of the river. A crowd of many hundred Irish emigrants had been landed during the present and former day and all . . . men, women and children, who were not confined to the sheds (which resembled cattle pens) . . . were employed in washing clothes or spreading them out on rocks and bushes to dry. The people appeared perfectly destitute of shame or a sense of common decency. Many were almost naked, still more partially clothed. We turned in disgust from the revolting scene. . . . Could we have shut out the profane sounds which came to us on every breeze, how deeply we should have enjoyed an hour amid the tranquil beauties of the . . . lovely spot.

That same summer, Catharine Traill passed the island and described in a letter home:

> There are several vessels lying at anchor close to the shore; one bears the melancholy symbol of disease, the yellow flag; she is a passenger ship and has the smallpox and measles among her crew. When any infectious complaint appears on board, the yellow flag is hoisted, and the invalids conveyed to the cholera hospital, a wooden building . . . surrounded with palisades and a guard of soldiers. There is also a temporary fort at some distance from the hospital, containing a garrison of soldiers, who are there to enforce the quarantine rules. The rules are . . . very defective and in some respects quite absurd. . . . When the passengers and crew of a vessel do not exceed a certain number they are not allowed to land, under penalty to both the captain and to the offender; but if . . . they should exceed the stated number, ill or well, passengers and crew must turn out and go ashore, taking with them their bedding and clothes, which are spread out on the shore, to be washed, aired and fumigated, giving the healthy every chance of taking the infection from the invalids near them.

The quarantine of the healthy and the sick together might be accepted as an example of the lack of appreciation of the means of transmission of cholera (since the vibrio responsible would not be discovered for another 50 years); however, its imposition was not the result of scientific ignorance at Grosse-Île but application of the practice of class distinction and scapegoating of the "wretched refuse" of Ireland. Indeed, cholera was felt by the English to be a vulgar and demeaning disease—a disease of the poor Irish—and this strengthened its repulsiveness and justified the quarantine. Quarantine shamed and blamed and contributed to the increased mortality on the island, estimated that summer at 2,000 deaths. By 1833, the number of sick immigrants arriving in Canada had dwindled, and in the intervening years up until 1846, all remained quiet at the Grosse-Île quarantine station.

The tragedy at Grosse-Île in 1847 stemmed from the famine in Ireland that began in 1845 and coincided with the arrival in Europe of the second cholera pandemic. Between 1845 and 1849, the time of the Great Irish Famine, the population of Ireland declined by over 2 million. Half died of starvation, disease, and malnutrition, while the other half emigrated. The United States was traditionally the route for Irish immigrants, but in 1847 the United States enforced an increase in the cost of passage, and ships that were overloaded were to be confiscated. This opened up new routes to the United States from Canada, as ship owners sought a cheaper option for their passengers. There was also an economic incentive for the English landlords to support the emigration of the Irish tenant farmers. It was reasoned that those tenants who remained on the land and were unable to pay their rents would become paupers. The cost of a landlord's maintaining a pauper in the poorhouse was 71 pounds per year, whereas the cost of passage was 31 pounds. The saving to the landlord in the first year was 40 pounds, and after that, the entire cost of pauper support was saved.

In 1847, fleeing the terrible failure of the potato crop and the rampant cholera in famine-stricken Ireland were hundreds of thousands of Irish who sailed for Quebec. Weakened by malnutrition and starvation, they were crowded aboard unsanitary sailboats unfit for transporting human beings. During the voyage of these "pest ships," those who died, along with their possessions, were hastily wrapped in canvas and thrown overboard as if they were dead birds or garbage.

On 15 May 1847, the curtain was raised on the drama of death that would play out on Grosse-Île. The first immigrant ship, Syria, arrived with 241 passengers. Fever and dysentery had broken out a few days after the ship had left the port of Liverpool (on 24 March). Nine passengers had died en route, and of those who landed, 84 were immediately hospitalized; another 24 would be admitted shortly thereafter. The first death in quarantine was 4-year-old Ellen Keane, and within a week 16 more were dead; before the end of May, 70 Irish had died. During that summer, 398 ships were inspected at Grosse-Île. Although ships usually took 45 days to cross the Atlantic Ocean, 26 of those that set sail in 1847 took over 60 days to reach Grosse-Île. Over 5,000 people died en route, and 5,424 were buried in a mass grave on the island. Other gravesites on the island replicated the lazy bed in which potatoes were grown in Ireland. Four physicians at Grosse-Île were aided by a crew of eight working from dawn until dark every day, digging trenches and burying the dead three deep. By August, dirt had to be imported to the rocky island to bury more bodies. Despite this, rats were coming off the ships to feed on the cadavers. All told, the number of deaths on the island probably exceeded 9,000, and many thousands more died elsewhere in colonial Canada during that "summer of sorrow."

The quarantine operation at Grosse-Île was characterized by haste, improvisation, and trial and error, and by physicians who lacked any understanding of what caused cholera, how it was transmitted, or how it might be treated. The island's facilities were ill equipped to accommodate and safely handle the large number of immigrants, especially those who were already ill. These problems were compounded by the effects of crowding and the unsanitary conditions aboard the ship during the long ocean crossing. No quarantine could meet this challenge and contain the spread of cholera. However, many historians feel that there were other factors that led to the death of those who were quarantined on Grosse-Île: they were Irish, they were Catholic, and they were subjects of an English government that considered them to be less than human for close to 1,000 years. In memory of that tragic summer, a Celtic cross was placed on a steep hill at Grosse-Île in 1909. Its inscription reads, in part, "Children of the Gael died in the thousands on this island having fled from the laws of foreign tyrants and an artificial famine. . . . God's blessing on them."

It has been suggested that pollution, especially sewage, has claimed more lives than smallpox and bubonic plague. "King Cholera," the most feared of all the sewage-related diseases, provoked global horror and terror, yet it also changed the way disease was looked at, how it was spread, and the manner by which human health could be preserved. In those countries where an adequate sewage and water supply system has been established, cholera has been eradicated. Cholera, however, remains endemic in areas of Asia, Africa, and in countries of South and Central America. Today, by quarantining and treating the sick, as well as preventing those who are infected from contaminating drinking water and food, cholera can be controlled. Provision of potable water almost always reduces epidemic spread.

Cholera, a disease that resulted in sanitary reforms and improvement of public health systems, has also been used to scapegoat those suspected of being carriers. The history of immigration provides many examples of massive movements of people resulting in transmission of disease across national borders. In large cities that attract a substantial number of immigrants, this can set the stage for dramatic shifts in civic and social organization and also for discrimination. At times, the public health organizations designed to protect the people at large became the "health police," and those who were already marginalized were labeled as bearers of disease and a threat to all of society. When some of the immigrants who arrived in developed countries became ill with cholera, this was sometimes extrapolated to all members of the ethnic or cultural minority to which the sick belonged, and the entire group was stigmatized. Naturally, the burden and blame fell, as it always does, on those without resources or political power—the immigrants and the urban poor. But the pandemics of cholera also had a positive effect: the fear and horror of cholera promoted the establishment of a public health system within many countries and inspired the formation of international bodies to monitor and control the global spread of all infectious diseases.

Figure 9.1 Mother and child infected with smallpox. (Courtesy of CDC/Dr. Stan Foster.)

Chapter 9

Smallpox, the Spotted Plague

In 1521, the subjects of the Aztec Empire numbered in the millions. Incredibly, Hernan Cortes, with fewer than 600 troops, was able to topple it. The Aztecs were militaristic and wealthy, having subjugated other indigenous Indian tribes and then extracted tribute from them. Cortes and his Spanish conquistadors set out to explore and claim Mexico for their king, Charles V. They landed in the Yucatan on the eastern coast of Mexico. With their armor plate, swords, horses, rifles, cannons, and attack dogs, they appeared to the Aztecs as a formidable fighting force. They moved toward the Aztec capital, Tenochtitlan (now Mexico City), without much incident, but at Tenochtitlan there were encounters that showed the limitations of guns, steel, and horseflesh. The outnumbered Spaniards lost one-third of their troops, and Cortes and his army were forced to retreat. Cortes expected a final and crushing offensive by the Aztecs, one that would result in their complete defeat. The attack never came. On 21 August, the Spaniards stormed the city, only to find that a greater force had ensured their victory. Bernal Diaz, witness to the scene, wrote,

> I solemnly swear that all the houses and stockades in the lake were full of heads and corpses. It was the same in the streets and courts. . . . We could not walk without treading on the bodies and heads of the dead Indians. I have read about the destruction of Jerusalem but I do not think the mortality was greater there than here in Mexico. . . . Indeed the stench was so bad that no one could endure it . . . and even Cortes was ill from the odors which assailed his nostrils.

Cortes's ally was a very efficient killer—smallpox (Fig. 9.1). The disease had preceded Cortes's arrival in Mexico. In 1520, an expedition led by Panfilo de Narvaez arrived from Spanish Cuba. Among the crew was a smallpox-infected slave. From this initial infection, smallpox spread from village to village throughout the Yucatan. So great was the die-off that there were not enough people to farm the land or protect the cities. Famine and havoc resulted. The paralyzing effect of the epidemic explains why the Aztecs did not pursue the demoralized Spaniards. It allowed them to rest and regroup, and in time permitted them to gather Indian allies. Smallpox was a devastatingly selective disease—only the Aztecs died; the Spaniards were left unharmed—and it so demoralized the survivors that the Amerindians had no doubt that the Spaniards were the beneficiaries of divine favor. It was this perceived superior power of the god the Spaniards worshipped that led the Aztecs and other Amerindians to accept Christianity. Without further resistance, they submitted to Spanish control. In the time of Cortes and his conquistadors, 3 million Amerindians, an estimated one-third of the total population of Mexico, were killed by smallpox.

Twelve years later (1532), another Spanish conquistador, Francisco Pizarro, led a group of 168 soldiers and captured Atahualpa, the leader of the Inca Empire. Pizarro held Atahualpa prisoner for 8 months, demanded a ransom of gold from his subjects for a promise of freedom, and, after the gold was delivered, reneged on the promise and had Atahualpa executed. Again, smallpox had made Pizarro's success in South America possible. Smallpox had arrived in Peru by land in 1526, killing much of the Inca population. When Pizarro landed on the coast of Peru in 1531, the stage for the Inca conquest had already been set; the smallpox epidemic that had preceded the conquistadors had weakened the Inca Empire, there was civil war, and Atahualpa's army was vulnerable and in disarray. Victory for Pizarro's conquistadors had been ensured by disease.

Smallpox was an Old World disease to which the New World Amerindians had never been exposed, and against which they had no immunity. Smallpox spread in advance of the Spaniards' entry into Mexico and Peru. In its wake followed death, devastation, and demoralization. Guns and steel were not the critical armaments the Spanish conquistadors used to destroy the Inca and Aztec empires. Disease moved ahead of them and, in effect, smallpox cleared the way.

Over the centuries, smallpox has killed hundreds of millions of people. In the 20th century alone, it killed 300 million people—three times the number of deaths from all 20th- century wars. Smallpox has been involved not only with war but also with exploration and migration. As a result, it has changed the course of human history. Smallpox is indiscriminate, with no respect for social class, occupation, or age; it killed or disfigured princes and paupers, kings and queens, children and adults, farmers and city dwellers, generals and their enemies, rich and poor. It has been suggested that in Europe the use of makeup began among wealthy survivors in order to hide smallpox-induced scars on their faces.

A Look Back

Smallpox was at one time one of the most devastating of all human diseases, yet no one really knows when smallpox began to infect humans. It is suspected that humans acquired the infectious agent from one of the pox-like diseases of domesticated animals, in the earliest concentrated agricultural settlements of Asia or Africa when humans began to maintain herds of livestock, some time after 10,000 BC. The best evidence of smallpox in humans is found in three Egyptian mummies, dating from 1570 to 1085 BC. One of these mummies is Pharaoh Ramses V, who died as a young man in 1155 BC. The mummified face, neck, and shoulders of the Pharaoh bear the telltale scars of the disease: pockmarks (Fig. 9.2). (The word "pox" comes from the Anglo-Saxon word *pocca*, meaning "pocket" or "pouch," and describes the smallpox fingerprint—craterlike scars.)

From its origins in the dense agricultural valleys of the great rivers in Africa and India, smallpox spread from the West to China, first appearing about 200 BC. Trade caravans assisted in the spread of smallpox, but at the time of the birth of Christ it was probably not established in Europe—the populations were too small and greatly dispersed. Smallpox was apparently known in Greece and Rome, but it does not seem to have been a major health threat until about AD 100, when there was a catastrophic epidemic called the Plague of Antonius (see p. 61). The epidemic started in Mesopotamia, and the returning soldiers brought it home to Italy. It raged for 15 years, and there were 2,000 deaths daily in Rome. Marcus Aurelius died of smallpox in AD 180. Smallpox, along with malaria, may have contributed to the decline of the Roman Empire.

There are records of smallpox in the Korean peninsula from AD 583. It had reached Japan by AD 585. In the western part of Eurasia, the major spread of smallpox occurred in the eighth and ninth centuries during the Islamic expansion across North Africa and into Spain and Portugal. As the disease

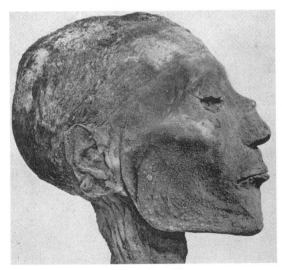

Figure 9.2 Pharaoh Ramses V and his pock-marked face.

moved into central Asia, the Huns were infected in either Persia or India. In the fifth century, when the Huns descended into Europe, they may have carried smallpox with them, but from other records it appears that by this time the disease had already established itself in Italy and France. In AD 622, a Christian priest, Ahrun, living in Alexandria 30 years before the Arab conquest of Egypt, wrote a clear description of the disease, differentiating it from measles, as did the Persian physician Al-Razi, living in Baghdad in AD 910. By AD 1000, smallpox was probably endemic in the more densely populated parts of Eurasia, from Spain to Japan, as well as in the African countries bordering on the Mediterranean Sea. The spread of smallpox was assisted by the caravans that crossed the Sahara to the more densely populated kingdoms of West Africa, and the disease was repeatedly introduced into the port cities of East Africa by Arab traders and slavers. The movement of people to and from Asia Minor during the Crusades in the 12th and 13th centuries helped reintroduce smallpox to Europe; in 1300, England and Germany suffered epidemics. By the fifteenth century, smallpox was in Scandinavia, and by the sixteenth century, it was established in all of Europe except for Russia. As Europe's population increased and people were crowded into the cities, smallpox epidemics appeared with increasing intensity and frequency.

Smallpox was a serious disease in England and Europe in the sixteenth century. As British and European explorers and colonists moved into the newly discovered continents of the Americas, Australia, and Africa, smallpox went with them. Besides the crucial role it played in the Spanish conquests in the New World, smallpox also contributed to the settlement of North America by the French and English. The English settlers who arrived in the American colonies in 1617 set off an epidemic among the Indians, thereby clearing a place for the settlers who came from Plymouth in 1620. The English used smallpox as a weapon of germ warfare. In the war of 1763 between England and France for control of North America, under orders of Sir Geoffrey Amherst (Commander-in-Chief of North America), the British troops were asked: "Could it not be contrived to send the smallpox among those disaffected tribes of Indians? We must on this occasion, use every stratagem in our power to reduce them." The ranking British officer for the Pennsylvania frontier, Colonel Henry Bouquet, wrote back: "I will try to inoculate the Indians with some blankets that may fall into their hands and take care not to get the disease myself." Blankets were deliberately contaminated with scabby material from the smallpox pustules and delivered to the Indians, initiating the spread of the disease among those highly susceptible people. This led to extensive numbers of deaths and ensured the Indians' defeat. The westward movement of smallpox among the Indian tribes moved far ahead of European and British settlements.

In West Africa, smallpox traveled with the caravans that moved from North Africa to the Guinea Coast. In 1490, it was spread by the Portuguese into the more southerly regions of West Africa. Smallpox was first introduced into South Africa by a ship docking in Capetown in 1713 that carried contaminated bed linen from a ship returning from India. Smallpox was again introduced into Capetown by a ship from Sri Lanka (1755) and later by one from Denmark (1767). The first outbreak of smallpox in the Americas was among African slaves on the island of Hispaniola in 1518. Here and elsewhere, the Amerindian population was so decimated that by the time of Pizarro's conquest it is estimated that 200,000 had died. With so much of the Amerindian labor force lost to disease, there was an increased need for replacements to do the backbreaking work in the mines and plantations of the West Indies, the Dominican Republic, and Cuba. This need for human beasts of burden stimulated, in part, the slave trade from West Africa to the Americas. Slaves also brought smallpox to the Portuguese colony of Brazil. Smallpox was first recorded in Sydney, Australia, in 1789, a year after the British established a penal colony there. The disease soon decimated many of the aboriginal tribes. As in the Americas, this destruction of the native population by smallpox paved the way for easy expansion by the British.

The Disease of Smallpox

The cause of smallpox is a virus—one of the largest viruses—and with proper illumination it can actually be seen with the light microscope. However, much of its detailed structure can be visualized only by using the electron microscope. The outer surface (capsid) of the smallpox virus resembles the facets of a diamond, and its inner dumbbell-shaped core contains the genetic material, double-stranded DNA (Fig. 9.3). The virus has about 200 genes, 35 of which are believed to be involved in virulence. Other poxviruses that can infect humans are monkeypox, cowpox, milker's node, tanapox, and chicken pox. Most of these viruses cause mild disease in humans, with the exception of the zoonotic monkeypox, which is virtually indistinguishable from smallpox.

Most commonly, the smallpox virus enters the body through droplet infection by inhalation. However, it can be acquired by direct contact or through contaminated fomites (inanimate objects) such as clothing, bedding, blankets, and dust. The infectious material from the pustules can remain infectious for months. In the mucous membranes of the mouth and nose, the virus multiplies. During the first week of infection, there is no sign of illness; however, the virus can be spread by coughing or by nasal mucus. The virus moves on to the lymph nodes and then to the internal organs via the bloodstream. Here the virus multiplies again. Then the virus reenters the bloodstream. Around the ninth day, the first symptoms appear: headache, fever, chills, nausea, muscle ache, and sometimes convulsions. The person feels quite ill. A few days later a characteristic rash appears (Fig. 9.4).

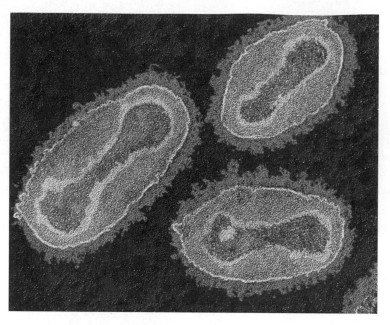

Figure 9.3   The smallpox virus as seen with the transmission electron microscope. The dumbbell-shaped structure is the core containing the genetic material (DNA). (Photo by Eye of Science/Photo Researchers, Inc. © 2005 Photo Researchers, Inc.)

Richard Preston in *The Demon in the Freezer* described it:

The red areas spread into blotches across [the] face and arms, and within hours the blotches broke out into seas of tiny pimples. They were sharp feeling, not itchy, and by nightfall they covered [the] face, arms, hands, and feet. Pimples were rising out of the soles of [the] feet and on the palms of the hands, too. During the night, the pimples developed tiny, blistered heads, and the heads continued to grow larger . . . rising all over the body, at the same speed, like a field of barley sprouting after the rain. They hurt dreadfully, and they were enlarging into boils. They had a waxy, hard look, and they seemed unripe . . . fever soared abruptly and began to rage. The rubbing of pajamas on [the] skin felt like a roasting fire. By dawn, the body had become a mass of knob-like blisters. They were everywhere, all over . . . but clustered most thickly on the face and extremities . . . The inside of the mouth and ear canals and sinuses had pustulated . . . it felt as if the skin was pulling off the body, that it would split and rupture. The blisters were hard and dry, and they didn't leak. They were like ball bearings embedded in the skin, with a soft velvety feel on the surface. Each pustule had a dimple in the center. They were pressurized with an opalescent pus.

The pustules began to touch one another, and finally they merged into confluent sheets that covered the body, like a cobblestone street. The skin was torn away . . . across . . . the body, and the pustules on the face combined into a bubbled mass filled with fluid until the skin of the face essentially detached from its underlayers and became a bag surrounding the tissues of the head . . . tongue,

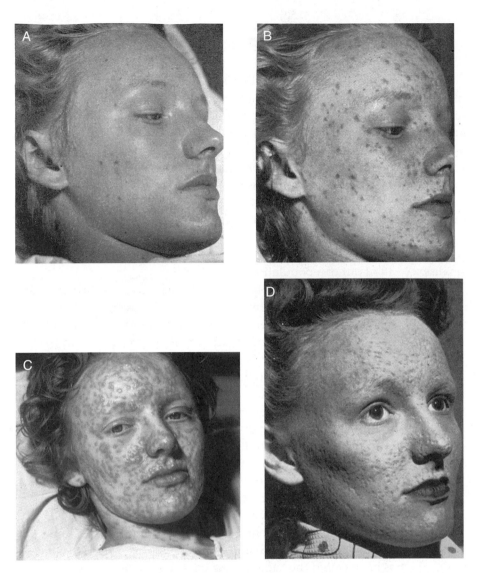

Figure 9.4   The progression of smallpox [From: C. W. Dixon, Smallpox (London: Churchill 1962)]. (A) Day 3; (B) day 4; (C) day 6; and (D) six months after infection with residual facial scarring and loss of eyebrows and eyelashes.

gums, and hard palate were studded with pustules . . . the mouth dry. . . . The virus had stripped the skin off the body, both inside and out, and the pain . . . seemed almost beyond the capacity of human nature to endure.

The individual is infectious a day before the rash appears and until all the scabs have fallen off. Many die a few days or a week after the rash appears. Not infrequently, there are complications from secondary infections. The infection results in a destruction of the sebaceous (oil) glands of the skin, leaving permanent craterlike scars in the skin—pockmarks.

There are two pathologic varieties of the smallpox virus: Variola major and Variola minor. Each can be distinguished from the other by differences in their genes. Variola major, the deadlier form, frequently killed up to 25% of its victims, although in naive populations, such as the Amerindians, the fatality rate could exceed 50%. Variola minor, a milder pathogen, had a fatality rate of about 2% and was more common in Europe until the 17th century, when it mutated to the more lethal form, perhaps as a result of reintroduction from the Spanish colonies in the Americas. In the 17th century, smallpox was Europe's most common and devastating disease, killing an estimated 400,000 each year. It caused one-third of all cases of blindness. England suffered most severely in the second half of the 17th century: between 1650 and 1699, there were 20 deaths a week. In 1660, in London alone, more than 57,000 died of smallpox (~10%) out of a population of 500,000.

Smallpox also played a role in the British royal houses, in particular, the succession of the House of Stuart by the House of Hanover. The pedigrees of the Houses of Stuart and Hanover show the royals who were afflicted with smallpox (Fig. 9.5). William of Gloucester, son of Queen Anne and heir to the throne, died of smallpox at age 11. This precipitated a constitutional crisis. The Act of Settlement that followed barred another Catholic sovereign and gave the English throne to the House of Hanover via Elizabeth, Queen of Bohemia. Succeeding Anne were George I, II, III, IV, William IV, and Victoria. Not only did smallpox alter the course of British history, but it also affected many of the other royal families of Europe. Indeed, in the 80 years before 1775, smallpox killed a queen of England (Mary II), William II of Orange, an Austrian emperor (Ferdinand IV), an emperor of Germany (Joseph I), a king of Spain (Balthazar), a tsar of Russia (Peter II), and a king of France (Louis XIV's son, Louis XV).

By the beginning of the 18th century, nearly 10% of the world's population had been killed, crippled, or disfigured by smallpox. The 18th century in Europe has been called the "age of powder and patches" because pockmarks were so common. The "beauty patch" (a bit of colored material) seen in so many portraits was designed to hide skin scars. In this way the pockmark set a fashion.

"Catching" Smallpox

Smallpox is a contagious disease of civilization that is spread from person to person, and there are no animal reservoirs, i.e., it is not a zoonotic disease. It can exist in a community only so long as there are susceptible humans. It has been estimated that a minimum population of the order of 100,000 persons is needed to ensure that there are enough susceptible persons born annually

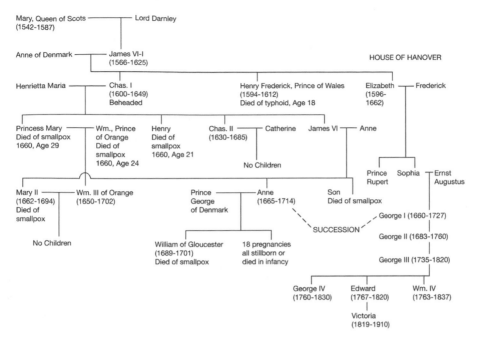

Figure 9.5   Pedigree of the Royal British Houses of Stuart and Hanover.

to sustain the chain of infection indefinitely; the R_0 value (i.e., the number of secondary infections that result from a single primary infection; see p. 13) has been calculated to be between 5 and 10 for epidemic smallpox. Smallpox spreads more rapidly during the winter in temperate climates and during the dry season in the tropics.

Smallpox caused large epidemics when it first arrived in a virgin community where everyone was susceptible, but afterwards, most individuals who were susceptible either recovered and were immune or died; then the disease died down. If the infection was reintroduced later, that is, after a new crop of susceptible individuals had been born or migrated into the area, then another epidemic would occur. However, after a period of time, the disease would reappear so frequently that only newborns would be susceptible to infection, older individuals being immune from previous exposure. In this situation, smallpox became a disease of childhood similar to measles, chicken pox, mumps, whooping cough, and diphtheria. Sometimes smallpox didn't disappear altogether, especially in larger communities, but remained at a low level. However, infection would break out as an epidemic every 5 to 15 years when enough susceptible individuals had accumulated. Smallpox affected both sexes and all ages. Mortality was highest among infants, and smallpox can induce abortions in all trimesters of pregnancy.

Variolation

Early treatments for smallpox involved prayer and quack remedies. In ancient Africa and Asia, there were smallpox gods and goddesses that could be enlisted for protection. In France, one could appeal to Saint Nicaise, the Bishop of Rheims, who recovered from the disease only to be killed by the Huns in 452. Because smallpox produces a red rash, a folklore remedy of "like cures like" arose in medieval times. In 1314, the Englishman John Gaddeson followed this recipe and recommended that smallpox victims could be helped by the color red—so those who were infected were dressed in red. In 1834, Niels Finsen, the Dane who pioneered light therapy, gave "like cures like" a pseudoscientific status by prescribing that those afflicted with smallpox be screened from ultraviolet light and exposed only to red light. Other charlatans suggested cure would come about by eating red foods and imbibing red drinks. These remedies, which did no good, persisted in some quarters until the 1930s! Today, prevention of smallpox is often associated with Edward Jenner, but even before Jenner's scheme of vaccination, techniques were developed to induce a mild smallpox infection. The educated classes called this practice of folk medicine "inoculation" from the Latin *inoculare*, meaning "to graft," since it was induced by cutting. It was also called "variolation" because the scholarly name for smallpox was "variola" from the Latin *varus*, meaning "pimple." The method of protection by variolation may have been inspired by the idea of "like cures like," since in some versions of this remedy, secretions from an infected individual would be inoculated into the individual to be protected from that same infection.

There were numerous techniques of variolation. The Chinese avoided contact with the smallpox-infected individual, preferring that the person either inhale a powder from the dried scabs shed by a recovering patient or place the powdered scabs on a cotton swab and insert this into the nostrils. In the Near East and Africa, material from the smallpox pustule was rubbed into a cut or a scratch in the skin. The first scholarly account of variolation (or immunization against smallpox) was written in 1675 by Thomas Bartholin, physician to King Charles V of Denmark and Norway. Forty years later, this work came to the notice of the physician to Charles XII, Emanuel Timoni, who was living in Turkey. He in turn communicated a description of a method of variolation to the Royal Society of London: material was taken from a ripe pustule and rubbed into a scratch or an incision in the arm or leg of the person being inoculated.

Mary Pierpont was a highborn English beauty who eloped with Edward Wortley Montague, a grandson of the first Earl of Sandwich. Soon after Edward became a member of Parliament, the couple became favorites at the court of King George I. In London in 1715, at age 26, Lady Mary Montague contracted smallpox. She recovered, but smallpox-induced facial scars and the loss of her eyelashes marred her beauty forever. Her 22-year-old brother was less fortunate, however, and died from smallpox. The following year, Lady Montague's husband was appointed ambassador to Turkey, and she accompanied him. There she learned of the Turkish practice of variolation. She was so impressed by what she saw that she had her son inoculated. Upon returning to England in 1718, she propagandized the practice, and she had her daughter inoculated against smallpox at age 4. She convinced Caroline the Princess of Wales (later Queen of England during the reign of George II) that her children should be variolated, but first the royal family required a demonstration of its efficacy. In 1721, the "royal experiment" began with six prisoners who had been condemned to death by hanging and were promised their freedom if they subjected themselves to variolation. They were inoculated, suffered no ill effects, and upon recovery were released as promised. Despite this, Princess Caroline wanted more proof and insisted that all the orphan population of St. James Parish be inoculated. In the end only six children were variolated, with great success. Princes Caroline, now convinced, had both her children (Amelia and Caroline) variolated. This, too, was successful. However, Lady Montague met with sustained and not entirely inappropriate opposition to the practice of variolation. The clergy denounced it as an act against God's will and physicians cautioned that the practice could be dangerous to those inoculated as well as to others with whom they came in contact. The danger of contagion was clearly demonstrated when the physician who had previously inoculated Lady Montague's children variolated a young girl who then proceeded to infect six of the servants in the house; although all seven recovered, it did cause a local epidemic. Although inoculation was embraced by the upper classes in England—Lady Montague's physician was called upon to variolate Prince Frederick of Hanover—it never caught on with the general population there or elsewhere. This had less to do with some of its demonstrated successes than it did with the fear of contagion, the risks involved in inoculation with smallpox, and the lack of evidence that the protection would be long lasting. Later, it was found that variolation did provide lifelong protection, but the danger from death remained at ~2%.

Variolation came into practice quite independently in the English colonies in America. Cotton Mather, a Boston clergyman and scholar, was a member of the Royal Society of London. Although he read the account of Emanuel Timoni, he wrote to the Society that he had already learned of the process of variolation from one of his African slaves, Onesimus. In April 1721, there was a smallpox outbreak in Boston. Half the inhabitants fell ill, and mortality was 15%. During this time, Mather tried to encourage the physicians in the city to carry out variolation of the population. The response from the physicians was negative save for one, Zabdiel Boylston. On 26 June Boylston variolated his 6 year-old son, one of his slaves, and the slave's three-year-old son without complications. By 1722, Mather, in collaboration with

Dr. Boylston, had inoculated 242 Bostonians and found that it protected them. His data showed a mortality rate of 2.5% in those variolated, compared with the 15 to 20% death rate during epidemics. Boylston's principal opponents, however, were not the clergy but his fellow physicians.

In the crowded cities of England during the 18th century, most adults had acquired some measure of immunity to smallpox through exposure during childhood. However, in the sparsely settled villages in the American colonies, there was little exposure and hence little immunity. With the outbreak of war in 1775, the most dangerous enemy the colonists had to fight was not the British but smallpox. Following the Battle of Bunker Hill in June of 1775, General Howe's forces occupied Boston, and Washington's troops were deployed on the surrounding hills. Howe was unable to attack because smallpox was so rampant among the Bostonians and to a lesser degree (perhaps because of their practice of variolation) among his troops. Washington, however, did not attack for fear that his troops would be decimated by smallpox. So he first ordered his troops to be variolated. When the British evacuated the city on 17 March 1776, Washington sent 1,000 of his men who were immune to smallpox to take possession of the city.

Earlier (in 1775), the American colonists feared that the British forces from Quebec would attack New York. The Americans sent 2,000 colonial troops under General Benedict Arnold to attack the poorly fortified garrison at Quebec. Although General Arnold's army was superior and better equipped, the colonials halted their advance on Quebec when they were suddenly stricken with smallpox. Over half of Arnold's troops developed the disease, and mortality was very high. Few of the others were well enough to fight. In contrast, the much smaller British forces were well variolated and fit, so they could hold out until reinforcements arrived. The colonials abandoned their siege of the garrison and retreated to Ticonderoga and Crown Point on Lake Champlain, where they continued to die at a high rate. The Americans never made another serious attempt to intrude into Canada. Ten years later, based on this experience, Washington ordered that his entire army be variolated, but by that time it was too late—Canada had been preserved for the British Empire.

Vaccination

Today, smallpox has been eradicated thanks to the development of the first vaccine by Edward Jenner (1749–1823), an English physician. Jenner did not know what caused smallpox or how the immune system worked; nevertheless, he was able to devise a practical and effective method of immune protection against attack by the lethal virus Variola major. Essentially, Jenner took advantage of a local folktale and turned it into a reliable and practical device against the ravages of smallpox. The farmers of Gloucestershire believed that if a person contracted cowpox, he or she was assured of immunity to smallpox. In milk cows, cowpox shows up as blisters on the udders that clear up quickly and produce no serious illness (Fig. 9.6). Farm workers who come into contact with cowpox-infected cows develop a mild reaction with the eruption of a few blisters on the hands or lower arm. Once exposed to cowpox, neither cows nor humans develop any further symptoms. It was noted that hardly a milkmaid or a farmer who had contracted cowpox showed any of the disfiguring scars of smallpox, and most milkmaids were reputed to have fair and almost perfect complexions as suggested in the nursery rhyme:

> Where are you going, my pretty maid
> I'm going a-milking sir, she said
> May I go with you, my pretty maid
> You're kindly welcome, sir, she said
> What is your father, my pretty maid
> My father's a farmer, sir, she said
> What is your fortune, my pretty maid
> My face is my fortune, sir, she said.

In 1774, a cattle breeder named Benjamin Jesty contracted cowpox from his herd, thereby immunizing himself. He then deliberately inoculated his wife and two children with cowpox. They remained immune even 15 years later, when they were deliberately exposed to smallpox. Edward Jenner, however, not Jesty, is given credit for vaccination because he carried out experiments in a systematic manner over a period of 25 years to test the farmer's tale. Jenner wrote:

> It is necessary to observe, that the utmost care was taken to ascertain, with most scrupulous precision, that no one whose case is here adduced had gone through smallpox previous to these attempts to produce the disease. Had these experiments been conducted in a large city, or in a populous neighborhood, some

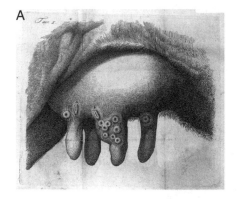 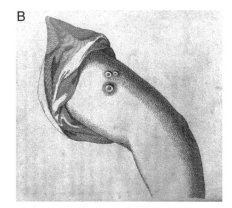

Figure 9.6 Cowpox pustules on cow's udder (A) and vaccination on human arm (B). (Courtesy of the Wellcome Library of Medicine.)

doubts might have been entertained; but here where the population is thin, and where such an event as a person's having had the smallpox is always faithfully recorded, no risk of inaccuracy in this particular case can arise.

As with the earlier "royal experiment" Jenner felt free to carry out human experimentation without any reservations, although it should be noted that all of his patients wanted the vaccination for themselves or their children. On 14 May 1796, he took a small drop of fluid from a pustule on the wrist of Sarah Nelms, a milkmaid who had an active case of cowpox. Jenner smeared the material from the cowpox pustule onto the unbroken skin of a small boy, James Phipps (Fig. 9.6). Six weeks later, Jenner tested his vaccine (from the Latin *vacca*, meaning "cow") and its ability to protect against smallpox by deliberately inoculating the boy with material from a smallpox pustule. The boy showed no reaction—he was immune to smallpox. In the years that followed, "poor Phipps" (as Jenner called him) was tested for immunity to smallpox about a dozen times, but he never contracted the disease. Jenner wrote up his findings and submitted them to the Royal Society; to his great disappointment, his manuscript was rejected. It has been assumed that the reason for this was that he was simply a country doctor and not a part of the scientific establishment. In 1798, after a few more years of testing, he published a 70-page pamphlet, "An Inquiry into the Causes and Effects of Variola Vaccinae," in which he reported that the inoculation with cowpox produced a mild form of smallpox that would protect against severe smallpox, as did variolation. He correctly observed that the disease produced by vaccination would be so mild that the infected individual would not be a source of infection to others—a discovery of immense significance.

The reactions to Jenner's pamphlet were slow, and many physicians rejected his ideas. Cartoons appeared in the popular press showing children being vaccinated and growing horns (Fig. 9.7A). But within several years, some highly respected physicians began to use what they called the "Jennerian technique" with great success. By the turn of the century, the advantages of vaccination over variolation were clear, and Jenner became famous. Parliament awarded him a prize of 10,000 pounds in 1802 and another 20,000 pounds in 1807. Napoleon had a medal struck in his honor, and he received honors from governments around the world.

By the end of 1801, Jenner's methods were coming into worldwide use, and it became more and more difficult to supply sufficient cowpox lymph or to ensure its potency when shipped over long distances. To meet this need, Great Britain instituted an Animal Vaccination Establishment in which calves were deliberately infected with cowpox, and lymph was collected. Initially, this lymph was of variable quality, but when it was found that addition of glycerin prolonged preservation, this became the standard method of production. The first "glycerinated calf's lymph" was sent out in 1895. With gen-

TRIUMPH OF DE-JENNER-ATION.

[The Bill for the encouragement of Small Pox awaits Third Reading in the Commons.]

1898.

Figure 9.7 The vaccination for smallpox has been described in both a humorous (A) and a frightening (B) way. (Courtesy of the Wellcome Library of Medicine.)

eral acceptance of Jenner's findings, the hide of the cow called "Blossom" that Jenner used as the source of cowpox in his experiments was enclosed in a glass case and placed on the wall of the library in St. Georges's Hospital in London, where it remains to this day.

Why does vaccination work? As mentioned earlier there are two kinds of smallpox virus in humans: Variola major, which can kill, and the less virulent, Variola minor. Variolation or inoculation with Variola minor obviously protects against the lethal kind of virus. But vaccination involves cowpox, and this disease is caused by a different virus named by Jenner *Variola vaccinae*. The exact relationship between the virus used in vaccines today and Jenner's cowpox is obscure. Indeed, in 1939 it was shown that the viruses present in the vaccines used were distinct from contemporary cowpox as well as smallpox. Samples of Jenner's original vaccine are not available, so it cannot be determined what he actually used, but current thinking is that the available strains of vaccine are derived neither from cowpox nor smallpox, but from a virus that became extinct in its natural host: the horse.

Variola vaccinae and Variola major are 95% identical and differ from each other by no more than a dozen genes. For there to be protection, there must be cross-reactivity between the two such that antibodies are produced in the vaccinated human that can neutralize the Variola major virus should the need arise. Fortunately, this does happen. However, although recovery from a smallpox attack gives lifelong immunity, vaccination ordinarily does not. The advantages of vaccination over variolation are twofold. First, the recipient of a vaccination is not infectious to others, and, second, death occurs very rarely after vaccination, whereas with variolation not only is the individual infectious, but the death rate is 1 to 3%.

Some consider smallpox a disease of historical interest because it was certified as eradicated by the World Health Assembly on 8 May 1980. This feat was accomplished 184 years after Jenner introduced vaccination. The program began in 1967. By 1970, smallpox had been eliminated from 20 countries in western and central Africa; by 1971, it had been eliminated from Brazil; by 1975, it had been eliminated from all of Asia; and in 1976, it was eliminated from Ethiopia. The last natural case was reported from Somalia in 1977. Eradication of smallpox was possible for three reasons: First, there were no animal reservoirs. Second, the methods of preserving the vaccine proved to be effective. Third, the vaccine was easily administered. Smallpox is the first and only naturally occurring disease to be eradicated by human intervention.

With eradication, there arose another question: what should be done with smallpox virus stocks? Should they be destroyed, or should they be transferred to a few reference laboratories? In 1979 the World Health Organization recommended transfer. Transfer, however, is hazardous to those working with the virus, and if stocks were obtained by terrorist groups or rogue governments, they could be used as an agent of biological warfare.

A nightmare biohazard scenario is that smallpox is released into a major city. The virus would spread rapidly through the entire unprotected population. The death rate in those without any immunity would be 30%. Anyone born after 1970, when most countries stopped vaccinating against smallpox, would be vulnerable. Imagine the worldwide catastrophe of millions of people dying from smallpox as a result of a bioterrorist attack. In the United States alone, the 120 million people who were born after the end of routine vaccination would be highly susceptible, and there could be 40 million deaths! What we have learned about biological weapons is this: they have a high destructive potential, they can be unpredictable in their consequences, they are cheap and easy to prepare with a minimum of scientific knowledge, they may be difficult to detect, and they will certainly have a devastating psychological impact. So perhaps destruction of the stocks of smallpox would be better—or would it? Destroying the stocks would result in a loss of scientific knowledge of smallpox pathogenesis and would prevent the design of new antiviral agents to prevent large outbreaks of smallpox should they ever recur. This is the dilemma. As yet, it remains unresolved.

Vaccination and Its Social Context

In 1799, news of Jenner's vaccine reached the American colonies. Benjamin Waterhouse (1754–1846), a physician and professor at the newly established Harvard Medical School, was one of the first to send for the vaccine, and he popularized its use in Boston. His first trial was in 1800 when he vaccinated his son; then he sent a pamphlet and list of successes to Vice President Thomas Jefferson. Jefferson enthusiastically endorsed the "Jennerian technique" and had his own family and his slaves vaccinated. Jefferson expected the public to follow his lead, but they did not. Indeed, it wasn't until the term of James Madison, the fourth president of the United States, who was familiar with the positions of Waterhouse and Jefferson, that a vaccine law was signed encouraging vaccination. But Waterhouse's success with Jefferson turned out to be grist for the mill of his adversaries. Waterhouse, a practicing Quaker, was a pacifist and opposed the Revolutionary War. Born in Rhode Island, he was considered an outsider to the conservative Boston community, and he further incurred the wrath of the Boston elite (who were supporters of Federalism) when he allied himself with the populist democracy espoused by Jefferson. His political enemies, consisting of a coalition of physicians from Harvard, the clergy, and others from the Boston area, arranged for his dismissal from the Harvard Medical School. In 1820, the vaccine law was repealed, and in 1822 Dr. James Smith, the federal agent for the distribution of cowpox, was dismissed from office. Politics was favored over good public health policy, and the result was that by 1840, epidemics and deaths from smallpox once again increased in the United States.

Some European states did make vaccination compulsory (e.g., Bavaria, in 1807), and by 1869 it was required in Germany that all citizens and especially soldiers be vaccinated. This proved to be of great benefit to Germany during the Franco-Prussian War (1870 to 1871) when the number of cases of smallpox in the German army was slightly more than 8,000, with fewer than 300 deaths, whereas among the French, who did not practice vaccination, there were 28,000 cases and 2,300 deaths. However, compulsory vaccination never became a federal policy in the United States; communities were free to vaccinate or not. But with the expansion of elementary and secondary education after the Civil War, and eventually the compulsory attendance of children, the most effective way of enforcing vaccination was for public schools to require a vaccination scar on the arm before any child could attend.

This should have solved the problem but it did not. Up until the 1900s, there were antivaccination societies whose members believed the practice of vaccination to be dangerous, ineffective, and a violation of their civil liberties (Fig. 9.7B). The strength of the movement was clearly demonstrated by the "Milwaukee riots" in 1890. Milwaukee, Wis., was home to an increasing number of German and Polish immigrants who worked in the factories and lived in the poorer sections. During a smallpox epidemic in 1894, the public health authorities pursued a strict policy of enforcement, in some cases actually removing children suspected of being infected from their homes and placing them in the city's isolation hospital. The immigrant community believed not only that their civil rights were being violated, but also that they were being discriminated against because they were poor immigrants. Some believed they were being poisoned, and others held to the belief that home care was better than hospitalization. They asked, if the city's smallpox hospital was so safe, why had it been built in the part of the city where the poorest people lived? When the health department ambulance came to collect a child suspected of having the disease, the people threw stones and scalding water at the ambulance horses, and the guards were threatened with baseball bats, potato mashers, clubs, bed slats, and butcher knives. In response, the ambulances withdrew. By the end of the epidemic, the health commissioner had been impeached and the health department had lost its police powers, which it never regained.

The vaccination controversy did not end in Milwaukee. In May 1901, there was an outbreak of smallpox in Boston. The fatality rate was 17%; there were ~1,500 cases in a population of ~550,000. By the fall of that year, the Boston Board of Health took steps to control the epidemic: a free and voluntary vaccination program was started. By December, 400,000 people had been vaccinated, but outbreaks of smallpox continued to be reported. The board of health established "virus squads" with the following orders: (i) all inhabitants of the city must be vaccinated or revaccinated, (ii) there will be a house-to-house program of vaccination, and (iii) vaccine should be administered to all. Persons who refused vaccination were fined $5 or sentenced to 15 days in jail. As is frequently the case, the disenfranchised of society were blamed for spreading the disease. In this instance, it was the homeless. Virus squads were dispatched to cheap rooming houses to forcibly restrain and vaccinate the occupants. The opponents of vaccination (Anti-Compulsory Vaccination League) swung into action, complaining that vaccination was a violation of civil liberties, calling the practice "De-Jenneration" (Fig. 9.7B). By January 1902, legislation was proposed to repeal the state's compulsory vaccination law. This led to a landmark legal case on the constitutionality of compulsory vaccination (*Johnson v. Massachusetts*). In 1905, the U.S. Supreme Court voted 7 to 2 in favor of the state, ruling that although the state could not pass laws requiring vaccination in order to protect an individual, it could do so to protect the public in the case of a dangerous communicable disease. The epidemic ended in March 1903.

The Boston epidemic of 1901 to 1903 illustrates the benefits of educating the public about the benefits of disease prevention, and the value of having a public debate on the pros and cons of public health policies. It also shows how well-meaning boards of health can institute policies that disregard civil liberties and ethical concerns.

Smallpox was endemic worldwide for more than 3,000 years. The fatality rate was particularly high in naive populations. The disease affected the course of British, European, and American history. It contributed to the exploitation of the Amerindians, helped to destroy their native culture, promoted their conversion to Christianity, accelerated the slave trade to the New World, affected the outcome of battles during the Revolutionary War, paved the way for European and British colonial settlements, and altered the succession of royal houses. Smallpox also provided the incentive for the development of protective measures (variolation and vaccination), affected the cultural responses to disease, and contributed to the establishment of more humane public health policies. The eradication of smallpox has demonstrated that through vaccination there is the possibility for the elimination of other infectious diseases that continue to plague humankind.

Figure 10.1   Prince Valiant rides to save the castle. Cartoon by Hal Foster. © King Features. (Courtesy of Manuscript Press.)

Chapter 10

Preventing Plagues

The year was 430 BC, and Athens was being devastated by a plague. The Greek historian Thucydides wrote,

> The infection first began, it is said in parts of Ethiopia above Egypt and thence descended into Egypt and Libya and into most of the king's country [i.e., Persia]. Suddenly falling upon Athens, it first attacked the population in Piraeus . . . and afterwards appeared in the upper city, when the deaths became more frequent. Fear of gods or laws of men there was none to restrain them. As for the first, they judged it to be just the same whether they worshipped them or not, as they saw all alike perishing. As for the latter, no one expected to live to be brought to trial for his offenses . . . even the most staid and respectable citizens devoted themselves to nothing but gluttony, drunkenness and licentiousness.

The Plague of Athens, described by Thucydides, raged for 4 years, during which time 25% of the population died. Many historians regard this plague as a "turning point" in the history of Western civilization. However, Thucydides' description does not allow for a clear distinction as to the specific agent that caused this epidemic. Some have argued it must have been typhus, because it broke out during wartime, its clinical course lasted for 7 to 10 days, and there was a rash, fever, and delirium. Hans Zinsser, an expert on typhus, was not persuaded that the plague's cause was typhus, since Thucydides makes no mention of lice or a cloudy mental state, and, furthermore, typhus usually does not have a fatality rate of 25%. Others argue that the disease was smallpox, because Thucydides describes skin pustules and hints that the rash spread from the face to the trunk; however, he does not mention pockmarks among the survivors. Its cause still remains a mystery. But Thucydides, who was stricken with this plague and recovered, wrote, "Yet still the ones who felt most pity for the sick and dying were those who had the plague themselves and recovered from it. They knew what it was like and at the same time felt themselves to be safe, for no one caught the disease twice, or if he did, the second attack was never fatal." There is no mystery here: those who survived this plague were immune.

What does it mean when we say someone is immune? The dictionary definition of "immune" states that it comes from the Latin and is derived from the time in ancient Rome when citizens were given freedom from some onerous duty owed to the empire. But in medicine, "immune" means protected from a disease.

Immune Defense

Imagine you are living in the time of the Crusades. You are a knight (Fig. 10.1) and a member of the royal court. Your lord resides in the castle and owns the land surrounding it, called the manor. The manor is worked by peasants, who are subjects of the lord and under his protection. Within the castle is a working community consisting of members of the royal family—lords and ladies, dukes and duchesses—an assortment of trades people, maids, cooks, children, horses, dogs, cows, chickens, pigs, and other livestock. The lord conferred knighthood on you and made you a professional soldier; in turn, you have sworn obedience to the lord and promise to protect him and his unarmed subjects because they are unable to defend themselves against an invading foe. When invasion is imminent, the subjects flee to the safe confines of the castle grounds. The castle has been designed to be virtually impregnable, and the expectation of the community is that any threat from invasion can either be repelled or prevented so that the inhabitants will not suffer from hostile outsiders. The outermost defenses of the castle include the water-filled moat and the close-fitting stones of the castle walls penetrated by windows and other very small openings. This first line of defense, consisting of a set of physical barriers, provides internal security for all members of the castle's community. If these barriers of moat and walls are threatened and partially breached by hostile invaders, it is possible to delay those who want to intrude by dumping boiling water, heated oil, or molten lead on them from the tops of the castle walls. Again, physical means are used to prevent invader access. If the walls are breached by the enemy, then the inhabitants, including you, will have to employ an additional set of defensive measures. In the earliest stages of battle, foot soldiers, without any specific directions from the lord, would engage the intruders. Then, to ensure victory, the lord could dispatch his archers to shoot arrows at the enemy. The lord might also have need to summon you to do battle. He instructs you and the other knights to suit up in armor and mount horses. The knights use their weapons—sword, mace, and axe—to fight off the enemy. Combat with the foreign invader is man to man, and the combatants are trained soldiers— knights and foot soldiers; the lord's subjects (nonprofessional soldiers) may also assist in the defense of the castle by shouting out alarms to alert others to enter the fray or using stones, sticks, pitchforks, hammers, pots, and pans to drive back the enemy. Once the invaders have been successfully repelled, the community can begin to repair the damage. To protect against another such attack, the archers and knights will remember the defensive strategies critical to defeat of the invading foe. It is their memory of a previous battle that will enable them to mount a swift and more effective protective response in the next encounter, should that occur.

Like the defense of the crusader castle, there are several levels of protective human immune mechanisms. The first involves the physical and physiological barriers of the body: the close-fitting epithelial cells of the skin ("castle walls") and secretions, such as tears containing the antibacterial lysozyme, mucus, sebum (the oil associated with hair follicles), and the acidic gastric juice of the stomach and urine ("the water-filled moat"). These tend to act as frontline defenses. But if the body's physical defenses are breached by an invading pathogen, there is a second level of immune responses: white blood cells capable of ingesting and digesting foreign substances. We call these "foot soldiers" of the immune system "phagocytes," or "eating cells," because they feed on, digest, and destroy foreign particles, especially bacteria, and they also clean up the "victims"—the dead and damaged cells. The remnants of living and dying bacteria, the healthy and wounded phagocytes, and the fluid that surrounds them constitute the material called pus. Pus is characteristic of the site where a battle is taking place: blood vessels dilate, the capillaries become more permeable, fluid leaks and produces swelling (edema), there is an influx of phagocytes (neutrophils, macrophages), histamine is released, and the battle area becomes heated up and reddened. This sequence of cellular events, called the innate inflammatory response, can be compared to the hand-to-hand combat in the castle that involved armed defenders—both professional and nonprofessional. We will revisit this immune mechanism subsequently.

Phagocytes were discovered in 1882 by the Russian Elie Metschnikoff (1845–1916) when he placed the thorn from a rose into the larva of a starfish. What he saw under the microscope was the tip of the thorn surrounded by a cluster of phagocytes that appeared to want to "eat" the intruding foreign object. Later, when Metschnikoff established himself at the Pasteur Institute in Paris in 1887, he noted that the phagocytic cells in a vaccinated animal were more active than those in animals that were not vaccinated. He formed the theory that all protective immunity was due to phagocytic activity. Much of his subsequent experimental work was directed toward establishing that phagocytes—including macrophages and their precursors, the blood-borne monocytes (found in higher numbers in infectious mononucleosis) are key elements in this process. His "solely the phagocyte" theory of immunity was flawed, but we shall find he was not entirely incorrect in his beliefs about the fundamental role of phagocytic cells in immunity.

Acquired Immunity

Let us turn our attention to another level of defense, the kind of immunity that develops as a consequence of exposure to a foreign intruder—acquired immunity. This is the immunity that Thucydides wrote about.

Diphtheria is a horrible childhood disease. The incubation period is ~7 days followed by fever, malaise, sore throat, hoarseness, wheezing, blood-tinged nasal discharge, nausea, vomiting, inflamed tonsils, and headache. The infected child (Fig. 10.2) is quite pale and has a sticky gray-brown membrane (made up of white cells, red cells, dead epithelial cells, and clots) that adheres to the palate, throat, windpipe, and voice box. This membrane impairs the child's breathing. If disturbed, it can bleed and break off and block the airways of the upper respiratory tract. Speech becomes thickened, and the breath is foul smelling. Within 2 weeks, this upper respiratory infection can spread to the heart, leading to inflammation of the heart, heart failure, circulatory collapse, and paralysis. During an epidemic, up to 50% of untreated children may die, but more commonly the death rate is 20%.

In 1883, two German scientists, Edwin Klebs and Friedrich Loeffler, working in Robert Koch's laboratory in Berlin, discovered the microbe of diphtheria in throat swabs taken from an infected child; they were able to grow the bacterium in pure culture and named it *Corynebacterium diphtheriae*. Then in 1888, Emile Roux and Alexandre Yersin of the Pasteur Institute in Paris found that the broth in which the *Corynebacterium* bacteria were growing contained a substance that produced all the symptoms of diphtheria when injected into a rabbit;

Figure 10.2   Detail from the painting *Diphtheria,* attributed to a follower of Goya (1802–1812). The father is attempting to remove the membrane in the throat that is blocking the breathing of the child. From the collection of Dona Carmen Maranon Vinda de Fernandez de Araoz, Madrid.

at high enough doses, it killed the rabbit. This killing agent was a poisonous toxin. In humans it is this toxin, a neurotoxin, that leads to heart muscle degeneration, neuritis, paralysis, and hemorrhages in the kidneys, adrenal glands, and liver. The toxin is produced because the diphtheria bacterium itself is infected with a virus. For the toxin to be synthesized, iron molecules must be present in its surroundings—body tissues or culture medium. Today, diphtheria is less of a problem because of childhood immunization (the diphtheria, pertussis, tetanus shot); if the infection is acquired, it can be treated with the antibiotics erythromycin or penicillin. Another childhood disease, whooping cough, was first described in 1578, and by 1678 its name, "pertussis" (from the Latin word meaning "intensive cough"), was in common use in England. In 1906, Jules Bordet and Octave Gengou identified the specific bacterium that causes whooping cough and named it *Bordetella pertussis*; its disease symptoms are produced by another kind of toxin, one different from that of diphtheria.

In 1885, Arthur Nicolaier discovered the tetanus bacillus *Clostridium tetani*, and in 1889, Shibasaburo Kitasato in Robert Koch's laboratory was able to grow it in pure culture and found a powerful toxin in the culture fluid. Then, in 1890, two critical discoveries related to protective immunity were made: Emil von Behring and Kitasato, both working with Robert Koch in Berlin, found that, if small amounts of tetanus toxin were injected into rabbits, the rabbit serum contained a substance that would protect another animal from a subsequent lethal dose of toxin. (Serum is the straw-colored liquid that remains when blood is allowed to clot and then the clotted material is removed.) They called the serum that was able to neutralize the toxin "immune serum." von Behring found that the same thing happened with diphtheria toxin. In effect, by toxin inoculation, the rabbit had been protected against a disease. This process was called immunization, although, in Jenner's honor, any immunization is called a vaccination. Further, the immune protection seen by von Behring and Kitasato could be transferred from one animal to another by injection of this immune serum. In other words, it was possible to have "passive immunity"; that is, immunity could be borrowed from another animal that had been immunized with the active foreign substance when its serum was transferred (by injection) to a nonimmunized animal. (Passive immunization, or "immunity on loan," can be life saving in the case of toxins such as bee or spider or snake venoms or tetanus, since it is possible to counteract the deadly effects of the poison by injection of the appropriate immune serum.) The implications of the research on immune serums for treating human disease were quite obvious to von Behring and others in the Berlin laboratory. Without waiting for the identification of the active ingredient in immune serum, the German government began to support the construction of factories to produce different kinds of immune sera against a wide range of bacterial toxins. Clinical trials were organized first in Germany, then in the rest of Europe and the United States.

von Behring and Kitasato called the substance in immune serum that neutralized these toxins "antitoxins," meaning "against the poison." Treatment of diphtheria victims with antitoxin serum (an antiserum produced in horses) was so effective that the death rate dropped precipitously. In some cases, mortality was halved. For his work on antitoxin therapy, von Behring received the first Nobel Prize in 1901. Initially, it was believed that all bacterial infections could be treated with antitoxin therapy, but, unfortunately, only the bacteria that cause diphtheria, pertussis, and tetanus and a few others produce and secrete toxins, so the therapeutic application of immune serum against most bacterial infections remains limited.

An antitoxin is one of many kinds of active materials found in immune serum after a foreign substance is injected into humans, rabbits, horses, chickens, mice, guinea pigs, monkeys, or rats; the general term for the active material is antibody. Simply put, antitoxin is one kind of antibody. At about this time, Paul Ehrlich (1854–1915) entered the scene in Koch's laboratory. Ehrlich established the principle that there was a quantitative (1:1) relationship between the amount of toxin neutralized by a specific amount of antitoxin. The potency or strength of an antibody is called its titer and is reflected by the degree of dilution required to make a specific amount of antigen that will bind to an equal amount of antibody. (The higher the titer, the greater the potency of the serum; put another way, a high-titer immune serum can be diluted to a greater degree to achieve the same neutralizing effect as that of a weaker immune serum.) Ehrlich's principle of titer was not only of theoretical interest, but it also enabled antitoxins (and other therapeutic biologicals, such as insulin and other hormones) to be provided in standardized amounts and strengths. Ehrlich also found that, although the diphtheria toxin lost its poisoning capacity with storage, it still retained its ability to induce immune serum. He called this altered toxin (usually produced by treatment with formalin) "toxoid." Even today, this is the standard means of inducing immunity to diphtheria, pertussis, and tetanus. Toxoids made from diphtheria, pertussis, and tetanus toxins are used to make a vaccine known as DPT. The Schick test for susceptibility to diphtheria also involves toxoid: a small amount of diphtheria toxoid is injected just under the skin; failure to react (i.e., failure to produce a red swelling—inflammation—at the site of injection) indicates a lack of protective immunity. Individuals who are immunosuppressed, such as AIDS patients, those on anticancer chemotherapy, and those receiving high doses of steroids, cannot mount an immune response and will get false-negative results with such tests.

Ehrlich went on to study other toxins and found that the serum made against one toxin did not protect against another toxin. Thus, the antitoxin made by an animal was specific. Since this specificity was true for almost any foreign substance, he concluded that antibodies were specific for the foreign material against which they were produced. Substances, such as a toxin or a toxoid, that were *anti*body *gen*erators, were called antigens. Antigens are any material that is foreign to the body, be it a bacterium or its toxin, a virus such as smallpox or human immunodeficiency virus (HIV), a piece of tissue from another individual, or a foreign chemical substance such as a protein, a nucleic acid, or a carbohydrate polymer, a polysaccharide.

Lock and Key

Paul Ehrlich also formulated a theory of how antigen and antibody interact with each other. It was called the "lock and key" theory. He visualized it in the following way: the tumblers of the lock were the antigen, and the teeth of the key were the antibody. One specific key would work to open only one particular lock. Another way of looking at the interaction of antigen with antibody is the way a tailor-made glove is fitted to the hand. One such glove will fit only one specifically shaped hand. Similarly, only when the fit is perfect can the antigen combine with the antibody. Yet, despite the fact that antigens are large molecules, only a small surface region is needed to determine the binding of an antibody to it. These antigenic determinants are called epitopes, and they can be thought of as a small patch on the surface of the much larger molecule of antigen. Earlier, when vaccination was discussed (p. 205), it was said, without explanation, that protection occurred because the antibodies produced against *Variola vaccinae* were able to neutralize the Variola major virus. Simply stated, it is because the epitopes of the cowpox antigens are so similar to those of smallpox that when the body produces antibodies to cowpox antigens, they are also able to fit exactly to the smallpox epitopes (i.e., the antibodies are said to be "cross-reactive"), and Variola major is prevented from multiplying.

Antibody

What is the chemical nature of an antibody? In the 1930s, techniques were developed to separate large molecules from one another and to identify them chemically. Antibodies were found to be made up of a large number of amino acids; that is, they were proteins. If one took serum and placed it in an electric field, the proteins would separate according to their charge and their size: the fastest-moving and smallest one was called albumin, and the three slower and much larger spherical proteins were called globulins. Of the three globulins, the fastest-moving one was called alpha, and the slowest was called gamma. In immune serum, the gamma globulin is increased, and that is where the antibodies were found. The gamma globulin that contains antibody is called immunoglobulin, because it is produced in response to the antigen, an immunogen. Simply put, antigens that elicit an immune response are called immunogens.

The Nature of the Immune System

Being immune means that the body is able to react specifically to a foreign material as a result of previous exposure. To be effective, the immune response has to be swift the second time around. So what are the essential properties of the immune system involved in protection against a foreign invader? First, it must be able to distinguish foreign substances—"not self"—from "self." Second, it must be able to remember a previous encounter, that is, there must be memory. Third, it must be economical; that is, antibodies should not be produced all the time, but they should be turned on and off as needed. Fourth, it must be specific for a particular antigen.

A key to understanding the protective mechanisms of the immune system is an appreciation of the role of the body's bloodstream and lymphatic system. Blood consists of a salty protein-containing fluid, the plasma, in which red blood cells and white blood cells float. (When blood plasma clots, the remaining fluid, which is deficient in clotting molecules, is serum.) The red cells carry oxygen and carbon dioxide and are not involved in immunity. The white cells, on the other hand, are a critical part of the immune system. White cells come in two varieties: (i) granule-containing granulocytes (called *eosin*ophils if they stain with *eosin*, an acidic red dye; *baso*phils if they stain with a *basic* blue dye; or neutrophils if they stain with both) and (ii) non-granule-containing cells (called monocytes and lymphocytes). The "plumbing" of our bloodstream does not exist as an entirely closed system of "pipes": it leaks fluid but not cells. This is because blood, headed for the tissues and organs, is pumped by the heart into the arteries. The outgoing arteries branch into smaller and smaller vessels, the farther they are from the heart; finally, they ramify into the smallest vessels, the capillaries.

Fluid, consisting essentially of plasma with much less protein and without blood cells, seeps out of the capillaries into the tissue spaces, bathing the cells. Cells are microscopic manufacturing plants: to accomplish their energy-demanding tasks, they contain an internal furnace that burns fuel (glucose) and dumps ashes (carbon dioxide and other waste products). The cell's bathing solution (called interstitial fluid) serves as a waterway for carrying fuel to the cells and removing ashes from them. But what happens to all the fluid that has leaked out into the tissues? How does it return to the bloodstream? The fluid cannot return in the direction from which it came, because the pumping of the heart creates an opposing hydrostatic pressure. The only means by which the fluid can return to the bloodstream is by an auxiliary drainage system that collects the seepage and returns it. (If the drainage is faulty or limb movement is limited, then fluid accumulates in the tissue spaces, especially in the lower limbs by gravity, and the result is swollen feet!) The drainage system consists of permeable "pipes" (lymph vessels) of the lymphatic system (Fig. 10.3). Just as the capillaries are widely distrib-

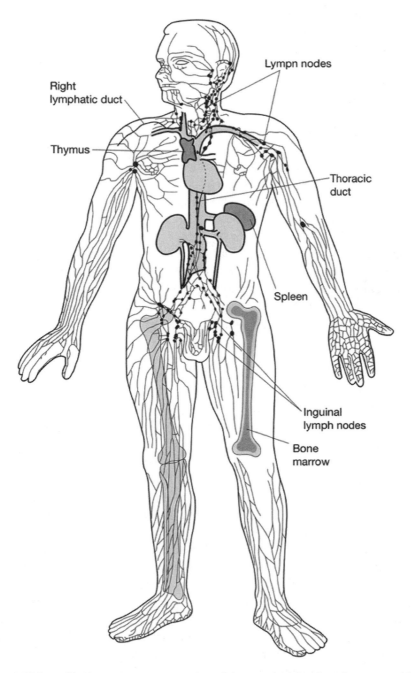

Right
lymphatic duct

Thymus

Lympn nodes

Thoracic
duct

Spleen

Inguinal
lymph nodes

Bone
marrow

Figure 10.3   The immune system consists of tissue containing lymphocytes and lymphatic vessels that transport the cells from the lymphatic tissues and the antibodies secreted by lymphocytes, returning them to the bloodstream via the lymphatic and thoracic duct in the neck region. Lymphocytes are manufactured in the bone marrow and multiply in the thymus, the spleen, and lymph nodes.

uted throughout all the tissues of the body, so too are the lymph vessels. Lymph vessels pick up the fluid from the tissue spaces (such movements are aided by muscular movements of the limbs) and return it to the blood circulation upstream from the heart by "connector pipes" (the thoracic duct and the right lymphatic duct which empty into the blood system in the region of the neck); in this way, there is a complete circulatory system.

Mammals have co-opted the lymphatic system for another purpose: immunological defense. This has been accomplished by placing clusters of immune tissues at "checkpoints" or "surveillance stations" along the lymphatic routes, where the fluids (now called lymph in the lymph vessels) are examined. The "surveillance stations" are called lymph nodes (Fig. 10.3). Lymph nodes are scattered throughout the body but tend to be clustered in the region of the groin, armpits, and chest. In these locations, the lymph nodes act as filters. The "filter" is principally composed of two kinds of white cells: macrophages and lymphocytes. Our body contains 10^{10} to 10^{12} lymphocytes, equivalent in mass to the brain or the liver! Lymphocytes also circulate in the bloodstream and are found in the spleen, a kind of lymph node. Now that we have an idea of the plumbing and surveillance properties of the immune system, let us consider how it works to protect us against disease.

Inflammation and Innate Immunity

The inflammatory reaction (mentioned on p. 213) is localized, swift, and characterized by redness, puffiness, tenderness, and warmth. It prevents invading bacteria and viruses from spreading through the body, and it is critical to our well-being until acquired immunity can develop—which may take days to weeks. How does inflammation work? There are two kinds of cells involved: macrophages and dendritic cells, each of which has 10 different kinds of receptors (or docking stations) on its outer surface. (Dendritic cells have a maze of long extensions; although they resemble nerve cells, they are more closely related to macrophages of the immune system.) The receptors on these cells are able to recognize about 20 different *p*athogen-*a*ssociated *m*olecular *p*atterns ("pamp"). Some "pamp" molecules are those in the cell wall of gram-positive bacteria, the viral coat protein, the cell wall of mycobacteria, and flagellin, a protein that makes up the bacterial flagellum. Once a "pamp" is fitted into the docking station of the dendritic cell or macrophage (that is, bound to the receptor), chemical messengers called lymphokines are released. Lymphokines make the blood vessels leaky, there is redness and swelling, and neutrophils and lymphocytes (of the T kind, which will be considered shortly) move out of the blood vessels and are attracted to the area where they begin to work. This painful, swollen, warm, and reddened site may become filled with pus (and is called a boil) and consists of an accumulation of live or dead infectious bacteria and live or dead neutrophils, the debris of battle. Neutrophils are the "foot soldiers" of the immune system that are activated immediately, and they fight and die on the inflammatory battlefield. Pus can be milky and thick in a staphylococcal infection, watery and blood stained in a streptococcal infection, and foul smelling when there are anaerobic bacteria present. In the good old days, a poultice (a word dating back to the 16th century), which consisted of a soft paste spread on muslin or some other cloth, was heated and placed on the boil or the inflamed area. The goal was to increase the blood flow and to enhance movement of the fluid containing white cells (neutrophils and lymphocytes) into the area and to produce pus.

Antibody-Mediated Acquired Immunity

Now imagine that the infected area on your hand is filled with bacteria called streptococci. The inflammatory reaction has been set in motion, but not all of the bacteria are destroyed, and some enter the lymph vessels that drain the inflamed site. These bacteria are carried to the regional lymph node, which enlarges and gets sore. (Recall the buboes of bubonic plague.) Sometimes the pathway is marked by a reddish streak of inflammation from the site where the pathogenic bacteria entered a break in the skin and have been carried to the regional lymph node. The bacteria contain foreign materials—"not self" antigens. Some of these antigens will now trigger the production of antibodies.

How are antibodies made? Antibodies are made by lymphocytes called B lymphocytes or B cells. Their discovery came about in an interesting way. It all began with a study of immunity in chickens. In the 1950s, Bruce Glick was a graduate student who sought to discover the function of a lymphoid organ, called the bursa of Fabricius, located at the lower end of the chicken's digestive tract. (The bursa is named after the 16th century anatomist Hieronymus Fabricius, who first described it.) Glick surgically removed the bursa and found no difference in the survival or health between chickens with a bursa or those lacking it. He left the project temporarily. Then another graduate student in Glick's department needed some chickens to demonstrate the production of antibodies. He used Glick's chickens: both those with a bursa and those without one. After the chickens were injected with antigen, he discovered that the chickens lacking a bursa made no antibody. Subsequently, it was found that the cells in the bursa that were involved in the production of antibody were lymphocytes; in honor of their location, they were named B cells after the first letter of "bursa." (In mammals, which lack a bursa, the B cells are made in the bone marrow, which fortunately also begins with a B.) The work was submitted for publication in *Science* magazine and rejected because it was "uninteresting." However, later it was published in the journal *Poultry Science*, where it went unnoticed for several years. Glick and his colleagues then made another interesting observation regarding the chickens that lacked a bursa: although they could not make antibody, they could reject skin grafts and were able to overcome a viral infection. This suggested that, although the bursa played a role in antibody-based immunity, there must be additional mechanisms that could protect against infection or foreign materials. As we shall soon find out, this additional immune response involved another kind of lymphocyte, called the T lymphocyte.

How do B cells make antibody? On the surface of the B lymphocyte is an antibody molecule that serves as a receptor for the foreign antigen. Once bound to the B-cell surface, the B cell (called a plasmablast) is triggered to divide asexually, and the population of identical offspring from such division is called a clone. The members of the clone are called plasma cells. Division to produce an adequate number of plasma cells may take several days, since the division rate is about 10 hours per division. Plasma cells are cellular factories that can manufacture large quantities of antigen-specific antibody, which they secrete into the blood plasma. A single plasma cell can secrete over 2,000 molecules of antibody per second, or about 10^6 molecules each hour, and it is possible for different plasma cells to synthesize 10^{15} different antibodies, each having a different specificity for the antigen that triggered its division in the first place. (Plasma cells, and the B cells that give rise to them, are analogous to the archers in the defense of the crusader castle; the arrows shot to kill the foreign invader are analogous to antibody molecules.) The specific antibody produced can protect in several ways: it can neutralize the antigen, or it can aggregate (clump) bacteria and free virus in the blood; it may lyse (disintegrate) virus-infected cells or bacteria; it may block virus entry into cells; or it may make the bacteria more attractive so the phagocytes readily eat the microbe. Because the antibody is in the liquid fraction (plasma or serum) of the blood, this is called humoral immunity after the four humors (black and yellow bile, phlegm, and blood) that the ancient Greeks and Romans believed were able to control our behavior. We still use these terms when we describe someone who is sluggish as "phlegmatic" or we say we are in a bad humor or in a good humor.

Not all B cells differentiate into plasma cells; some remain quiescent and persist as memory cells. When there is another encounter with the same antigen, the memory cell undergoes rapid division, plasma cells are produced, and antibody is synthesized and released. This rapid response is due to memory of a past antigenic invasion and is the basis of immunization with booster shots.

Clonal Selection Theory for Antibody Formation

Now you might ask, how does the body learn to distinguish its own cells and its own antigens from those that are foreign or "not self"? In the 1950s, Macfarlane Burnet, an Australian immunologist, proposed a theory to explain how the immune system is able to distinguish "us" from "them." He called it the clonal selection theory. According to Burnet's theory, the lymphoid organs of each embryo contain immature progenitor lymphocytes, and each is capable of interacting with a different antigen, including the body's own antigens—"self antigens." However, in the embryo, interaction of a "self antigen" with an immature lymphocyte specific to it results in the death of that particular cell. As a result, by the time of birth, the immune system is purged of the lymphocytes that can react with self-antigens. In short, the lymphocytes that react with "self" have been selected against. The remaining lymphocytes, which are considered to be mature, may react to an assault by a foreign or "not-self" antigen because they are able to recognize only "non-self" antigens. When such an encounter with a "non-self" antigen does take place, the binding of antigen to the antibody molecules on the surface (now called a receptor) of the B lymphocyte causes the B cell to multiply, a clone of plasma cells is produced, and the plasma cells synthesize and secrete antibody specific for that particular antigen into the blood. Thus, only the antigen-selected cells multiply to make an antibody, and the levels of that specific antibody increase in the blood as a result of the expansion of the plasma cell population. Antibody levels usually rise over a 2- to 4-week period following exposure to antigen and may last for quite a long time. A secondary exposure to antigen (a booster) produces an even more rapid response, because now there are more plasma cells making that particular antibody. Prolonged exposure to antigen is no longer required. Furthermore, the binding ability of an antibody improves with time because the antigen "selects" for the multiplication of those plasma cells that carry the receptor with the better match. That is, the key "selects" the lock with the best fit according to a very specific arrangement of its tumblers. In effect, the immune system "learns." In 1960, Burnet received the Nobel Prize for his work on the clonal selection theory, which was fully substantiated by further experimental work over the course of the next decade and beyond.

The immune system has as its prime function surveillance; that is, it is constantly "looking about" for foreign substances, whether these be cancer cells or bacteria or viruses or toxins. However, the system is not perfect; things can go wrong, especially when an antibody reacts with a foreign antigen that resembles a "self" antigen. This is a case of mistaken identity in an antigen. Such wayward antibodies can result in what are called autoimmune diseases, literally "immunity against oneself." Rheumatic fever is an example of an autoimmune disorder. Rheumatic fever, a disease almost unknown today, thanks to the effectiveness of the antibiotic penicillin, starts as an inflammation in the throat, pharyngitis. It may, however, result in heart valve destruction with permanent scarring so that the individual can suffer from a heart murmur for life. The cause of the pharyngitis is a hemolytic streptococcus (bacterium). One strain of this streptococcus causes a benign infection of the skin called pyoderma or impetigo, but others—especially those heavily encapsulated by hyaluronic acid and M protein—are invaders of the throat. However, it is not the direct action of the streptococci that results in heart damage. Instead, the M protein antigens are swallowed in large amounts during pharyngitis, and this triggers an immune response, producing antibodies that cross-react with heart valves as well as autoimmune cell-mediated cytotoxic cell reactions (see below). These all act to destroy the heart valves. Other autoimmune diseases are insulin-dependent diabetes, rheumatoid arthritis, multiple sclerosis, lupus erythematosus, myasthenia gravis, and Addison's and Graves' diseases.

Cell-Mediated Acquired Immunity

Thus far, we have considered the humoral (antibody) or B-cell component of the immune system involved in protective immunity. Now let us look at the other part, the one that Glick and his colleagues found was not bursa-related. This alternative immune pathway involves another set of lymphocytes, the T variety, and is concerned with nonhumoral immunity. In the 1960s, it was discovered that the bone marrow and thymus are the "master organs" of our (and other mammals') immune system. The bone marrow is the site where all the different cell types of the blood and the immune system are made, and it is here that the blood stem cells also occur. Stem cells are primitive, unspecialized cells, and they have no function other than to divide and to make other cells. But when a stem cell divides to produce two daughter cells, the offspring are not alike: one daughter cell is an identical replica of the stem cell, but the other can grow and become a very specialized cell. In this way, stem cells are both self-renewing and give rise to a variety of specialized cell types, including B and T lymphocytes. A T lymphocyte arises in the bone marrow but migrates via the bloodstream from the bone marrow to a gland in the neck called the thymus (Fig. 10.3). The thymus was so named because Galen in the second century, in an imaginative mood, thought the shape of the gland resembled the leaves of the thyme plant. The thymus glands of young calves and lambs have another name: butchers and restaurants call them sweetbreads. These are prized by gourmets and are delectable, with a very delicate flavor.

The thymus is made of lymphoid tissue. In children, the gland is rather large. It is located just above the heart. As we age, it tends to atrophy and shrink in size. The thymus is the place where the T lymphocytes from the bone marrow establish residence for some time and where they mature (or become educated) and multiply. It is because of their maturation in the thymus that they are called "thymus-educated" or T cells. While in the thymus, some T cells are selected to become either helper T cells (also called T4 cells) or killer T cells (also called T8 cells). The killer cells are also referred to as "cytotoxic T lymphocytes," or CTLs. (In the crusader castle analogy, these would be the knights on horseback.) Some of the progeny of these T cells leave the thymus and by way of the bloodstream reach and settle down in other lymphoid organs—the spleen, the tonsils, and the variously distributed lymph nodes—or they circulate in the blood. The CTLs patrol the body in search of altered cells (cancer), foreign cells (grafts of tissues), or cells compromised by an internal pathogen such as a virus or a rickettsia. Because they are capable of attacking and destroying any cell in the body that appears to be "not self," they are aptly called "killers." But in the attack, the killer does not direct its action against the virus itself but rather the cell (no longer recognized as "self") in which the virus is hiding. Killing may occur in one of two ways: when a foreign cell is encountered, the CTLs attach to it and punch in a "security code" that then triggers a self-destruct mechanism in the "nonself" cell. In effect, the CTL does not have to shoot the invader with a "silver arrow," it simply tells the pathogen-infected cell to die. The other way is for the CTL to punch a hole using a pore-forming molecule (analogous to the knight's sword) and then inject it with a lethal cocktail of destructive enzymes. The CTL response is strong within 5 days after exposure to a foreign antigen, peaks between 7 and 10 days, and then declines over time. (Antibody formation, when it does occur, usually follows the CTL response.)

What roles do the T cells play in protective immunity? The first line of defense in cell-mediated immunity is the macrophage or "big eater" described by Metschnikoff more than a century ago. Macrophages, the "foot soldiers" of the immune system, are also the cellular "vacuum cleaners" or "filter feeders" of the immune system. The immune system cannot respond to an entire microbe, be it a virus or a bacterium, but it can handle the fragments of these parasites. The microbe is fed into the system, both literally and figuratively, when the macrophages ingest, digest, and then regurgitate bits and pieces of the foreigner (virus or bacterium) or even the dead and dying parasite-containing cells they have eaten. In this way, the macrophages are able to display the broken antigenic bits on their outer surface in association with another molecule called the *m*ajor *h*istocompatibility *c*omplex (MHC) II. This combination of MHC II and foreign antigen is information that is offered to a T-helper (T4) lymphocyte. Once the T4 lymphocyte binds to the antigen-presenting cell and thus receives information, it is triggered to divide, forming a clone. However, instead of secreting antibody, the T cell secretes chemical messengers called lymphokines. (Using the crusader castle analogy, this T cell would be the lord who conferred knighthood by "dubbing.") The chemical messages then activate other T lymphocytes ("knights") or macrophages ("foot soldiers") or can activate the B cells ("archers"), which are also MHC II, so they can begin the process of antibody synthesis and secretion ("shooting their arrows"). In addition, as with B cells, there are memory T cells left after an encounter with a foreign antigen, so the response to a second encounter is swift and specific. The T4 cell is the keystone ("lord" of the crusader castle and manor) that maintains the connection between the two branches of the immune system: cell-mediated and humoral. The humoral branch produces antibodies that prevent foreign invaders from infecting cells, while the members of the cell-mediated branch search out and destroy infected or foreign cells. The central role of the T-helper cell in protective immunity is most clearly seen in AIDS: destruction of the T-helper cell by HIV results in a collapse of the immune system, after which the uncoordinated and defenseless body can be assaulted by opportunistic pathogens and cancer, the syndrome of AIDS.

CTLs are also critical in organ transplantation. If the graft is "not self," then the CTLs kill it and the graft is rejected. Graft rejection does not involve antibody. Other examples of cell-mediated immunity are seen in delayed-type hypersensitivity, such as reactivity to tuberculin—an inflammatory response that takes 24 to 72 h. Contact dermatitis, which occurs, for example, after a person touches poison ivy or poison oak, is also cell mediated, but because this reaction occurs in a matter of hours, it is called immediate-type hypersensitivity. Immediate-hypersensitivity reactions are hay fever, asthma, hives, eczema, food allergies, sulfonamide and penicillin and bee-venom sensitivities, and transfusion reactions; all are cell mediated and not due to antibody.

Why doesn't a CTL destroy our own cells or the T-helper cells? Because all the cells in our body (except for red blood cells) express MHC I. This allows infected cells to signal their plight to the CTL, which is also MHC I. Once this information is received by the CTL, the infected cell is killed. But because the T-helper cell is MHC II, it is either not recognized by the CTL or the information received tells the CTL that it is "self," and killing does not occur. Not all MHC molecules are alike. It is the many different kinds of MHC I molecules on the surface of our body's nucleated cells that are used as the basis for matching tissues (called tissue typing) to be used in grafting. When a proper match is made, compatibility is ensured and the graft will not be rejected.

Attenuation and Immunization

In 1875, Louis Pasteur attempted to induce immunity to cholera. He was able to grow the cholera bacterium that causes death in chickens and to reproduce the disease when healthy chickens were injected with it. The story is told that he placed his cultures on a shelf in his laboratory that was exposed to the air and went on summer vacation; when he returned to his laboratory, he injected chickens with an old culture. The chickens became ill, but to Pasteur's surprise they recovered. Pasteur then grew a fresh culture of the bacterium with the intention of injecting another batch of chickens, but the story goes that he was low on chickens and used those he had previously injected. Again he was surprised to find that the chickens did not die from cholera; they were protected from the disease. The prepared mind of Pasteur recognized that aging and exposure to air (and possibly the heat of summer) had weakened the virulence of the bacterium and that such an attenuated or weakened strain might be used to protect against disease. In honor of Jenner's work a century earlier, he called the attenuated strain a vaccine. Pasteur extended these findings to other diseases and in 1881 was able to protect sheep against anthrax (Fig. 10.4). In 1885, he administered the first vaccine (prepared from dried spinal cords from rabies-infected animals) to a human, a young boy, Joseph Meister, who had been bitten by a rabid dog. The boy lived, and Pasteur proved vaccination worked. Attenuation was also used in the development of *Mycobacterium bovis* BCG (bacillus Calmette-Guerin), the vaccine for tuberculosis; it took 231 subcultures of a bovine tubercle bacillus in an ox bile-containing medium every 3 weeks over a period of 13 years before the bacterium was attenuated. But some see a problem with attenuation: how can it be ensured that the benign form will not, in time, revert to its virulent state and, instead of giving protection, cause disease?

It is difficult to underestimate the contribution immunization has made to our well-being. It has been estimated that were it not for childhood vaccinations against diphtheria, pertussis, measles, mumps, smallpox, and rubella, as well as protection afforded by vaccines against tetanus, cholera,

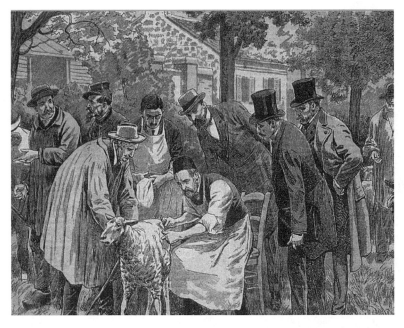

Figure 10.4   Louis Pasteur vaccinating sheep against anthrax using an attenuated form of the bacillus. All 25 sheep that had been vaccinated were protected, whereas those that did not receive the vaccine died.

yellow fever, polio, flu, hepatitis B, bacterial pneumonia, and rabies, childhood death rates would probably hover in the range of 20 to 50%. Indeed, in those countries where vaccination is not practiced, the death rates of infants and young children remains at that level.

After Pasteur's success in creating standardized and reproducible vaccines at will, the next major step came in 1886 from the United States. Theobald Smith and Edmund Salmon published a report on their development of a heat-killed cholera vaccine that immunized and protected pigeons. Two years later, these two investigators claimed priority for having prepared the first killed vaccine. Though their work appeared in print 16 months before a publication by Chamberland and Roux (working in the laboratory of Louis Pasteur) and with identical results, the fame and prestige of Pasteur was so widespread that the claim of "first killed vaccine" by Smith and Salmon was lost in the aura accorded those who worked in Pasteur's Institute in Paris.

In the period 1870 to 1890, killed vaccines for typhoid fever, bubonic plague, and cholera were developed, and by the turn of the century, there were two human attenuated virus vaccines (*V. vaccinae* and rabies) and three killed ones. Although there were other successes, such as a formalin-inactivated diphtheria toxin (toxoid) in 1923 and BCG (1927), the golden age of vaccine development really began in 1949 when viruses could be grown in chick embryos and in human cell cultures. This led to the first live polio vaccine (1950), the Salk formalin-inactivated polio vaccine, attenuated Japanese encephalitis, measles and rubella vaccines, and an inactivated rabies vaccine. In the 1970s and 1980s, bacterial vaccines were made by using the capsular polysaccharides from pneumococcus, meningococcus, and *Haemophilus influenzae*. The majority of vaccines being developed today use new technologies: DNA vaccines and subunit vaccines (purified proteins or polysaccharides), as well as genetically engineered and virus-transmissible antigens. The older methods continue to be useful, however, as in the case of the live flu virus vaccine. A prominent vaccinologist said, "If these new technologies succeed, the golden age will turn to platinum." And just think, all of this began when a perceptive English country doctor, Edward Jenner, turned a folk belief into a practical method to protect humans from their own diseases.

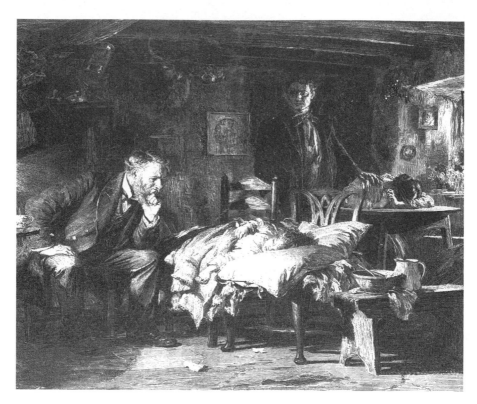

Figure 11.1 *The Doctor* by Sir L. Fildes (1843–1927). Commissioned by Henry Tate in 1887 for his new National Gallery of British Art. The painting shows the artist's son attended by Doctor Murray, who, though he showed care, could do little to cure the dying child. (Courtesy of The Wellcome Library of Medicine.)

Chapter 11

The Plague Protectors

As he watched his beloved wife, Virginia, dying from tuberculosis, Edgar Allan Poe wrote,

> "The Red Death" had long devastated the country. No pestilence had ever been so fatal, or so hideous. Blood was its Avatar and its seal—the redness and the horror of blood. There were sharp pains, and sudden dizziness, and then profuse bleeding from the pores, with dissolution. The scarlet stains upon the body especially upon the face of the victim were the pest band which shut him out from aid and from the sympathy of his fellow-men. And the whole seizure, progress and termination of the disease, were the incidents of half an hour.

For Poe's wife and millions of others, the catastrophic effects of disease were clear, but there was little that could be done to protect or treat or cure (Fig. 11.1).

Although some of the prehistoric diseases that afflicted our ancestors are recorded in their bones, little or no evidence exists to document the treatments used for the illnesses they suffered. Based on present-day primitive cultures, however, it is clear that the medical treatments were inseparable from religion and magic. At the core was the priest or healer, who was given the task to restore the ill to health by driving out evil spirits, luring back the good ones, or begging forgiveness from an offended god. On occasion, there was a direct attack on sickness: sucking, bleeding, cupping, fumigating, steam bathing, drugging to deaden the senses or to reduce fever, and doing surgery. Most of these treatments were based on trial and error, few had a rational or pharmacological basis, and for the most part the benefits to the infirm were psychological. Indeed, the greatest ally these "medicine men" had was the body's ability to cure itself.

Secular medicine, which developed in parallel to the religious approach, is usually dated from the time of the Greeks in the fifth century BC and most often is personified by the physician Hippocrates of Cos. However, it should be recognized that Western medicine did not spring forth fully formed from the forehead of Zeus, as Athena did, but was dependent on knowledge accumulated over hundreds and even thousands of years. Let us consider the healing arts and the science of medicine that have been, and continue to be, used to protect us from plagues and to understand how attitudes toward disease have influenced prevention, treatment, and cure.

Barbers, Bloodletting, and Antisepsis

Archeological records show that cleanliness was important to the ancients. Their motivation, however, was not prevention of disease but the belief that being clean would appease the gods and prevent them from sending down plagues. The word "hygiene," meaning "healthy," comes from the name of the Greek goddess Hygeia, who was the daughter of the god of medicine, Asclepios. Asclepios is depicted in paintings and sculpture (Fig. 11.2) as an older, bearded wise man holding a staff around which is wrapped a serpent

Figure 11.2   Asclepios, the god of medicine. Wrapped around his staff is a serpent, signifying death and suggesting that Asclepios was able to ward off illness and prevent death. The tiny figure standing by his right foot is Telesphorus, the child god of convalescence. From a ceremonial ivory diptych carved in the late Roman style.

(signifying death), suggesting that he is able to ward off illness. His daughter Hygeia is represented as a young, beautiful woman, and another daughter, Panacea, represents treatment. The most famous of the healing temples dedicated to Asclepios were those in Epidauros and Pergamon; these temples were not hospitals but were a mixture of a religious shrine and a health spa. The teachings of healthy living, first by the Greeks and later by the Romans, produced aqueducts to bring clean water, bathhouses, toilets, and sewage systems. However, since effective remedies for disease were almost non-existent, the "cure" provided in the temples of Asclepios and the bathhouses was faith, not medicine.

Hippocrates, often regarded as the father of modern medicine, believed there was a rational basis for disease. His theory on the mechanism of disease was based on an imbalance in the body of the four humors (fluids). Key to his theory was that the body contained varying amounts of blood (warm and moist), phlegm (cold and moist), yellow bile (warm and dry), and black bile (cold and dry). Illness was due to poor diet, bad air, and injuries that disturbed the balance. Cure, he believed, could be effected by restoring the balance. At the time, it was accepted as fact that blood was made from food in the liver and that overeating led to an overproduction of blood. This excess of blood resulted in a hot and wet condition (fever, sweating, coughs, headaches, rheumatism, and heart disease). To cure such a patient, purging the humor causing the condition was required. In the case of blood overproduction, bloodletting was undertaken. Bloodletting would, it was claimed, eliminate the overbalance of blood and result in cure. Hippocrates said, "What cannot be cured with medicines is cured by the knife; what the knife cannot cure is cured with the searing iron, and whatever this cannot cure must be considered incurable." The grisly art of bloodletting (based on Hippocrates' theory) flourished for the thousand years that encompassed the Dark Ages (476 to 1453). Its practice would last well into the 19th century. Indeed, a bimonthly bloodletting was, if one could afford it, a preventive medicine. George Washington died from a throat infection in 1799 after being drained of nine pints of blood in a 24-hour period!

During the first millennium of the Christian era, few people were able to read or write. The majority of people were treated by the literate clergy, especially the monks (who became physicians), women who were acquainted with medicinal plants, and lay surgeons who treated wounds. The physicians enlisted barbers (from the Latin *barba*, meaning "beard") to act as assistants in bloodletting. In 1163, when the Council of Tours ruled by papal decree that it was sacrilege for the clergy to draw blood from humans, the barbers became the only individuals capable of performing bloodletting. The practice was so successful that barber-surgeons began to thrive throughout Europe. Soon the royals as well as commoners went to the barbers to be shaved, to have a haircut and to have their illnesses cured or prevented through bloodletting. Barbers also used their razors for lancing boils and were able to set fractures and amputate limbs. Later, the barbers expanded on their reputations as surgeons and began practicing dentistry.

Bloodletting involved slicing small cuts in the flesh of the arm through which blood was allowed to drain into a bowl. To encourage blood flow, the barber-surgeon commonly gave his patient a pole to hold on to and squeeze. Bleeding would continue until the patient fainted. The barber-surgeons also used a strip of cloth as a tourniquet, which was later applied as a bandage when the operation was complete. Afterwards, the bloody bandage was rinsed and hung on the pole outside, flapping in the breeze. As it dried it twisted around the pole in the spiral pattern that we now associate with the sign of a barbershop, the barber pole.

In the middle of the 13th century, the barbers of Paris, also known as the Brotherhoods of St. Cosmos and St. Damian, established the first school for instruction in the practice of surgery; later this was expanded and served as a model for other surgical schools in the Middle Ages. The oldest barber organization in England (the Worshipful Company of Barbers) was established in London in 1308. For the most part, barber-surgeons were unskilled and uneducated and lacked not only an appreciation of contagious diseases but also a knowledge of how they might be prevented; as a consequence, in their hands, postoperative infections became more and more common. Until 1416, barbers had a free hand in the practice of surgery, but complaints of barber-surgeon incompetence and the sicknesses their cutting induced led the mayor of London in that year to forbid them from taking care of anyone "in danger of death or maiming unless within 3 days after being called in they presented the patient to one of the masters of the Barber-Surgeon Guild." To facilitate improvement in their surgical skills, Henry VIII (in 1540) allowed a merger of the barbers and surgeons and decreed that barbers were to receive the bodies of four criminals yearly for the purposes of dissection. The merger, however, was an uneasy one, and in 1745 a bill was passed to separate the barbers and surgeons. The Company of Surgeons was formed with its own guildhall close to Old Bailey (the courts) and Newgate Prison (where the bodies of criminals to be used in anatomical dissection could be obtained), and in 1800 the Company of Surgeons was granted a royal charter to become the Royal College of Surgeons of England. Barbers, on the other hand, began to decline in stature as medical practitioners, and by the late 1800s they had lost all rights to perform surgery and were restricted to cutting hair, shaving beards, and fashioning wigs.

Although attention to cleanliness in surgery can be traced as far back as 600 BC to the Hindu surgeon Sursuta, as well as to the writings of Hippocrates, who advised that surgeons thoroughly cleanse their hands before operating and that boiled water be used to irrigate a wound, these recommendations were later largely forgotten. Indeed in Rome, Galen (AD 130–ca. 200) proposed a "laudable pus theory," claiming that pus was an essential part of the healing process. As a result, wounds were encouraged to become contaminated and filled with pus. It may seem inconceivable to us today to think that microbe-induced rotting and pus-filled flesh would be accepted as a part of wound recovery, but because Galen was regarded as the ultimate medical authority, sepsis (literally "putrefaction" or "decay") was considered for more than a thousand years as an indicator of a process that would eventually lead back to health.

The disease puerperal fever (also called childbed fever) was known since the time of Hippocrates, who described it this way: "The disease usually commenced with a diarrhea; the uterus became hard, dry and painful . . . pain in the head and sometimes a cough. On opening the bodies, curdled milk was found on the surface of the intestines, a milky . . . fluid in the stomach, the intestines and the uterus when carefully examined, appear to have been inflamed." In 1772 in Paris, Vienna, and other European cities outbreaks of puerperal fever were common and 20% of all delivered mothers died from it. Alexander Gordon (1752–1799), a Scottish surgeon, was the first to suggest its cause, writing, "I will not venture positively to assert that puerperal fever and erysipelas (a streptococcal skin infection causing redness and swelling and also known as St. Anthony's fire) are precisely the same in specific nature (but) that they are concomitant epidemics I have unquestionable proof. They began at the same time and afterwards kept pace together; they arrived at their acme together and they both ceased at the same time. After delivery the infectious matter is readily and copiously admitted by numerous . . . orifices, which are open to imbibe it, by separation of the placenta from the uterus." Dr. Gordon established that in 77 cases of the disease, 28 patients had died. The cause, he claimed, was not a noxious constituent in the atmosphere but was transmitted from one patient to another by doctors or midwives as carriers, as they themselves were not affected by the disease. He argued that "the cause of puerperal fever . . .was contagion of infection altogether unconnected with . . . a miasma." He suggested, "Nurses and physicians who have attended patients affected with puerperal fever ought to carefully wash themselves and to get their apparel properly fumigated before it be put on again." Yet, despite these observations and remedies, the standards of care in hospitals remained poor. One surgeon in 1801 wrote, "In hospital . . . the patient sinks almost inevitably under the formation of pus of a compound fracture . . . and some surgeons continue to believe that the formation of pus is one of the essential processes of healing—a legacy from the days of Galen." By 1840, 3,000 mothers in Britain were dying every year from puerperal fever, yet doctors remained unconvinced that they carried the disease from one patient to another.

Oliver Wendell Holmes (1809–1894), father of the great American jurist, was not a surgeon but a professor of anatomy and physiology at Harvard

University. He collected anecdotal evidence on puerperal fever, describing two cases:

> A practitioner opened the body of a woman who had died of puerperal fever, and continued to wear the same clothes. A lady whom he delivered a few days afterwards was attacked with and died of a similar disease; two more of his lying in patients, in rapid succession, met with the same fate; struck by the thought that he might have carried the contagion in his clothes he instantly changed them and met with no more cases of the kind.
>
> Dr. Campbell . . . assisted in the post-mortem examination of a patient who died with puerperal fever. He carried the pelvic viscera in his pocket to the classroom. The same evening he attended a woman in labor without previously changing his clothes; this patient died. The next morning he delivered a woman with forceps; she also died . . . and within a few weeks, three shared the same fate in succession.

A mid-19th century operating theatre and the surgeon's clothes are shown in Thomas Eakins' 1875 painting *The Gross Clinic* (Fig. 11.3).

In 1843, Holmes summarized his findings in an essay, "The Contagiousness of Puerperal Fever," in which he clearly stated the case for communication from one person to another, directly or indirectly. Holmes became so engrossed by the disease that he attended several women who developed puerperal fever; he died of a blood infection he had contracted from one of them. Despite Holmes's persuasive contagionist view, his work attracted little attention save for the obstetrician Ignaz Semmelweisz.

Semmelweisz (1818–1865) was a Hungarian acting as an assistant in the obstetrics department of the General Hospital in Vienna, Austria. The department had two clinics: the First Clinic was attended by medical students, and the Second Clinic was attended by midwives. The overall death rate of pregnant women who had been admitted to the hospital was between 5 and 10%. Semmelweisz found that for the period 1841 to 1846, the death rate for the two clinics was quite different, as shown in Table 11.1.

In 1846, a colleague of Semmelweisz, Jakob Kolletschka, professor of forensic medicine, age 43, died of septicemia from a scalpel wound he received during a postmortem exam. The knife had slipped, the skin of his finger had been pierced, and infection had set in. He developed a severe inflammation in his hand, followed by fever, delirium, a rapid pulse, difficulty in breathing, stiffness in the neck, peritonitis, and multiple abscesses over his body. Semmelweisz described his colleague's death: "Professor Kolletschka . . . frequently participated with his pupils in the performance of medico-legal autopsies; during such an exercise, he was stuck in the finger by a student with a knife . . . used during the post-mortem . . . [he] became ill with lymphangitis and phlebitis . . . died . . . of . . . pleuritis, pericarditis, peritonitis, meningitis. . . ." Semmelweisz noted that the cause of Kolletschka's death was similar to that of women who had died from puerperal

Figure 11.3   *The Gross Clinic* by Thomas Eakins. © Courtesy Jefferson Medical University, Philadelphia, Pennsylvania.

fever. He was reminded of how the medical students did their autopsies and then with only a quick wipe of a towel attended the women in delivery. He reasoned that an agent—a hand-borne cadaveric particle—was transferred from the dead by the particles clinging to the knife in the case of Kolletschka or on the hands of those physicians who examined pregnant women. He set up an experiment based on his observations: "In order to destroy the cadaveric particles adhering to the hand . . . I began to use Chlorina liquida with which I and every student were obliged to wash our hands before making

Table 11.1 Death rates in two clinics in General Hospital, Vienna, Austria, 1841 to 1846

Year	First clinic		Second clinic	
	No. of births	Death rate (%)	No. of births	Death rate (%)
1841	3,036	7.7	2,442	3.5
1842	3,287	15.8	2,659	7.5
1843	3,060	8.9	2,739	5.9
1844	3,157	8.2	3,241	2.0
1845	3,492	6.8	3,241	2.0
1846	4,010	11.4	3,754	2.7
Total	20,242	9.92	17,791	3.35

an examination. After some time I abandoned Chlorina liquida because of its high price and changed to the considerably cheaper chlorinated lime." The death rate in the First Clinic was 12/100 in April, but 2 months later, after Semmelweisz had instituted the chlorine hand washing, the death rate dropped to 2/100.

Semmelweisz concluded that decaying animal organic matter was conveyed to the individual from external sources. "These are the cases represented as epidemics of childbed fever; these are the cases that can be prevented. The source of decaying-organic matter can be a corpse of any age, either sex . . . whether the corpse is a pregnant woman or not. Decaying matter is carried by examining fingers, operating hands, instruments, bed linen, sponges, basins and attendants. In other words, anything that is contaminated . . . and then comes in contact with the genitals of the patients." Semmelweisz also wrote, "Puerperal fever is caused by conveyance to the pregnant woman of putrid particles derived from living organisms through the agency of contaminating fingers. Consequently must I make my confession that God only knows the number of women whom I have consigned prematurely to the grave." Semmelweisz did not identify the cause of childbed fever (that was Pasteur, who in 1879 identified it as a streptococcus), but he did develop a method for preventing it—asepsis, literally "without putrefaction or decay." Semmelweisz encountered strong opposition from the hospital officials in Vienna, and in 1850 he left for the University of Pest, Hungary. As professor of obstetrics at the hospital, he enforced aseptic practices, and the death rate from puerperal fever was reduced to 0.85%. His findings and publications, however, were resisted by hospital and medical authorities in Hungary and elsewhere. Semmelweisz, a manic-depressive, was admitted to a lunatic asylum, where a minor cut on his right hand, contracted during a gynecological operation, was noted. On 1 August 1865, he died of septicemia from the cut.

Semmelweisz is virtually unknown today, and much of the credit for aseptic technique is credited to Joseph Lister (1827–1912), whose name is familiar to us from the mouthwash Listerine. Lister actually used another method to prevent infection—antisepsis—and this occurred 18 years after Semmelweisz's chlorine hand-washing experiment and 35 years before the establishment of the germ theory of disease. Earlier, John Pringle, a British Army surgeon, coined the term "antisepsis" to describe substances, such as mercuric chloride, that stopped the rotting of wounds. Although Lister lived at a time when it was already recognized that microbes could be responsible for infectious diseases, he was unaware of the work of Pasteur on fermentation and putrefaction, and he developed his techniques of antisepsis independent of these ideas on infection.

In 1861, Lister was made head of the new surgical ward in Glasgow's Royal Infirmary and set out to prevent putrefaction in a wound by chemically destroying the offending microbes. Lister, however, erroneously believed that such germs were carried in the air, and so he decided to attack these before they entered the body through the open wound. In his search for antiseptics, he found that carbolic acid (creosote or phenol) was quite effective. On 12 August 1865, Lister treated an 11-year-old boy, James Greenlees, whose leg had been crushed by a cart. He covered the wound with a bandage soaked in linseed oil and carbolic acid and enclosed it in tinfoil to prevent evaporation. The wound did not become infected, the leg fracture healed perfectly, and the boy left the clinic 6 weeks later. After success in treating the wounds of several other patients, Lister published his findings in the British medical journal *The Lancet*, stating that germs were the cause of infection and that infection and pus formation were neither beneficial nor a normal part of the healing process. But Lister remained concerned that exposure of the wound to "septic air" while the dressing was changed could result in infection, so he used a hand-pump to spray a 2.5% solution of carbolic acid during bandage changes (Fig. 11.4). Lister was an expert practical surgeon and insisted on hand washing, cleansing of the wound with carbolic acid, and spraying of carbolic acid during surgery. In effect, Lister used both antisepsis (killing the infective agents already in the wound) and asepsis (preventing bacteria from getting into the wound).

Lister's techniques were put to the test during the Franco-Prussian War (1870 to 1871) when Prussian military surgeons used his carbolic acid treatment for battle wounds. Their success in reducing infection was far greater than that of the French, who completely rejected Lister's methods, and Lister was hailed as a hero in Germany. Later, Lister developed a steam-powered spray that was able to fill the entire operating theatre with carbolic acid; because British surgeons hated the smell as well as being drenched with carbolic acid, the method was soon abandoned. By 1880, Lawson Tait replaced Lister's antiseptic method by a thorough cleansing of the wound area with soapy water;

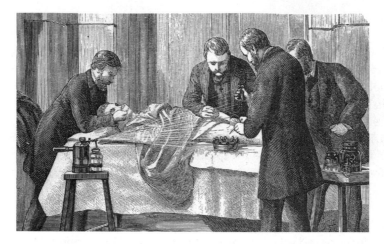

Figure 11.4   Lister using carbolic acid (phenol) as an antiseptic. (Courtesy of the Wellcome Library of Medicine.)

instead of soaking his hands in carbolic acid, Tait washed his hands with soap and hot water. During surgery, Tait wore a clean coat, used clean towels and instruments, and instead of treating sponges with carbolic acid to mop up the blood and pus, he used washing soda. For suturing, he used silk sterilized by boiling instead of Lister's carbolic acid-treated catgut. And he replaced Lister's carbolic acid-impregnated bandage with clean, dry dressings. In 1886, steam sterilization of instruments was introduced, and this allowed surgery to move toward a germ-free environment from the very beginning of an operation. By 1918, few "old guard Listerians" still believed in antiseptic surgery. Today, aseptic surgery is used—that is, surgery utilizing germ exclusion rather than germ destruction; disinfection of the operating theatre, the patient, and the surgical team; and sterile instruments, gowns, masks, and gloves.

Surgery, Disease, and Anesthesia

During the Middle Ages and until the 19th century cleanliness was a priority neither of medicine nor of people in general. Communal wells provided water to the streets but there were few facilities for removing sewage. Wastes polluted the wells. The streets were dumping grounds for garbage and human wastes and domestic animals including pigs roamed the streets. Uneaten food from the table was thrown on the floor to be devoured by the dog and cat or to rot . . . and draw swarms of flies from the stable. The cesspool in the rear of the house would have spoiled . . . the appetite . . . if the sight of the dining room had not. (H.W. Haggard, *The Doctor in History*)

At this time (and even before), surgery was an unsafe practice because control of infection (by asepsis and antisepsis) and anesthesia were unknown.

Surgery was regarded as a manual art and was looked upon by the public with fear and hatred because of the suffering (Fig. 11.5) and almost certain death from infections. In general, surgery was restricted to surface cutting; opening of the body cavity was avoided because it usually meant disaster. There were two kinds of surgeons: "true" surgeons, who carried out major operations such as removal of tumors, repair of fistulas, operations on the face, and amputations, and the barber-surgeons, who would assist surgeons by carrying out bloodletting, extraction of teeth, resetting of bone fractures, and removal of skin ulcers. Gunshot wounds and injured limbs, frequently gangrenous, were amputated and then treated with cautery or boiling oil. The settings for early practitioners of the "cutting art" were not the hospital or operating theatre but the kitchen table, the battlefield, or below deck on a ship. Prior to the 1840s, hospitals were not institutions for the care of the sick as they are today, but existed primarily to provide a refuge for the poor, orphans, and cholera victims. Hospitals—dirty, poorly ventilated, and overcrowded—were perpetrators of disease, and they were more like prisons than places where health might be restored.

The methods for minimizing pain during surgery were few—opium, nightshade, strapping, compression of nerve trunks, distracting noises, and

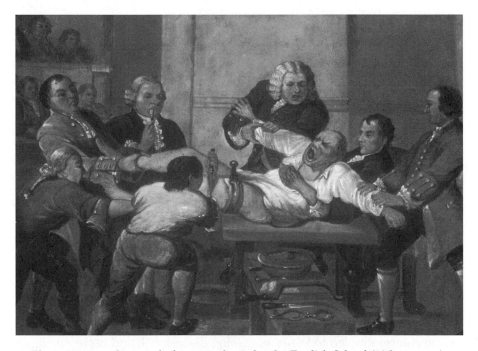

Figure 11.5   Surgery before anesthesia by the English School (18th century). Royal College of Surgeons, London, United Kingdom. The scene is in the operating theatre of St. Thomas' Hospital in Southwark London where a surgeon is beginning to amputate a leg. (Courtesy of the Bridgeman Art Library.)

hypnotism. Among the surgeons themselves, the notion existed that anesthesia was a threat to their profession. As V.S. Thatcher wrote in his book *History of Anesthesia*, surgeons felt "that it was unworthy to attempt to try by artificial sleep to transform the body into an insensitive cadaver before beginning to use the knife." Some even claimed that the pain endured during surgery was critical to recovery. The "best" surgeons were speed artists. Indeed, one of the most famous was the London surgeon Robert Liston (1794–1847). Liston removed a 45-lb scrotal tumor in 4 minutes, but in his enthusiasm he cut off the patient's testicles as well. He amputated a leg in 2.5 minutes, but the patient died shortly thereafter from gangrene (as was often the case in those days); during the same operation, he amputated the fingers of his young assistant, who also died later from gangrene, and he slashed the coat tails of a distinguished surgical spectator, who was so terrified that the knife had pierced his vital organs that he dropped dead in fright. So ended the only operation in history with 300% mortality! However, in the early 1840s, attitudes toward pain began to change, but the change came from dentists, not surgeons. Why were dentists so inclined to use agents that were able to produce insensibility to pain? Some have suggested that, unlike surgeons, who would have contact with their patients only once (and sometimes with fatal consequences), dentists would have to minimize pain to keep their patients coming back.

The first dental anesthetic to be used was nitrous oxide (N_2O), "laughing gas." In 1798, Sir Humphrey Davy performed research on nitrous oxide, which had been identified by Joseph Priestly in 1772. He found that severe pain from dental inflammation would be relieved by breathing the gas, and he correctly observed "that it is capable of destroying pain, [and] it may probably be used with advantage during surgical operations." Yet very few European surgeons used it. They believed that the pain endured during an operation was beneficial to the patient. However, in the United States, nitrous oxide was used to amuse; demonstrations accompanied lectures for the purpose of entertainment. The American dentist Horace Wells attended one of these lecture-demonstrations. He saw the possibilities, borrowed some of the gas, and had a fellow dentist extract one of Wells's own teeth painlessly during inhalation. There followed dozens of extractions using "painless dentistry," and with the help of another dentist, William T. G. Morton (1819–1868), Wells arranged to demonstrate the anesthetic properties at Harvard University. One of the Harvard students volunteered to have his tooth removed, but because the bag of laughing gas was removed too early, he screamed from excruciating pain. This failure and others, as well as his addiction to chloroform, led to Wells' becoming discredited, and in the spring of 1848 he committed suicide during imprisonment in the Tombs in New York City. The practice of nitrous oxide anesthesia was abandoned until J. H. Smith, another dentist, anesthetized several hundred patients and removed

2,000 teeth painlessly. By 1886, nitrous oxide was in common use in dentistry in the United States and is still used today. Nitrous oxide inhibits N-methyl-D-aspartate receptors, which play a role in learning, memory, and pain perception. Another drug that affects these receptors is phencyclidone—also known as PCP and angel dust—a powerful hallucinogen.

Ether (named by Froebenius in 1730) is a distilled mixture of sulfuric acid and alcohol. In 1605, its properties were described by Paracelsus: "if it is taken even by chickens . . . they fall asleep from it . . . but awaken later without harm. It quiets all suffering without any harm and relieves all pain and quenches all fevers. . . ." In 1818, the effects of inhaling ether were reported by Michael Faraday, who was working in Davy's laboratory. He found that after inhalation of ether vapors, the lethargic state was followed by a great depression of spirits and a lowered pulse—conditions that might threaten one's life. It found no use in England, but in the United States, ether inhalation became a part of social entertainment: there were "ether parties." In 1842, Crawford Long (1815–1878), a physician who had used ether for amusement when he was a student at the University of Pennsylvania, successfully employed it for the painless surgical removal of a tumor of the neck after the patient (a student) inhaled ether. A second patient, a child who required amputation of two severely burned fingers, was given ether, but when he awoke prematurely and was in great pain, he had to be tied down to complete the operation. Under popular pressure, Long was forced to abandon his use of ether. In England, ether was derided as the "Yankee dodge." However, in the late 1840s, American dentists, especially Morton, who had previously employed ether for tooth extractions, promoted its use in surgery. The first public demonstration of its successful use took place at the Massachusetts General Hospital on 16 October 1846 (Fig. 11.6). A large vascular tumor of the neck was removed painlessly by the surgeon John Collins Warren as Morton administered ether using a special but simple device: a sponge soaked in ether inside a glass vessel with a spout that was placed in the patient's mouth. Morton called his drug "letheon" and attempted to patent it, but he was later forced to reveal that it was nothing other than ether. Finally, Charles Jackson, a Boston physician who had advised Morton to use ether, claimed to have played a large part in its discovery and pressed for credit all the way to Congress, but it held that Morton was the true discoverer. Dentists and surgeons in England soon began to catch on to the benefits of ether anesthesia, and on 21 December 1846, Liston (the surgical speed artist) performed the first painless major operation (a leg amputation).

Chloroform, a highly volatile substance with an agreeable taste and smell, was described simultaneously by three chemists in 1831. James Simpson (1811–1870), an Edinburgh obstetrician, had experimented with chloroform at a "chloroform frolic party," attempting to produce the same effects as nitrous oxide—but when all of the guests fell asleep, he realized the potential

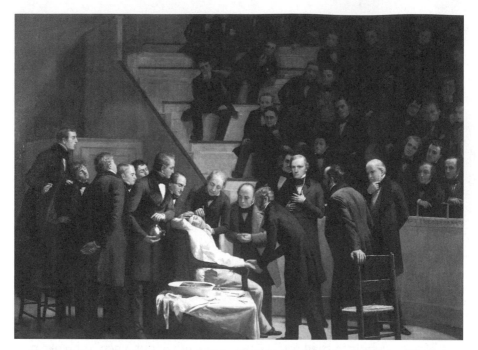

Figure 11.6   *First Operation under Ether* by Robert C. Hinckley. Courtesy of Boston Medical Library in the Francis A. Countway Library of Medicine.

of the substance as an anesthetic (Fig. 11.7). On 4 November 1847, Simpson used chloroform to relieve the pain of labor. Other successes in both Britain and the United States soon followed Simpson's public announcement in 1848. However, the Scottish Church was vehemently opposed to this use of anesthesia to relieve pain during childbirth, because the Bible says, " in sorrow thou shalt bring forth children." Simpson responded in kind with, "And the Lord caused a deep sleep to fall upon Adam; and he slept; and He took one of his ribs, and closed up the flesh instead thereof."

The suffragists knew that the discrimination against painless childbirth came from the male-dominated medical establishment and the clergy and promoted the use of chloroform and ether. In 1850, Simpson used ether to deliver Queen Victoria's seventh child, Arthur. The London physician John Snow used chloroform during the birth of Queen Victoria's eighth child, Leopold, in 1853, and Queen Victoria described it as "the blessed chloroform." When chloroform was used at the birth of her ninth child, Beatrice, in 1857, Victoria said, "We are having a child and we are having chloroform." Snow anesthetized 77 obstetric patients with chloroform; in 1858, he wrote a treatise, "Chloroform and Other Anesthetics," in which he wrote, "The most important discovery that has been made in the practice of medicine since the introduction of vaccination is undoubtedly the power of making persons

Figure 11.7 James Simpson experiments with chloroform. Simpson and friends did not inhale chloroform vapors; instead, they drank liquid chloroform. (Courtesy of the Wellcome Library of Medicine.)

perfectly insensible to the most painful surgical operations, by the inhalation of ether, chloroform and other agents of the same kind." He also described 50 deaths during chloroform administration and recommended means for prevention.

The history of the development of anesthesia is filled with success as well as tragedy: Wells committed suicide at age 33, Long died impoverished and unrecognized at age 63, Morton died of a stroke at age 49, Jackson became an alcoholic and was committed to an insane asylum where he died at age 75, and Simpson died at the rather early age of 58.

Anesthetics were medicine's original wonder drugs. Surgeons used ether and chloroform 40 years before Louis Pasteur made them realize that washing hands and boiling instruments might prevent disease and death. Semmelweisz developed a practical method for asepsis before the germ theory was established. Sanitation, antisepsis, asepsis, and anesthesia prevented disease from needlessly destroying human life, but perhaps the most important point of these advances in health and medicine was their influence on the social climate and the greater acceptance and recognition of scientific truths.

Disease, Dyes, and Drugs

Today, we take for granted that when we become sick there is, with rare exception, a drug to cure us. How did it happen that the drug salvarsan was developed for the treatment of syphilis or that AZT (zidovudine) was designed to cripple the human immunodeficiency virus, and what provided the foundation for the discovery of antibiotics such as penicillin and streptomycin?

It all began with the German dye industry and the microbe hunter Paul Ehrlich, who said, "My dear colleague for seven years of misfortune I had one moment of good luck!"

In the first 3 decades of the 19th century, studies of the chemistry of dyes began. This was the result of the large-scale commercial manufacture of illuminating gas (ethylene) for lighting purposes. In England, illuminating gas was obtained by distillation from coal; however, it was not the gas itself that the English were after, it was the tar. Tar was needed by the British navy as a result of the loss of their American colonies, which had supplied it. Unwilling to be an economic captive to the colonies, the British attempted to become self-sufficient and to make the tar from coal. For several decades, illuminating gas, the first by-product of coal distillation, was of little value; however, after 1812 the importance of the gas outweighed that of tar. As a consequence of the stepped-up production of illuminating gas, great quantities of tar would have accumulated had not the chemists discovered a new use for it. By boiling and distilling the tar, it was possible to obtain light oils and heavier oils, called creosote and pitch. The creosote was used for preserving wood, and the pitch was sold for asphalt. A sample of the light oil was sent to Germany for analysis by a brilliant young chemist, August Hoffmann. He found that it contained benzene and aniline. Hoffmann made a variety of dyes by starting with aniline. By 1850, owing to the efforts of Hoffmann and his students, Germany had become the center of the European dye industry, and aniline dyes in all shades of the rainbow could be had. It was therefore natural that another German, Paul Ehrlich, would become interested in dyes and their uses.

Ehrlich was born in 1854 in the city of Strehlen, Silesia, then Germany, now a part of Poland. He studied at the University of Strasbourg, where his tutor, Professor Waldeyer, indulged his experimentation with various kinds of dyes, and "little Ehrlich," as Waldeyer called him, discovered a new type of cell—the mast cell. (Later, it would be found that mast cells secrete the chemical histamine, making nearby capillaries leaky and in some cases provoking an allergic response. Drugs designed to blunt this effect are called antihistamines.) Ehrlich graduated as a doctor of medicine at age 24. His thesis was on "The value and significance of staining tissues with aniline dyes." The fact that certain dyes stained only certain tissues and not others suggested to Ehrlich that there was chemical specificity of binding. This notion became the major theme in his scientific life and led to a search for "magic bullets"—drugs (derived from dyes) that would specifically strike and kill parasites.

Ehrlich wrote that

> curative substances—a priori—must directly destroy the microbes provoking the disease; not by an "action from distance," but only when the chemical compound is fixed by the parasites. The parasites can only be killed if the chemical has a specific affinity for them and binds to them. This is a very difficult task because it is necessary to find chemical compounds, which have a strong

destructive effect upon the parasites, but which do not at all, or only to a minimum extent, attack or damage the organs of the body. There must be a planned chemical synthesis: proceeding from a chemical substance with a recognizable activity, making derivatives from it, and then trying each one of these to discover the degree of its activity and effectiveness. This we call chemotherapy.

Ehrlich's first chemotherapeutic experiments were carried out in 1904 with mice infected with the trypanosomes that cause African sleeping sickness (see chapter 15). He was able to cure these mice of their infection by injecting a red dye he called trypan red. This aroused some interest, but because the drug was inactive in human sleeping sickness, he turned his attention to arsenicals. He began with the compound atoxyl—an arsenic compound that was supposedly a curative for sleeping sickness—but he found the drug was useless: it destroyed the optic nerve, so when patients were treated with the drug, they not only were not cured of sleeping sickness but became blind. In 1906, he prepared compound number 418, arseno-phenyl-glycine, which killed trypanosomes, and then he prepared the 606th derivative, a compound he called 606, a dioxy-diamino-arseno-benzene. In the spring of 1909, Professor Kitasato of Tokyo sent Ehrlich his pupil, Dr. Hata; shortly thereafter, Hata successfully treated syphilis in rabbits with compound 606. So, in 1909, they had in hand a drug for treating syphilis in humans. It was dangerous to make 606, because the ether vapors used in its preparation could cause fires and explosions, and it was unstable—a trace of air changed it from a mild poison to a lethal one. There were problems with side effects. and, at times, the syphilis had progressed so far that 606, now renamed salvarsan, was not effective. But it was the best available treatment for 40 years. Although Ehrlich shared the 1908 Nobel Prize with Elie Metschnikoff for his research on immunity, his greatest contribution was the development of a magic bullet, salvarsan.

How did humans come to use dyes in the first place? Humans, even from prehistoric times, have been fascinated with color. For example, the discovery of red ochre (a pigment from iron ore) in ancient burial sites and in the cave paintings at Lescaux in France reveals that color was used for aesthetic purposes more than 15,000 years ago. The art of dyeing cloth goes back to China in about 3000 BC; at about this time, the principal dyes were obtained from the roots and other parts of plants as well as from insects or shellfish. But despite their use for thousands of years, only a dozen natural dyes have found a practical use. Why, one might ask, are humans so fixated on color? It lies in our biology. Perhaps if we were color-blind, that is, if we lived in a black-and-white world as dogs, cats, horses, cows, pigs, and most mammals do, then colored objects would have little appeal—but primates (apes, monkeys, humans) and birds are not color blind and do see the colors of the rainbow.

Natural dyes come in the colors red, yellow, blue, purple, and black. By mixing some of them, such as blue and yellow, one can get green. The red dyes have two principal sources: madder and cochineal. Madder is obtained from the roots of herb madder; the dye is also called alizarin because *alizari* means "root." Coats and trousers dyed red using madder and alizarin have been used by the armies of Britain and France for several centuries. Indeed, the British were called redcoats after their red jackets. It may be that the color red was selected for military clothing to minimize the visual effect of a wounded soldier seeing his own blood. The red dye cochineal comes from the female scale insect *Dactylopius coccus,* which lives on the prickly pear cactus. It takes 200,000 insects to produce 2 lb of dye.

Yellow dyes, though the largest group of natural dyes, are distinctly inferior to the reds, blues, and blacks, because they have weak coloring properties and they bleach out or fade easily. The most important yellow dye in the Middle Ages was weld obtained from the seeds, stems, and leaves of *Reseda luteola,* a plant called Dyer's rocket. Other sources of yellow dyes are saffron from the anthers of the plant *Crocus sativus,* safflower petals, and the inner bark of the black oak (*Quercus velutina*). This last dye bleaches and fades more easily than weld. Blue dyes come from the leaves of the indigo plant, *Indigofera tinctoria,* also called woad. This plant probably originated in India and is used for coloring blue jeans. Weld in conjunction with woad gives Lincoln green, a color made famous by Robin Hood and his Merry Men. Related to the blue dyes are the purple dyes. The most famous is that obtained from the sea snail *Murex,* called Tyrian purple because it was widely distributed by the Phoenicians out of the city of Tyre (now in Lebanon). It is said that the Romans paid $300 for 2 lb of wool dyed with Tyrian purple, so at the time only kings and priests could afford it. Thus, the phrase "born to the purple" signifies people of wealth or position. The only black dye of any importance comes from the heartwood of the tree *Hematoxylin campechiancum,* found in Central America; it gives a deep color when combined with metallic salts. Obtaining dyes from natural sources was slow, inefficient, wasteful, and very labor-intensive. Natural dyes were usually impure, and the proportions of mixtures were so variable that reproducibility was hard to come by. Another shortcoming was that few natural dyes are fast (fade resistant). These limitations of natural dyes set the stage to begin a search for synthetic dyes.

After 1700, a complex series of social and economic events took place in England and Germany, much of it based on the development of the technology for using steam-powered engines. Along with this, coal was used for powering these engines as well as for smelting iron ore into iron and steel. Coal came into increased demand because woodlands in Britain had been depleted, and supplies of charcoal were scarce. At this time, the coke oven was developed for heating coal to high temperature in the absence of air to produce a dense, smokeless, high-temperature-yielding fuel. By the 18th century, the textile industry, once dependent on human power for weaving, had begun to use machines. In Britain, the rising wool and cotton textile industries characterized the industrial revolution. All this brought a sudden increase in interest on the part of the chemical industry, since textiles required bleaching, and natural processes such as sunlight, rain, and urine were too slow. This need for coloring textiles was responsible for the development of the synthetic dyestuffs industry, and the synthesis of aniline dyes paralleled the early advances in organic chemistry. It is doubtful one would have succeeded without the other.

The founding father of the synthetic dye industry was the Englishman William Perkin (1838–1907), who at age 18 produced the purple dye mauveine. However, the story actually begins in Germany with the appointment of Justus Liebig (1803–1873) as professor of chemistry at Giessen, Germany, in 1824, the same year that coal tar (a viscous by-product of coke production) was discovered. This discovery received little notice until a brilliant pupil of Liebig, August Hoffmann (1818–1892), was appointed director of the Royal College of Chemistry in London in 1845. Hoffmann's main research interest was aniline. Perkin entered the Royal College in 1854 at age 16, and in 1856 Hoffmann gave him the task of making quinine from aniline by oxidation with potassium dichromate in sulfuric acid. If Perkin and Hoffmann had known the chemical structure of quinine, they certainly would have abandoned this route, but at that time it was unknown. Perkin oxidized aniline with dichromate and got a black precipitate. He boiled the black sludge with ethanol, and it gave a striking purple solution, which he called aniline purple or mauveine. This he recognized as a valuable dye, and it was successful in dyeing silk. From this discovery, the synthetic dyestuffs industry began. Perkin left school at 18, patented his discovery, built a factory, and retired in 1873 (age 35) a rich man. He devoted his "retirement" to basic science studies at Glasgow.

The magazine *Punch* wrote the following about Perkin's discovery:

> There's hardly a thing a man can name
> Of beauty or use in life's small game,
> But you can extract in alembro or jar,
> From the physical basis of black coal tar.
> Oil and ointment, and wax and wine,
> And the lovely colors called aniline:
> You can make anything from salve to a star
> (If you only know how) from black coal tar.

Ironically, the Royal College of Chemistry lost its financial support because it could not show a contribution to agriculture, so in 1865 Hoffmann returned to Berlin, Germany, where he held a chair in chemistry. Thus, when the center of the dyestuffs industry moved from Britain to Germany, it was Paul Ehrlich's luck to be on the scene. By World War I, about 90% of all dyestuffs were manufactured in Germany.

The era between 1890 and 1939 might be called the "age of chemotherapy," because at that time, drug development consisted of the discovery of chemical compounds that could be used therapeutically. This was a great advance over the use of disinfection and antisepsis, which involved the practice of preventing microorganisms from entering the body. Magic bullets—therapeutic drugs—enabled physicians to treat disease after the parasite had already established itself. Among the important drugs that made landmark contributions to the elimination of infectious disease agents and their vectors are quinine, atebrine, chloroquine, and primaquine for malaria; sulfonamides for bacterial infections; arsenicals for syphilis; tetrachlorethylene for hookworm; metronidazole for *Giardia*; and DDT for fleas, mosquitoes, flies, and lice.

The next era of magic bullets was the development of antibiotics, which began in 1939. A review of how one of the more recent antibiotics, sulfa drugs, was discovered shows how its development was linked to aniline dyes. One of the German dye companies, IG Farben, had a principal goal of producing dyes for the textile industry. In 1929, IG Farben established a new research institute for bacteriology and pathology with the objective of finding medicinal uses for their dyes. One of the scientists at IG Farben was Gerhard Domagk (1895–1964), who was provided with a dye that was useful for dyeing silk and wool. The Farben chemists had started with a yellow dye called chrysoidin and then added a sulfur group to produce sulfo-amino-chrysoidin with a rich ruby red color. In 1932, Domagk put this dye into mice that had been infected with streptococci, and the mice were cured. However, if the dye was added directly to a culture of the bacteria, the streptococci lived. The red dye was given the trade name Prontosil.

This red dye apparently did not follow Ehrlich's rule: for a chemotherapeutic agent to be effective, it must bind to the pathogen. Something must have happened to Prontosil in the mouse that made it a magic bullet. What happened in the mouse body was that Prontosil was chemically reduced by enzyme action, splitting the molecule into two separate molecules, sulfanilamide and tri-amino-benzene. The actual magic bullet, sulfanilamide, was formed in the body from Prontosil. For this work (which he published in 1935 after many clinical trials), Domagk was awarded the Nobel Prize in 1939, but Hitler prevented him from accepting it at the ceremonies in Sweden.

Sulfa drugs, AZT, and the like are sometimes called antibiotics because they inhibit or kill other microbes, but strictly speaking, a true antibiotic is a substance produced by one living organism that is deleterious to the growth of (or kills) another living organism. The development of true antibiotics began with a chance discovery made by Alexander Fleming (1881–1955). In 1929, Fleming found that one of his petri dishes had a lawn of streptococci with a zone of inhibition around a contaminating green mold, *Penicillium*. Although Fleming coined the term "antibiosis"—literally "against life"—for the phenomenon of bacterial growth inhibition, his subsequent work using "mold juice" produced inconclusive results, and he discontinued further research. A decade later, a British group at Oxford University, headed by a brash Australian, Howard Florey, and a temperamental Jewish refugee, Ernst Chain, took up the project. Despite working under wartime conditions, and with limited equipment and supplies, they eventually purified and tested the active material, named penicillin. Mice were infected with a fatal dose of streptococci; within hours, those mice that had been given penicillin improved, whereas the penicillin-deprived mice were dead. The curative possibilities for this wonder drug were obvious, and by the end of 1943, penicillin production was the second highest priority in the U.S. War Department. Because the British Medical Research Council took the position that patenting medicines was unethical, American companies patented the production techniques they had been using. Fleming, Florey, and Chain shared the 1945 Nobel Prize for their work on penicillin; however, accolades and media attention were disproportionately showered on Fleming. Regrettably, the contributions of Florey and Chain were soon forgotten, and Fleming alone came to be considered responsible for penicillin.

By the late 1940s, penicillin, now available in amounts suitable for the treatment of human disease, became the drug of choice for the treatment of syphilis and other bacterial diseases. Only later would its mode of action be discovered: penicillin blocks the synthesis of the bacterial cell wall. In 1943, Selman Waksman (1888–1973) isolated another antibiotic from the mold *Streptomyces*. It was called streptomycin. By 1944, because penicillin was found to be ineffective against tuberculosis, streptomycin became the drug of choice for treatment.

Microbes Fight Back

Just as there is resiliency in the human species, there is also resiliency in pathogens that enable these species to survive in the presence of drugs. This phenomenon is called drug resistance. How does resistance develop? Mutations occur in the organism that permit that organism to survive, and this capacity is passed on to the offspring. The presence of the drug acts as a selective agent—a sieve, if you will—that culls out the sensitive organisms and allows only the resistant ones to pass through to the next generation. There are three basic mechanisms that allow organisms to develop or promote resistance. (i) They become impermeable to the drug or pump out the drug from the cell so that toxic levels are not reached within the target organism. (ii) They develop an altered enzyme, which has a lower affinity for the drug. (iii) They manufacture excessive amounts of enzyme, thus counteracting the drug, a phenomenon called gene amplification. In some organisms, such as insects, there may be behavioral resistance—the insect avoids coming into contact with the insecticide. Drug resistance can develop without exposure to the drug, but once the drug is present, natural selection promotes the survival of resistant individuals. The problem of drug resistance is that it can blunt our ability to eradicate pathogens and suppress virulence.

The 19th century ushered in measures for protection from and treatment of plagues. The germ theory of disease, anesthesia, antisepsis, and the discovery of magic bullets worked together to alter the way disease was looked upon and how hygienic practices could be harnessed to improve the public health; they also provided a stimulus for the rational development of therapeutic measures. By the 20th century, wonder drugs such as the antibiotics penicillin and streptomycin, as well as chemotherapeutic agents such as AZT and chloroquine, and insecticides such as DDT and malathion were available. In concert with improvements in immunization, these plague protectors—imperfect as they were and as they continue to be—benefited the health of many people, but regrettably not all of humankind. There is much work to be done, but our greatest hope is that the tragedies of coming plagues can be mitigated by the rational use of drugs, insecticides, and vaccines.

Figure 12.1   A scene from the play *Miss Evers' Boys.* (Courtesy Mark Taper Forum, Los Angeles, Calif.)

Chapter 12

The Great Pox Syphilis

It was the fall of 1932, and syphilis was rampant in small pockets of the American South. The U.S. Public Health Service began a study of the disease and enlisted 399 poor, black sharecroppers living in Macon County, Ala., all with latent syphilis. Cooperation was obtained by offering financial incentives such as free burial service on the condition that they agreed to an autopsy; the men were also given free physical exams, and a local county health nurse, Eunice Rivers, provided them with incidental medications such as "spring tonics" and aspirin whenever needed. The men (and their families) were not told they had syphilis. Instead, they were told they had "bad blood" and that annually a government doctor would take their blood pressure, listen to their hearts, obtain a blood sample, and advise them on their diet so they could be helped with their "bad blood." However, these men were not told they would be deprived of treatment for their syphilis (which they did not know they had), and they were never provided with enough information to make anything like an informed decision. The men who were enrolled in the Tuskegee Syphilis Study (as it was formally called) were denied access to treatment for syphilis even after penicillin came into use in 1947. They were left to degenerate under the ravages of tertiary syphilis. By the time the study was made public, largely through James Jones's book *Bad Blood* and the play *Miss Evers' Boys*, 28 men had died of syphilis, 100 others were dead of related complications, at least 40 wives had been infected, and 19 children had contracted the disease at birth (Fig. 12.1).

The Tuskegee study was designed to document the natural history of syphilis, but it came to symbolize racism in medicine, ethical misconduct in medical research, paternalism by physicians, and government abuse of society's most vulnerable—the poor and uneducated. On 16 May 1997, the surviving participants in the study were invited to a White House ceremony, where President Bill Clinton said, "The United States government did something that was wrong—deeply, profoundly, morally wrong. It was an outrage to our commitment to integrity and equality for all our citizens. Today all we can do is apologize but you have the power. Only you have the power to forgive. Your presence here shows . . . you have not withheld the power to forgive. I hope today and tomorrow every American will remember your lesson and live by it."

A Look Back

In 1996, the 500th anniversary of the arrival of syphilis in England was "celebrated." Syphilis—"the Great Pox," as the English called it—was a disease that from 1493 onward swept over Europe and the rest of the world, including China, India, and Japan. The claim was made that this new disease was brought to Naples by the Spanish troops sent to support Alphonso II of Naples against the French king Charles VIII. Charles VIII launched an invasion of Italy in 1494 and besieged Naples in 1495. During the siege, his troops, consisting of 30,000 mercenaries from Germany, Switzerland, England, Hungary, Poland, and Spain as well as those from France, fell ill with the Great Pox, and this forced their withdrawal. It is generally believed that, with the disbanding and dispersal of the soldiers of Charles VIII, who themselves had been infected by the Neapolitan women, the pox spread rapidly through Europe. In the spring of 1496, some of these mercenaries joined Perkin Warbeck in Scotland and, with the support of James IV, invaded England. The pox was evident in the invading troops. Within 5 years of its arrival in Europe, the disease was epidemic: it was in Hungary and Russia by 1497 and in Africa and the Middle East a year later. The Portuguese carried it around the Cape of Good Hope with the voyage of Vasco de Gama to India in 1498. It was in China by 1505, in Australia by 1515, and in Japan by 1569. European sailors carried the Great Pox to every continent save for Antarctica. Syphilis was so ubiquitous by the 19th century that it could be considered to be the AIDS epidemic of that era. As we shall see, the parallel between AIDS and syphilis does not end there.

The French called this pox "the disease of Naples," blaming the Italians for it, and the Italians called it "the French disease" (Fig. 12.2). The basis for these various names is that the undisciplined troops of King Charles VIII, during their retreat from Italy, carried the disease back to their homelands in many parts of Europe; shortly thereafter, it came to be called after the national origins of those people who were disliked and considered unclean— the Russians called it "the Polish disease"; in Japan it was referred to as "the Chinese disease," and in England it was "the Spanish disease."

Victims of syphilis suffered with fevers, open sores, disfiguring scars, disabling pains in the joints, and gruesome deaths, leading Joseph Grunbeck (1473–1532) of Germany to write in the late 15th century: "In recent times I have seen scourges, horrible sicknesses and many infirmities affect mankind from all corners of the earth. Among them has crept in, from the western

Figure 12.2 This etching by the Belgian artist Felicien Rops (1833–1898) shows the destructive effects of syphilis on an old prostitute.

shores of Gaul, a disease which is so cruel, so distressing, so appalling that until now nothing so horrifying, nothing more terrible or disgusting, has ever been known on this earth."

But if the story of the outbreak of this Great Pox in the army of Charles VIII is accurate, how did his men contract this disease? There are two main theories, called the Columbian and the pre-Columbian (or, by some, the anti-Columbian). Let us consider the Columbian theory first. Christopher Columbus (1451–1506) visited the Americas and on 12 October 1492 arrived in San Salvador. He set sail for home 3 months later on 16 January 1493 and arrived in Spain in March of 1493, carrying with him several natives of the West Indies. Upon arrival in Spain the crew of 44 disbanded, and some of them joined the army of Charles VIII. The first mention of the disease occurs in an edict by Emperor Maximillian of the Holy Roman Empire at the Diet of Worms in 1495, where it is referred to as "the evil pox." Twenty-five years later, in a book published in Venice and authored by Francisco Lopes de Villalobos, it was claimed that syphilis had been imported into Europe from the Americas. Favoring this idea was the severity of the outbreak—indicative of a new import. Indeed, from 1494 to 1516, the first signs were described as genital ulcers, followed by a rash, and then the disease spread throughout the body, affecting the gums, palate, uvula, jaw, and tonsils and eventually destroying these organs. The victims suffered pains in the muscles and there was early death—an acute disease. From 1516 to 1526, two new symptoms were added to the list: bone inflammation and hard pustules. Between 1526 and 1560, the severity of symptoms diminished, and thereafter its lethal effects continued to decline, but from 1560 to 1610 there was another sign: ringing in the ears. By the 1600s, "the Great Pox" was an extremely dangerous infection, but those who were afflicted did not suffer the acute attacks that had been seen in the 1500s.

In the 1700s, syphilis was a dangerous but not an explosive infection. By the end of the 1800s, both the virulence of the pathogen and the number of cases declined. Even so, the numbers were by no means trivial: by the end of the 19th century it was estimated that 10% of the population of Europe was infected, and by the early 20th century mental institutions noted that one-third of all patients could trace their neurological symptoms to syphilis. Clearly, either the people were developing an increased resistance, or the disease's pathogenicity was changing.

What produced this dramatic outbreak of syphilis? Some have suggested that it was a new disease introduced into a naive population and that the increased rate of transmission by sexual means transformed what once had been a milder disease into a highly virulent one (such as happened with human immunodeficiency virus [HIV]). Another hypothesis was that European syphilis was derived from yaws, and initially infection was by direct contact; one simple means might have been by mouth-to-mouth kissing, as this was the more common practice of greeting (rather than a handshake) in Tudor England. In addition to kissing, it was suggested, infection could have been passed by shared drinking cups. In this case, the infectious canker (chancre) occurred on the lips and tongue and either went unnoticed or was mistaken for more benign diseases, such as cold sores or impetigo. Another reason for the rapid spread of syphilis may have been that precautions against transmission were not observed, since the later stages would not have been recognized as being associated with the earlier and more-infectious stages. Even the early stages—chancre and rash—might have been considered nothing more than minor, self-healing skin disorders.

Treatments for "the Great Pox" varied with the times. George Sommariva of Verona, Italy, tried mercury for the treatment of "the French pox." By 1497, mercury was applied topically to the suppurating sores, or it was taken in the form of a drink. The treatment came to be called "salivation" because the near-poisoning with mercury salts tended to produce copious amounts of saliva. Another treatment was guaiacum (holy wood) resin from trees (*Gauiacum officianale* and *G. sanctum*) indigenous to the West Indies and South America. In fact, the resin was a useless remedy, but it was popularized probably to lend further credence to the American (Columbian) origins. However, few adherents of the Columbian theory paid attention to the fact that guaiacum was introduced as a treatment in 1508 fully 10 years before the first mention of the West Indian origin of syphilis. Further, the claim for the Columbian origin of syphilis would seem to be weakened by several written reports that the crew of Columbus and the Amerindians were healthy. There are others, however, who suggested (but never showed) that some of Columbus' sailors did have syphilitic lesions. Perhaps the best evidence for the Columbian origin of syphilis lies in the bones (Fig. 12.3) and teeth: bone lesions—scrimshaw patterns and saber thickenings on the lower limbs of adults and notched teeth in children—are diagnostic of syphilis and have been found in the skeletal remains of Amerindians; until recently, no such characteristic bone lesions were found in skeletons in Europe or China that dated from before the 1500s, or in Egyptian mummies. Hippocrates and Galen made no mention of the disease, so it appears that syphilis was not present in ancient Greece or in the Roman Empire. But in 2000, there was a report on 245 skeletons that were

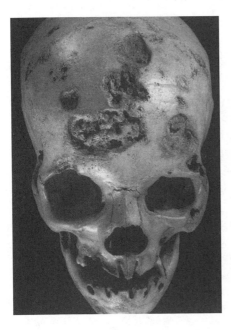

Figure 12.3   The destruction of the skull bones (called caries sicca) by bone gummas. Courtesy of Mütter Museum, College of Physicians of Philadelphia.

unearthed from a medieval monastery, known as Blackfriars, in Hull, England. Carbon dating of the bones established the date of death to be between 1300 and 1420—at least 70 years before Columbus' voyage. Some paleoanthropologists believe that these Blackfriar skeletons show the telltale signs of syphilis, whereas others dispute the evidence and contend that the bone lesions are more like those that result from a related (but nonvenereal) disease, yaws. However, this refutation seems not to hold up in light of the notched teeth found in the remains of individuals from the port city of Metaponto, Italy, and dated to 600 BC, as well as in those who died in Pompeii during the eruption of Mount Vesuvius (AD 79). Some believe that this is proof that syphilis has been present in Europe for thousands of years.

Port cities with their characteristics of high sexual activity and prostitution provide the locale for the possibility of high transmission rates of sexually transmitted diseases. Immigration contributes to the introduction of foreign illnesses at a greater frequency, and urbanization facilitates disease spread. It has been hypothesized that when syphilis became a disease of cities and there were changes in social habits, clothing, sharing of eating utensils, and hygiene, the propagation of the milder form was reduced, and this allowed transmission of only the more virulent form. According to this view, syphilis would have existed in Europe and Asia in a milder form prior to 1493, but then it changed and became virulent. Exactly when and how venereal syphilis arose continues to be debated. Indeed, the solution to the question of the origin of "the Great Pox" may come only when the bones of Columbus and members of his crew are found and can be shown to contain, by DNA analysis, the genetic signature of the germ of syphilis.

Spirochete Discovered

What is the causative agent of syphilis? Early observers believed that syphilis was God's punishment for human sexual excesses. Public bathhouses were closed, and there was distrust between friends and lovers. One result was that Platonic love emerged as a social cult. Others believed that syphilis had an astrological basis. In 1484, Mars, Jupiter, and Saturn were in conjunction with Scorpio, the constellation most commonly associated with sexuality. But as early as 1530, Girolamo Fracastoro (1483–1553) recognized that the disease was contagious, and he called it syphilis after a fictitious shepherd, named Syphilis, who got it by cursing the gods. Fracastoro described the earliest stages of syphilis as small ulcers on the genitals, followed by a skin rash; the pustules ulcerated and the person suffered with a severe cough that eroded the palate. Sometimes the lips and eyes were eaten away, or ulcerated swellings appeared, and there were pains in the joints and muscles. Fracastoro theorized that syphilis was a result of "seeds of contagion," but for 400 years no one saw the "seeds" that caused the symptoms of the Great Pox.

Then, in 1905 Fritz Schaudinn (1871–1906) and Erich Hoffmann (1869–1959) in Germany identified a slender, spiral-shaped bacterium, a spirochete (*spiro* meaning "coiled" and *chete* meaning "hair"), in the syphilitic chancres. When Hideyo Noguchi isolated the same bacterium from the brains of patients suffering with insanity and paresis, it became clear that all stages of the disease were linked to one kind of spirochete (see Fig. 1.2B). Because of its shape, Schaudinn and Hoffmann called the microbe *Treponema* (from the Latin *trep* meaning "corkscrew" and *nema* meaning "thread"), and because it stained so poorly, they named its species *pallidum* (meaning "pale"). *T. pallidum* has no spores and cannot be cultured on bacteriological medium, but it can be grown in laboratory animals such as rabbits and guinea pigs. However, human beings are the only natural host for *T. pallidum*. It divides slowly (in ~24 hours) and is quite fragile, requiring a moist environment. Other spirochetes related to *Treponema* cause relapsing fever and Lyme disease (*Borrelia burgdorferi*).

One of the reasons it took so long for microbe hunters to identify the cause of syphilis was that it was confused (and associated) with another venereal disease, gonorrhea. Even the great anatomist and physician John Hunter, of St. George's Hospital in London, could not solve the puzzle.

It was 1748, and London was the center of the universe . . . they talked of . . . Voltaire, Blackstone was a judge, Chippendale a furniture maker. Samuel Johnson wrote . . . of the permanent and certain characteristics of the mind. The pox was sexually transmitted. The pox, if you were unlucky, could rot the organ of manhood. Some said the pox was one disease, others said it was really two similar diseases. In John Hunter's mind . . . it wasn't transmitted by miasmas at all. The pox must be caused by . . . a putrid liquid of some sort. John Hunter had the pox all figured out. In his opinion there were two forms. If it began as a pimple on the penis, it took one specific course. The other form was seen when . . . the interior of the urethra was affected . . . and produced a liquid discharge. A patient with the pox stood before him . . . a tiny drop of yellowish liquid hung from the tip of his penis . . . it was a classic case . . . of the pox. To prove that the pox was a single disease all he needed to show was that the pus from this patient, a "wet" case, could produce a chancre, or a "dry" case in another penis. John only had one healthy penis handy, so he used it. While the patient stood before him, he took the droplet on a lancet and transferred it to the glans of his own penis. Then, through the droplet, he stabbed himself with the lancet . . . and again. He squeezed the small cuts open, so that the liquid . . . of the pox could take hold. He gave the man a bit of mercury . . . to rub . . . into his thighs. That Sunday he confided in his notebook that his penis began tingling . . . there was also a redness where the lancet had pierced the skin. By the following Tuesday there were two pimplelike chancres. . . . A runny disease had produced a dry disease, proving, once and for all . . . the pox was but one disease. He rubbed mercury into his thighs and the chancres disappeared. Three months later he got a skin rash and . . . he rubbed a massive dose of mercury in his thighs and the symptoms, he wrote in his notebook, disappeared for good. The inevitable happened in 1793 . . . he died. John Hunter had been wrong. But it would be a century before microscopes focused on the gonococcus bacterium and syphilis spirochete. Only then did it become clear that a patient could have both at once and pass on both . . . or only one. They would discover that the form of the pox that seemed the worst to John Hunter, the wet one . . . was really less dangerous. It was gonorrhea. The other one that produced the chancre was syphilis. They would learn that syphilis had not two stages, but three. In its third stage it could attack a variety of organs, including the heart, brain, spinal cord—and aorta. John Hunter had died of tertiary syphilis—and by his own hand.

The Disease Syphilis

In the past, diseases such as syphilis were called venereal diseases, or VD. "Venereal" refers to the Roman goddess of Love, Venus, because under most circumstances these diseases are transmitted by sexual contact. However, since love and sex are not equivalent, the classic venereal diseases—gonorrhea, genital herpes, lymphogranuloma inguinale, HIV, and syphilis—are now referred to as STDs, or sexually transmitted diseases.

Before considering the pre-Columbian (or anti-Columbian) theory for the origins of syphilis, let us first describe its clinical characteristics. Since there is enormous variation in the disease symptoms, syphilis has been called "the great imitator." How do we know about the clinical course of a syphilitic infection? Between 1890 and 1910, 1,404 untreated Swedish patients were studied as part of the Oslo study. Seventy-nine percent of those with late-stage syphilis developed neurosyphilis, 16% had destructive ulcers (gummas), 35% had cardiovascular disease, and syphilis was the primary cause of death in 15% of the men and 8% of the women. Between 1917 and 1941, the Rosahn study of cadavers of those who died from syphilis found that late-stage syphilis was evident in 39%; neurological complications were found in 9%, and cardiovascular disease was found in 83%. Between 1932 and 1972, the now infamous Tuskegee Syphilis Study tracked untreated syphilis in poor, rural, uneducated black males in Macon County, Ala., where the overall infection rate was 36%. The study compared 399 men with latent syphilis with a control group of 201 healthy men. Signs of cardiovascular disease were found in 46% of the syphilitics and 24% of the control subjects, and 24% of the syphilitics but only 5% of the control subjects had an inflammation of the aorta. Bone disease was found in 13% of the syphilitics but only 5% of the controls. The greatest differences were seen in the central nervous system: of the syphilitics, 8% had signs of disease, compared with 2% of the controls. After 12, 20, and 30 years, the death rate in the syphilitics was 25, 39, and 59% compared with 14, 26, and 45% in the control subjects, respectively.

The chancre stage is the earliest clinical sign of disease. After initial infection, that is, ~21 days (range, 3 to 90 days) after initial contact, a painless, pea-sized ulcer—a chancre—appears at the site of spirochete inoculation. The chancre is a local tissue reaction and can occur on the lips, fingers, or genitals. At this stage, the infection can be spread by kissing or touching a person with active lesions on the lips, genitalia, or breasts and through breast milk. If untreated, the chancre usually disappears within 4 to 8 weeks, leaving a small, inconspicuous scar. Despite the fact that the individual may not notice the chancre (or the scar), the presence of a lesion larger than that of smallpox gives the disease the name the Great Pox.

The secondary or disseminated stage usually develops 2 to 12 weeks (mean, 6 weeks) after the chancre stage; however, this stage may be delayed for more than a year. *T. pallidum* is present in all the tissues but especially in the blood, and there is a high level of syphilis antigen. Results of serological tests (such as the Wassermann, VDRL [Venereal Disease Research Laboratory], and RPR [rapid plasma reagin]) are positive. There is now a general tissue reaction—headache, sore throat, mild fever, and, in 90% of cases, a skin rash. The skin rash may be mistaken for measles or smallpox or some other skin disease. The highly infectious secondary stage does not last very long. Then the patient enters the early latent stage, in which he or she appears to be disease-free, i.e. there are no clinical signs. Indeed, the most dangerous time is during the early latent stage, because the infected individual can still transmit to others. Transmission can also occur by blood transfusion, but this is rare, because the spirochetes do not survive longer than 24 to 48 hours under blood bank storage conditions.

The infection continues to progress, and after about 2 years, the late latent or tertiary stage develops. In tertiary syphilis, there are still spirochetes present in the body, but the individual is no longer infectious. To all intents and purposes, individuals are not infectious through sexual contact 4 years after initial contact. Tertiary syphilis develops in one-third of untreated individuals 10 to 25 years after initial infection. The disease then becomes chronic. In about 20% of cases, destructive ulcers (gummas) appear in the skin, muscles, liver, lungs, and eyes. In 10% of cases, the heart is damaged and the aorta is inflamed. In severe cases, the aorta may rupture, causing death, as it did in the case of John Hunter. In 40% of untreated cases, the spinal cord and brain become involved, causing incomplete paralysis (paresis), complete paralysis, and/or insanity, accompanied by headaches, pains in the joints, impotence, and epileptic seizures. The remaining cases have asymptomatic neurosyphilis. Most untreated patients die within 5 years after showing the first signs of paralysis and insanity.

The symptoms of syphilis have been set to rhyme:

> There was a young man from Back Bay
> Who thought syphilis just went away.
> He believed that a chancre
> Was only a canker
> That healed in a week and a day.

But now he has "acne vulgaris"—
(Or whatever they call it in Paris);
On his skin it has spread
From his feet to his head,
And his friends want to know where his hair is.

There is more to his terrible plight;
His pupils won't close at night.
His heart is cavorting,
His wife is aborting,
And he squints through his gun-barrel sight.

Arthralgia cuts into his slumber;
His aorta's in need of a plumber;
But now he has tabes,
And sabre-shinned babies
While of gummas he has quite a number.

He's been treated in every known way,
But his spirochetes grow day by day;
He's developed paresis
Has long talks with Jesus,
And thinks he's the Queen of the May.

"Canker" means a mouth sore; "acne vulgaris" is a rash; "arthralgia" means pains in the joints; and "gun-barrel sight" is a loss of peripheral vision due to destruction of the optic nerve.

"Catching" Syphilis

At the primary (chancre) stage, syphilis can be spread by kissing or touching a person with active lesions on the lips, genitalia, or breasts. During the secondary stage, which usually does not last very long, the skin lesions render the individual infectious. In the early latent stage, there are no clinical signs; however, the individual remains infectious.

Syphilis can be transmitted from the mother to the developing fetus via the placental blood supply, resulting in congenital syphilis; this is most likely to occur when the mother is in an active stage of infection. If the mother is treated during the first 4 months of pregnancy, the fetus will not become infected. Fetal death or miscarriage usually does not occur until after the fourth month of pregnancy, at the earliest. Repeated miscarriages after the fourth month are strongly suggestive of syphilis (but not unequivocal proof). A diseased surviving child may have the same symptoms as the mother, or there may be deformities, deafness, or blindness. Of particular significance is that some offspring who are congenitally infected may have Hutchinson's triad: deafness, impaired vision, and a telltale groove across peg-shaped teeth, first described in 1861 by the London physician Jonathan Hutchinson.

The Pre-Columbian Origin of Syphilis

All of the treponemes that cause the human diseases yaws, pinta, and syphilis are identical in their morphology and differ very little genetically. Indeed, their genomes contain only 1,000 genes, and none has been found to be specifically associated with virulence. The species are differentiated from one another by their clinical manifestations. The *Treponema* family of spirochetes, with its varying clinical picture, forms the basis of the theory of a pre-Columbian origin of syphilis. This theory suggests that human treponemes may have come from animals and that an infection similar to pinta (*Treponema carateum*), localized in the skin, arose about 15,000 BC in Africa and then with human migration passed across the Bering Straits to become isolated in the tropics of the Americas, places where the humidity is high. Pinta is restricted to the skin, is disseminated by introduction into skin lesions, and is usually found in persons 15 to 30 years of age. Associated with the disease are pigment changes in the skin. Pinta is found in tropical Central and South America where hygiene is poor and conditions are crowded.

It has been hypothesized that about 10,000 BC, the spirochete-causing pinta mutated into a disease very similar to yaws (*Treponema pallidum* subsp. *pertenue*) and was restricted to tropical areas of Africa and the associated landmasses. This mutation may have been triggered by the tropical climate. Later, it was brought to the Americas by the slave trade. Yaws exists where there is a warm, moist climate, especially in places characterized by poor hygiene. It is transmitted primarily in children by skin contact and has probably existed in Africa since prehistoric times. Yaws is not a benign disease; it can spread throughout the body via the blood to cause disfigurement of the face and bones. Yaws occurs in Africa, South America, Southeast Asia, and Oceania. Today, the number of cases has been reduced worldwide thanks to antibiotic treatment.

As populations of *Homo sapiens* began to penetrate into temperate and drier regions, the presence of a cooler climate and the wearing of clothing provoked (or so it is believed) the spirochetes to mutate again, allowing them to invade the mouth and throat and to produce skin lesions similar to those of endemic syphilis or bejel (*Treponema pallidum* subsp. *endemicum*). This infection can be destructive to tissues such as the skin and bones. Transmission occurs via contaminated objects—drinking vessels, kitchen utensils, saliva, and contaminated fingers—as well as by mouth-to-mouth contact. It has been known to exist in Africa for centuries. Today, it is found in prepubescent children living under seminomadic conditions in the northern Sahara, southwest Asia, and Australia. In pinta, endemic syphilis, and yaws, the primary chancre may resemble impetigo (a skin disease caused by strep-tococci) or herpes simplex (cold sores).

According to the pre-Columbian theory, true venereal transmission from spirochetes in the vagina and the penis occurred as a result of a third muta-tion, coinciding with the emergence of cities about 3000 BC in the Middle East. From this region it spread through the Mediterranean and into Europe, and by about the first century BC it was present in the Mediterranean in a mild form; there it remained an endemic disease for centuries. Finally, a fourth mutation in the spirochete occurred in Europe in the 15th century. This muta-tion, triggered by environmental conditions and behavioral patterns associ-ated with city life in Europe, allowed the virulent, venereal form of syphilis to emerge.

Syphilis and Its Social Context

The highly pathogenic form of *T. pallidum* caused devastating effects from the 1500s to the 1700s. However, over a 200- to 300-year period, it changed from an acute, lethal infection to a chronic, corrosive disease. Cerebral damage due to syphilis caused the deaths or affected the lives of Charles VIII and Francis I, Pope Alexander Borgia, Benvenuto Cellini, Henri Toulouse-Lautrec, Hein-rich Heine, Franz Schubert, Peter the Great, Catherine the Great, Florence Nightingale, John Keats, Guy de Maupassant, and Al Capone. Randolph Churchill, the father of Winston, was a syphilitic, prompting the question, would Winston Churchill have been so driven had he not felt compelled to revive and restore his father's name? Syphilis may have also caused Ludwig von Beethoven's deafness.

Ivan, Grand Duke of Muscovy, was born in 1530 and ascended the throne upon the death of his father. For years, he ruled wisely and humanely. In 1552, his wife Anastasia gave birth to a son Dimitri, who died at age 6 months, probably of congenital syphilis. Nine months later, Anastasia gave birth to another son, Ivan, and in 1558 a third son, Fedor, was born. It is sus-pected that Ivan was infected with syphilis prior to his marriage, when he was a notorious womanizer. When Anastasia died in 1560, Ivan drowned his grief in drink and debauchery. Ivan remarried in 1561, and his wife gave birth to a son, Vassili, who lived 5 weeks. By 1564, there is evidence that Ivan was suffering from neurosyphilis. From 1565 to 1584, Ivan engaged in a reign of terror. He tortured, flogged, burned, and boiled those he considered to be his enemies. Claiming a conspiracy, he had thousands of citizens of Nov-gorod flogged to death, roasted alive, or drowned under the ice; public exe-cutions were also held in Moscow. Ivan and his son raped the widow and daughter of Prince Viskavati, whom they had hanged, and in 1581, during a fit of rage, he stabbed his own son to death, the czarevitch Ivan. Ivan died a gibbering idiot at age 54. This left the throne to the congenital idiot Fedor, who was incapable of ruling, so the throne fell to Boris Godunov. After the death of Godunov in 1605, chaos gripped Russia, and order was not restored until the election of the first Romanov in 1613.

On the basis of the offspring of several of his six wives, Henry VIII (1491–1547) is suspected of being a syphilitic. His first wife, Catherine of Aragon, gave birth to a child who died within a few days, and she had three more stillborns, all in the seventh and eighth months of pregnancy. Anne Boleyn miscarried one child at 6 months and another at three-and-a-half months. Jane Seymour had one son, Edward, born in 1537, who died at age 15, possibly of tuberculosis or poisoning, but he also had a suspicious skin rash shortly before his death. Elizabeth I, born of Anne Boleyn, lived to age 69 but was nearsighted, and so was Mary Tudor, who was also deaf and had a rather large, flat nose that discharged foul-smelling pus. Henry became sterile or impotent in his late forties, suggesting that he may have been syphilitic. In 1527, Henry's character, like that of Ivan the Terrible, began to change. He suffered from headaches, insomnia, sore throats, and an ulcer on his leg. He had a gumma on his nose. Any person who slandered his marriage to Anne Boleyn was guilty of treason and was sentenced to death by hanging and quartering while still alive. He was bent on terror and slaughtered the Lollards, Lutherans, Anabaptists, and Catholics. He executed the Prior of the Charterhouse and all his monks, and beheaded Thomas Moore and Bishop John Fisher in 1535. Although he could have divorced Anne, he preferred to put her to death and to declare her daughter a bastard. He dissolved the monasteries and hung monks and abbots who resisted or delayed establishment of the Anglican Church. Henry never decayed mentally but became profoundly obese. His failure to produce a male heir was the beginning of the end of the Tudor line. The firm and efficient Tudor dynasty gave way, perhaps as a result of syphilis, to the absolutism of the much weaker Stuarts, and England was wracked by civil war. Did Henry VIII actually have syphilis? We do not know for sure, but syphilis was certainly common during his reign, and the disabilities of his offspring and his behavior make it likely that he did not escape infection. The historian William H. McNeill has aptly noted, "The inability of royal and aristocratic leaders to give birth to healthy children accelerated social mobility making more room at the top than there would have been otherwise"—thanks to syphilis!

Diagnosing Syphilis

In the chancre stage, syphilis is rarely diagnosed by growth and isolation of *T. pallidum* itself. Instead, examination of fluid from the lesion by dark-field microscopy and clinical findings are the basis for disease determination. The Wassermann test is a serological reaction that originally used extracts of tissues infected with *T. pallidum*; however, when it was found that uninfected tissues reacted positively with syphilitic sera, these antibodies (called reaginic) were considered nonspecific. The nonspecific reaction appears to be due to the external waxy cell wall of the spirochete. Despite this, these serological tests, which are cheap, quick, and easy to use, continue to play a role in screening patients—especially those in high-risk populations—as well as in evaluating adequacy of treatment. The most commonly used tests are the VDRL and the RPR tests. Both become negative 1 year after successful treatment of primary syphilis and 2 years after successful treatment of secondary syphilis. Patients who are infected with both HIV and syphilis commonly have a serological response with very high titers. Fluorescent antibody absorbed tests and hemagglutination assays are used to confirm a positive nontreponemal (nonspecific) test.

Treating Syphilis

In 1909, Paul Ehrlich (1854–1915) developed an arsenical derivative, salvarsan, that was effective in reducing the severity of the disease. Salvarsan was not without its problems—it was toxic, required years of treatment, and was difficult to administer. Only 25% of patients ever completed the therapy. In her book *Out of Africa*, Isak Dinesen tells how she was diagnosed with syphilis, contracted from her husband, and was then treated (apparently successfully) with salvarsan. Salvarsan remained the drug of choice until 1943, when penicillin was introduced. Penicillin remains the drug of choice in the treatment of syphilis.

Syphilis and the Social Reformers

Although the advent of drug treatment has helped to significantly reduce the incidence of syphilis, the disease has by no means been eradicated. A great deal of the failure can be attributed to social attitudes toward syphilis in particular and sexually transmitted diseases in general. When salvarsan treatment became available in the early 1900s, syphilis became the focus of social reformers. To them the disease was a clear indicator of the breakdown of the home, morality, and marital fidelity and could be ascribed to promiscuity, prostitution, and immigration. So, they said, something had to be done about ridding society of the disease and its carriers.

Social reformers focused on the impact of sexually transmitted diseases on the family and the person who was infected—generally, a man or a woman who had strayed from the path of moral virtue and consorted with prostitutes or people of ill repute. Such individuals in turn infected the innocent spouse and the children. The moral code at the time (the Victorian era) held that only sex within marriage was socially sanctioned, and therefore the victims of syphilis were those who "deserved" the disease and had brought it on themselves by immoral behavior. The disease was considered a fitting punishment. This attitude led public health officials in the early 1900s to spread the idea that simple contact was sufficient to spread syphilis, that one could get it from a drinking cup; touching doorknobs, pencils, or pens; or sitting on a toilet seat. This was of course not true, but saying that so-called "innocent" transmission was possible allowed people to be socially "pure" even if they had contracted the disease. Physicians and politicians also used ~uch information as a further caution to keep individuals from engaging in socially unacceptable behavior and as a means of reducing immigration into the United States, because immigrants were supposedly spreading syphilis in America through either prostitution or casual contacts.

By the time of World War I, concern about syphilis had reached an unprecedented high. The military draft and consequent medical examinations had revealed that 13% of those drafted were infected with either syphilis or gonorrhea, and this touched off a vigorous anti-VD campaign. The focus of the campaign was on prostitutes, who were seen as the source of infection of military personnel. Posters were used by the Army to discourage immoral behavior; the idea was that, as one federal official stated, "To drain a red-light district and thereby to destroy the breeding ground of syphilis and gonorrhea is as logical as to drain a swamp thereby destroying the breeding place of malaria and yellow fever." As a result, thousands of women living and working near military training sites were quarantined during the war.

Those in charge of the VD campaign wanted not only to stop the spread of disease but also to change sexual and social attitudes. It wasn't just that soldiers should be cured of syphilis or that they should adopt precautions to avoid getting it whether they wanted to have sex or not. They were supposed to live a morally pure life and avoid extramarital sex under any circumstances; in doing so, they would avoid contracting a sexually transmitted disease. Latex condoms were available during the time of World War I, but the military refused to provide them to the troops because it was believed that this would contribute to their moral decline and encourage extramarital sexual behavior. The treatment for syphilis was made painful, and this was intended as a deterrent to contracting the disease. The U.S. Army also ruled that sexually transmitted diseases were injuries not incurred in the line of duty, so those who became infected lost their pay. The whole attitude was punitive: the infected soldier or sailor or marine had done something wrong, and the next best thing to telling him not to do it in the first place was to punish him afterwards.

As one might expect, the campaign was unsuccessful. Rates of VD continued to stay at high levels or rose during and after World War I, and efforts to control syphilis lagged during the 1920s. In the 1930s, President Franklin D. Roosevelt appointed Thomas Parran as Surgeon General. Parran avoided taking a moral stance on the disease and instead concentrated on how to eradicate syphilis—an achievable goal, but one that was unpopular in some quarters. Parran developed a 5-point program based on successful programs in Scandinavia. (i) Find the cases. To this end, diagnostic centers were established where there would be confidentiality and free testing to identify cases. (ii) Treat infected individuals promptly to prevent disease spread and to reduce virulence. (iii) Break the chain of transmission by tracing and treating sexual contacts. (iv) Institute mandatory premarital testing using the Wassermann test. (To get a marriage license, individuals would have to be Wassermann negative. While in principle this was a reasonable idea, Wassermann testing for premarital screening turned out not to be effective, first because of false-positive results [i.e., for individuals who did not have syphilis but had a positive Wassermann test result], and, second, because even a positive blood test did not indicate an active infection. Further, the tests were directed at the lowest-risk group, and this requirement assumed that those who wanted to marry would not do so if they tested positive and that after marriage no one would acquire an infection. By the 1980s, mandatory premarital Wassermann testing was dropped.) (v) Institute a massive public education campaign to inform people about syphilis, how it is contracted, its symptoms, and where to get treatment when infected. The heart of the campaign was provision of information: people were not told what to do or what not to do, and a moral position was not taken. The plan met with some opposition from the public and private sectors, since VD was something polite people were not supposed to discuss in public. Indeed, in 1934, when Parran was to make a radio broadcast about VD, he was allowed to do so only if he did not use the words syphilis or gonorrhea. Parran and Roosevelt eventually prevailed, and the National Venereal Disease Control Act was passed in 1938. Congress allocated money for VD clinics and testing, and treatment was provided to those who could not afford it. During World War II, antisyphilis efforts once again intensified when it was found that ~5% of the recruits had syphilis. Unlike in World War I, troops were provided with condoms, education, prophylaxis, and rapid treatment. If a soldier or sailor did not report the disease, however, it was a court martial offense.

Over the past 50 years there have been recurrent epidemics of syphilis in the United States. These cycles—with a periodicity of approximately 10 years—have generally been attributed to changes in human sexual behavior. For example, the epidemic of the 1970s has been linked to the sexual revolution and the gay liberation movement, and that of the 1980s was claimed to be due to poverty, urbanization, and crack cocaine prostitution. However, using mathematical modeling and data reported to the Centers for Disease Control between 1941 and 2002, it appears that the cycling of syphilis incidence is not attributable to these factors but is the result of prolonged partial protective immunity following infection. According to this model, cycles occur because, when there is a depletion of the number of susceptible individuals in the at-risk group (as a result of partial immunity), the epidemic fades, but over time, when the at-risk population builds up, there is sufficient "fuel" for the next outbreak. Consistent with the 10-year oscillation of syphilis epidemics is the calculated R_0 value of 3 (see page 13).

Distribution and Incidence of Syphilis

The Centers for Disease Control (CDC) reported in 1990 that the syphilis rate had risen from 11.5 cases per 100,000 in 1985 to 20 cases per 100,000 in 1990. By 2003 the rate had declined to 2.5 cases per 100,000. In the U.S. over 34,000 cases of syphilis (primary, secondary, latent, and congenital) were reported to the CDC in 2003—amounting to nearly 100 new cases every day. The highest syphilis rates were in the American South, which accounted for 44% of all cases.

In 2003, syphilis occurred primarily in males aged 35 to 39 and in women aged 20 to 24 years. The incidence among men was five times that in women (11.8 cases per 100,000 population versus 2.4 cases per 100,00 population). Commercial sex workers and persons who exchange sex for drugs had the highest incidence of syphilis. The rate of primary and secondary syphilis in African-Americans (7.8 cases per 100,000 population) was five times greater than the rate among non-Hispanic whites (1.5 cases per 100,000 population). In 2003, the overall rate of congenital syphilis in the United States remained steady at 10 cases per 100,000 live births.

Although there has been substantial progress in the United States in the elimination of syphilis the disease remains an important problem, particularly in the American South, in some urban areas, and among African-Americans. Some fundamental societal problems, such as poverty, inadequate access to health care, drug use, prostitution, lack of education, and male homosexuality remain associated with disproportionately high levels of syphilis in certain populations.

Vaccines against Syphilis

What about a vaccine against syphilis? So far, there is none. However, an early disastrous attempt to produce a vaccine is instructive. Joseph Auzias-Turenne (1812–1870), a French physician, drew inspiration from Jenner's smallpox vaccine and mistakenly considered soft chancre to be an attenuated form of hard chancre. So, as with smallpox, he took part of a soft chancre and prepared it for use as vaccine. However, soft chancre is caused by another bacterium, *Haemophilus ducreyi*, and there is no cross-protection to syphilis. Had this scheme prevailed in 19th-century France, thousands would have mistakenly believed they had been protected against syphilis; instead, they would have gotten chancroid!

Syphilis and AIDS in the Social Context

Public attitudes toward the AIDS epidemic are reminiscent of that toward syphilis a century ago. Although AIDS differs from syphilis in many respects, there are some important similarities. Historically, there have been two main approaches on how a sexually transmitted disease should be fought: one, the moral approach, contends that the best way to prevent infection is to advocate a social and sexual ethic that makes it impossible to acquire an infection—sexual abstinence until marriage. This, proponents believe, can be achieved through education and the suppression of prostitution. The other approach attempts to divorce itself from any particular judgment about sexual behavior and suggests that individuals should be provided with the means for protecting themselves from infection should they choose to engage in sexual behavior, and treatment should be nonpunitive so as to encourage infected individuals to seek help. If there is to be effective protection of the public at large, both approaches will be needed.

Syphilis is more than a corrosive, infectious disease. The history of syphilis is replete with discrimination against those with differing lifestyles and those who are marginalized in society—the urban poor, the uneducated, immigrants crowded into the cities, and those with alternative lifestyles such as sexual orientation. The terror of syphilis promoted the search for new drugs. Drug treatments, however, even those that have been shown to be successful, cannot be relied upon for complete eradication of diseases such as syphilis and gonorrhea. Syphilis has galvanized public health authorities to stem its spread through education, treatment, avoidance of moral judgments, and attempts to blunt stigmatization. The incidence of syphilis, however, is a reflection of multiple factors, cultural beliefs and practices as well as economic and political forces. Its control will require comprehensive programs that are able to fit within the social and cultural dynamics of this sexually transmitted disease.

Figure 13.1   Movie poster for the 1936 tragic romance *Camille*. Courtesy Popcorn Posters.

Chapter 13

The People's Plague: Tuberculosis

Disease and death are represented by Violetta in Giuseppe Verdi's opera *La Traviata* (1853) and by Mimi in Giacomo Puccini's opera *La Boheme* (1895). The young heroines are tall, thin, and pale-faced, with cherry-red lips and flushed cheeks, and their voices are like those of the nightingale. But Mimi and Violetta are also mysteriously ill with a disease called consumption (from the Latin *con* meaning "completely" and *sumere* meaning "to take up"). To those living in the 19th century, it seemed natural to link artistic talent to consumption, and Verdi and Puccini were well acquainted with this connection. Other composers and writers, such as Keats, Shelley, the Brontes, Chopin, and Schiller, were also consumptives. Consumption was characterized in an 1853 medical text as having the following features: nostalgia, depression, and excessive sexual indulgence. Indeed, at the time it was believed that mental activity and artistic talent were stimulated by the poisons of this wasting disease.

Today, consumption is more commonly called tuberculosis, or TB, but the pallor and emaciation of those afflicted gave it another name: "the White Plague." In *Illness as a Metaphor*, Susan Sontag wrote, "TB was—still is—thought to produce spells of euphoria, increased appetite, exacerbated sexual desire. . . . Having TB was imagined to be an aphrodisiac, and to confer extraordinary powers of seduction." In the 1800s, when epidemic TB reached its peak in Western Europe, persons with tuberculosis were considered both beautiful and erotic: extreme thinness, long neck and hands, shining eyes, pale skin, and red cheeks. Yet such "beauty" had its price: a painful death by drowning in one's own blood. Because it was neither recognized nor understood that TB was a chronic infectious disease, it was romanticized.

The operas *La Traviata* and *La Boheme* were both based on Alexandre Dumas' 1848 semiautobiographical novel *The Woman of the Camellias* (made into the classic 1936 movie *Camille* with Greta Garbo and Robert Taylor [Fig. 13.1]). In the Dumas work, the heroine (a courtesan) coughs blood onto her white handkerchief that recalls the red and white flower colors of the camellia and symbolically represents her sexual availability at various times during the menstrual cycle. (The spitting up of blood was sanitized in the operatic versions.) A more accurate and less romantic description of the consumptive includes incessant coughing, which made talking and eating almost impossible and breathing painful; weight loss that prevented walking; and pain that required opium and whisky to ameliorate. By the time of death, emaciation was so complete that the individual resembled a cadaver.

The romantic notion of TB in *La Traviata* (literally "the fallen woman") can be explained in part by the fact that the opera was composed before the cause of TB was discovered in 1882. The themes of artistic genius and eroticism persist in *La Boheme* despite its later date, largely because "consumptive decline" was considered to be due to a hereditary predisposition or specific living habits, such as the debauched bohemian lifestyle, poverty, or sexual promiscuity. Although today it is known that this romanticized and morbid fascination with consumption was entirely without scientific basis, the linkage between disease, creativity, and eroticism persists: the 1996 rock musical *Rent* transplants Puccini's beloved bohemians to New York City, and despite giving them AIDS, rats, and roaches, composer Jonathan Larson passionately describes their hopes, losses, striving, death, and climactic resurrection.

A Look Back

Tuberculosis is an ancient disease that has plagued humans throughout recorded history and even before. Tuberculosis of the lungs (called pulmonary TB) is the form of the disease we are most familiar with, giving rise to the slang word "lunger." (In the United States the lung is the primary site of infection in 80 to 85% of cases.) When localized to the lungs, tuberculosis can run an acute course, causing extensive destruction in a few months—so-called galloping consumption. It can also wax and wane with periods of remission—mistaken in some cases for chronic bronchitis with spitting up of blood. In 1839, Harriet Webster, the daughter of Noah Webster (author of Webster's Dictionary), wrote, "I began to cough and the first mouthful I knew from the look and feeling was blood. . . . I concluded to lay still and try what perfect quiet could do—swallowed two mouthfuls of blood and became convinced that if I could keep from further coughing I should be able to wait until morning without disturbing anyone. As soon as morning arrived, I looked at the contents of my cup. Alas, my fears were realized." She died 5 years later.

Tuberculosis can affect organs other than the lungs, including the intestine and larynx; sometimes the lymph nodes in the neck are affected, producing a swelling called scrofula (Fig. 13.2). (Scrofula comes from the Latin word for "pig" because the shape of the swollen neck looks like a little pig.) Tuberculosis can also produce the fusion of the vertebrae and deformation of the spine, called Pott's disease after Sir Percival Pott, who described the condition in 1779. This may lead to a hunchback, and it may also affect the skin (when it is called lupus vulgaris), and the kidneys. TB of the adrenal cortex destroys adrenal function and results in Addison's disease; this is probably what killed Jane Austen.

The microbes that cause TB (and also leprosy) are called mycobacteria. Their free-living relatives inhabit the soil and water, where they fix nitrogen and degrade organic materials. Mycobacteria have a protective cell wall that is rich in unusual waxy lipids, such as mycolic acid, and polysaccharides such as lipoarabomanan and aribinogalactan. Three mycobacteria, *Mycobac-*

Figure 13.2   Scrofula in a young man. The enlarged lymph glands in the neck resemble a piglet ("scrofus" is "pig" in Latin).

terium tuberculosis, *M. leprae*, and *M. avium*, are human pathogens that respectively cause TB, leprosy, and a pulmonary disease with swollen glands in the neck. *M. avium* is an opportunistic infection (page 101) found in some immunocompromised people (e.g., those with acquired immunodeficiency syndrome [AIDS]) with under 50 T4 cells per ml^3; its symptoms can include weight loss, fevers, chills, night sweats, abdominal pains, diarrhea, and overall weakness. Closely related to *M. tuberculosis* is a parasite of cattle, *M. bovis*. Although *M. bovis* can infect people, it does so infrequently and with great difficulty. *M. bovis* grows under conditions where the oxygen levels are low; when it does infect people, it is not associated with lung disease. *M. tuberculosis* grows best when oxygen is plentiful, and it is associated with pulmonary TB, probably because the lung has high levels of oxygen. But TB of the spine is associated with *M. bovis* and results from a blood infection that spreads to the spine via the lymph vessels. It has been hypothesized that *M. bovis* arose from soil bacteria, and humans first became infected with *M. bovis* by drinking milk. *M. tuberculosis*, on the other hand, is specific to humans and spreads from person to person through droplets of saliva and mucus. This airborne mode of transmission was clearly favored when humans settled down and established towns and cities. Genetically, *M. bovis* and *M. tuberculosis* have been shown to be >99.5% identical, so differences in their pathogenic nature are still to be explained.

Evidence of TB is found in bony remains that predate human writing. Pott's disease has been described in Egyptian mummies dating from 3700 BC to 1000 BC. One of these mummies, from the XXI Dynasty (100 BC), discovered near the city of Thebes in 1891, is that of a priest who died with extensive destruction of the bones of the spine (Fig. 13.3). Curiously, there is no evidence of pulmonary TB in any of the mummies from this period; however, in the tomb of a high priest of Ramses II, the body of a small boy was found whose lungs had been stored, and these show the telltale signs of pulmonary TB. The burial site has been dated to be from 1000 BC to 400 BC. Thus, it appears that tubercular disease of the lung is more recent than that of the bones. Based on this, it has been suggested that *M. tuberculosis* evolved from *M. bovis* after cattle were domesticated between 8000 and 4000 BC. Before this time, the epidemic form of TB as we know it today did not occur. TB, it is believed, then spread to the Middle East, Greece, and India via nomadic tribes (Indo-Europeans) who were milk-drinking herdsmen who had migrated from the forests of central and eastern Europe ~1500 BC. A clay tablet in the library of the Assyrian King Ashurbanipal (668–626 BC) describes the disease: the patient coughs frequently; his sputum is thick and sometimes contains blood; his breathing is like a flute; his skin is cold. The Greek physician Hippocrates (ca. 470–380 BC) called the disease *phthisis*, meaning "to waste," and noted that the individual was emaciated and debilitated, and had red cheeks and that it was a cause of great suffering

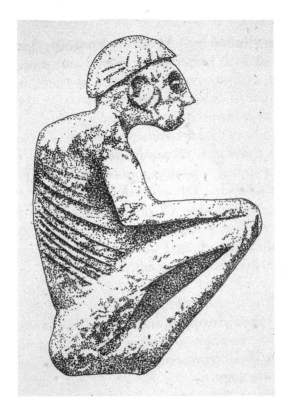

Figure 13.3   Tuberculosis of the spine (Potts' disease) in a reconstruction of an Egyptian mummy.

and death. Although Hippocrates believed the disease was due to evil air, he did not consider it contagious. Aristotle (384–322 BC), however, suggested that it might be contagious and due to "bad and heavy breath." However, by the time of Galen (AD 130–ca. 200), the theory of contagion of phthisis came to be accepted in the Roman Empire. But the contagious agent could not be found.

During the Middle Ages (AD 500 to 1500), a feudal system developed in Europe by which a small elite (the nobility) ruled the rest of society, their subjects. Royalty claimed that their right to rule and their talents were of divine origin, and they publicized this through claims of royal supernatural powers to heal disease, specifically scrofula. Kings and queens were able to heal those afflicted with scrofula by simple touching (Fig. 13.4). Clovis of France (481–511) and Edward the Confessor of England (1042–1066) were the first kings endowed with this gift. Edward I (1272–1307) touched 533 of his subjects in one month; Philip VI of Valois (1328–1350) touched 1,500 in a single ceremony; and Charles II (1660–1682) touched 92,102 people during his reign. On 14 June 1775, Louis XVI "ritually touched 2,400 stinking sufferers from scrofula" (Daniel and Daniel).

Figure 13.4   The royal touching to cure scrofula. Queen Anne is touching young Samuel Johnson to cure him of the "king's evil." (Courtesy of Bridgeman Art Library.)

In Shakespeare's *Macbeth* (act 4, scene 3), an accurate description of the ceremony is given:

'Tis called the evil:
A most miraculous work in this good king;
Which often, since my here-remain in England,
I have seen him do. How he solicits heaven,
Himself best knows: but strangely visited people,
All swoln, and ulcerous, pitiful to the eye,
The mere despair of surgery, he cures,
Hanging a golden stamp about their necks,
Put on with holy prayers; and 'tis spoken,
To the succeeding royalty he leaves
The healing benediction. With this strange virtue,
He hath a heavenly gift of prophecy,
And sundry blessings hang about his throne,
That speak him full of grace.

During the ritual ceremony, the king or queen touched the sufferer, made the sign of the cross, and provided the afflicted with a gold coin. In England this practice was known as the Kings Evil or the Royal Touching, and it persisted until the early 18th century. One of the last sufferers of scrofula to be touched was the English writer, critic, and lexicographer Dr. Samuel Johnson, who was infected with tuberculosis by his wet nurse. Johnson was brought to Queen Anne (1702–1714) at age 2 and touched; from all evidence, it appears he was not cured by the touching. But the crowd of sufferers anxious to be cured rushed to be touched, and several were trampled to death. Today we know that royal touching had nothing whatsoever to do with curing scrofula; cure occurred in some cases because of the natural remission of the disease.

The word "tuberculosis" is of recent vintage and refers to the fact that in the lung there are characteristic small knots or nodules called "tubercles." These were first described by Franciscus Sylvius in 1679; he also described their evolution into what he called lung ulcers. However, almost all the great pathologists of his time believed that the disease was due to tumors or abnormal glands rather than an infection. The first credible speculation on the infectious nature of TB was that of Benjamin Marten, who in 1772 proposed that the cause was an "animalcule or their seed" transmitted by the "Breath [a consumptive] emits from his Lungs that may be caught by a sound Person."

In the 16th century, the present epidemic wave of TB began in England, reaching its peak in about 1780. At that time it was estimated that 20% of all deaths in England were due to consumption. From there it spread to the rest of Western Europe, reaching a peak in the 1800s when *La Traviata* and *La Boheme* were written. Peaks of TB also occurred in Eastern Europe between 1875 and 1880, and by 1900 it had reached North America.

The cause for this rise in TB may have been the demographic shift from rural to urban living as well as the creation of "town dairies." These dairies were wooden buildings in the town center that housed dairy cows that had formerly been pastured; now the tubercular cows were within the town, which provided ideal circumstances for animal-to-animal transmission as well as animal-to-human (zoonotic) transmission of TB. The result: a sharp rise of scrofula in the 17th century. Later, with a resurgence of trade, the walled towns of England continued to provide the means for human-to-human transmission. This was especially so when the textile industry became mechanized; this development led to a shift from a rural cottage industry to the more urban riverside sites where waterpower was available. Towns grew in population size to become cities; as peasants streamed into these urban centers, people were more and more crowded together. These conditions are vividly shown in the paintings of Pieter Bruegel (1525–1569) and William Hogarth (1697–1764). Further, the practice in England of taxing a building partly based on its number of windows tended to affect building design to minimize the number of windows. This enhanced the rebreathing of exhaled air of those living and working in the crowded, airless rooms. The increased density of people provided ideal conditions for the aerial transmission of *M. tuberculosis* and pulmonary TB. By the 19th century, epidemic TB had raged for more than 2 centuries, and it was feared by some that tuberculosis might bring about the collapse of industrialized Europe as well as the end of all civilization.

John Bunyan wrote, "The Captain of all these men of death that came against him to take him away, was the consumption; for it was that that brought him down to the grave." And an early 20th century journalist described it this way: "Tuberculosis is a Plague in disguise. Its ravages are insidious, slow. They have never roused people to great, sweeping action. The Black Plague in London is ever remembered with horror. It lived one year; it killed fifty thousand. The Plague Consumption kills this year in Europe over a million; and this has been going on not for one year but for centuries. It is the Plague of all plagues—both in age and power—insidious, steady, unceasing."

To those living during the Victorian Age (1837 to 1901) when the British Empire was at its height, TB was not only romantic, but it was also attractive because the disease produced no obvious repulsive lesions, as did leprosy, its mycobacterial cousin. To the Victorians, the blood in the sputum blended metaphorically with menstrual blood—and so in a strange way sickness and death were blended with eroticism and procreation. There is an eminent gallery of victims of tuberculosis, including Baruch Spinoza (1633–1677), Johann Wolfgang Goethe (1749–1832), Friedrich Schiller (1758–1805), Fedor Dostoevsky (1821–1881), Anton Chekhov (1860–1904), Sir Walter Scott (1771–1832), D.H. Lawrence (1885–1930), Percy Bysshe Shelley (1792–1822), John Keats (1795–1821), Alexander Pope (1688–1744), Samuel Johnson (1709–1784), Jean Antoine Watteau (1684–1721), Niccolo Paganini (1782–1840), Elizabeth Barrett Browning (1806–1861), Igor Stravinsky (1882–1971), Robert Louis Stevenson (1850–1894), Edgar Allan Poe (1809–1849), Franz Kafka (1883–1924), Amadeo Modigliani (1884–1920), Frederic Chopin (1810–1849), Henry David Thoreau (1817–1862), George Orwell (1903–1950), Eleanor Roosevelt (1884–1962), and Vivien Leigh (1913–1967). Keats died at age 21, and all six children of Reverend Bronte and his wife Maria (including Emily and Charlotte), as well as Shelley, died of "galloping consumption " before the age of 40. Some, having looked at this list of writers, composers, and artists, believed that TB sparked genius. Arthur Fishberg wrote, "Tuberculosis patients particularly young talented individuals . . . display enormous intellectual capacity of the creative kind. Especially is this to be noted in those who are of artistic temperament, or who have a talent for imaginative writing. They are in a constant state of nervous irritability, but despite the fact that it hurts their physical condition, they keep on working and produce their best works." However, there is little scientific evidence to show that tuberculosis had any real effect on the brain or on creativity.

The deaths of Vivien Leigh and Eleanor Roosevelt were less romantic and more tragic, because they occurred after the disease agent had been identified and drugs for the cure were available. Vivien Leigh developed TB in 1945, was hospitalized for a brief time, and recovered, but her physician recommended that she seek further treatment at a sanitarium. She refused. Although streptomycin and isoniazid became available to treat TB in the 1940s and 1950s, she again refused to avail herself of these cures. She died of TB at age 54. Eleanor Roosevelt was infected as a young woman and developed active TB at age 12; then the disease went into remission. In the last 2 years of her life, her health began to deteriorate, and she developed miliary TB (so called because the small tubercles in the lungs look like millet seeds and spread throughout the body via the bloodstream). It was too late for treatment, and she died from disseminated TB at age 75.

Some contend that TB was brought to the Americas by European explorers and settlers; however, this appears not to be true. An Incan mummy of an 8-year-old boy who lived about AD 700—more than 700 years before Columbus arrived in the Americas—shows clear evidence of Pott's disease, and the lesions in the spine contain mycobacteria, most probably *M. bovis*. Domestic herds of cattle as well as wild herbivores probably served as the source of infection. In 1999, investigators found *M. bovis* in bone tissue removed from a 17,000-year-old bison that had fallen to its death in the Natural Trap Cave of Wyoming, providing clear evidence that TB was already present in prehistoric America waiting for new human hosts. With urbanization, immigration of infected individuals, higher population densities, and poor hygiene, the conditions for the spread of TB in the New World was favored. Its first peak was in the early 19th century. TB was especially prevalent in the cities along the Atlantic coast. Mark Caldwell, in his book *The Last Crusade*, described it this way: "Though its crowds and bad air made the city a crucible of TB, it was also the place where the poor congregated, where oppressive economic conditions prevailed, where filth and ugliness constantly assaulted the senses." In 1804 in New York City, a quarter of all deaths were due to consumption, and between 1812 and 1821, Boston had a similar fatality rate. TB did not affect all segments of the U.S. population equally. In 1850, African Americans in Baltimore and New York City had higher death rates than did whites. In Baltimore, above the age of 15, the number of female deaths was twice that of males, but in New York and London it was the reverse. The reasons for this are not clear. TB was not strictly an urban disease—it was also present in rural areas. The critical element was found to be not the total population but the size of the household. In colonial America in the 18th century, irrespective of the size of the town or city, a typical dwelling had 7 to 10 inhabitants. Such crowding facilitated household transmission. Inefficient heating of the home usually led to a sealing of the windows and doors in winter to keep out the cold, so transmission was enhanced. Behavioral patterns also contributed to the spread of pulmonary TB: caretakers of the sick frequently slept in the same bed as their patients, and physicians advised against the opening of windows in rooms where there were those sick with consumption.

Inadequate ventilation was also a contributing factor in the spread of urban tuberculosis. For example, during the 19th century, most tenements were built with little concern for proper ventilation. As more and more impoverished immigrants arrived in the United States (especially during the 1830s and 1840s), they were forced to live crowded together in these miserable tenements. In 1800, a Boston apartment typically had an average of 8 people, and by 1845 the figure was over 10. The generally filthy conditions and lack of ventilation characteristic of tenement housing surely played a key role in the rise in the incidence of pulmonary TB in the United States.

In the urban centers of New York and Boston, consumption came to be regarded as "a Jewish disease" or the "tailors' disease," because so many young immigrant Jews who were consumptives earned their living in the garment industry, cutting, sewing, and stitching. Despite the fact that TB was peculiar neither to Jews nor to those who worked in tailoring, they continued to be stigmatized as carriers of tuberculosis. During the 1920s, American nativists became concerned over the millions of emaciated and undersized Jewish immigrants who were lacking in the physical vitality so characteristic of the sturdy and robust Anglo-American stock and who were now pouring into the cities along the Atlantic seaboard. It was contended that because Jews were highly susceptible to TB, these sickly Jews were not only racially inferior, but they would soon become public charges. Data gathered by the Jewish community refuted the claim that Jews were especially prone to TB. However, it was a fact that TB did flourish among Jews (as it did among Gentiles) who lived in crowded, unhealthy conditions in the poorest parts of the city. Although statistical analyses showed that the rates of consumption among Jews in both Europe and the United States were not significantly higher than that of non-Jews, this did little to change the public mind. In effect, TB was used as a tool of anti-Semitism that justified the ostracism and persecution of Jews.

In the penitentiaries of the United States, approximately 10 to 12% of white inmates died of TB between 1829 and 1845, whereas during this time the annual mortality due to TB in eastern cities of the United States was less than half a percent. Even today, there is a higher incidence of TB in U.S. prisons. The increased incidence is due to several factors: prevalence is higher among new prisoners than in the general population, because there is a preponderance of prisoners from the lower end of the socioeconomic scale; close living arrangements make transmission more likely; and prisoners are at a higher risk for TB owing to their higher incidence of human immunodeficiency virus (HIV) infections.

In the 1900s, poorer people tended to have a higher mortality from TB; the greatest number of deaths occurred between the ages of 15 and 45, but there was also a minor peak between 5 and 10 years of age. The American Indians, who were highly susceptible to TB, were virtually killed off by being herded together on reservations. Indeed, between 1911 and 1920, 26 to 35% of all deaths occurring among American Indians were due to TB, and an examination of 600,000 American Indians in 1920 showed that 36% had TB. TB was virtually unknown in sub-Saharan Africa until the 20th century. Indeed, TB was a disease of Europe and North America and was absent from those countries in Africa where there had been little or no European immigration.

During the 19th century, some believed that TB was an act of God against which there was no defense. Others, however, were convinced that the disease was a result of bad air present in the crowded and dirty cities. Many consumptives sought refuge in warmer or milder climates. In the United States, there was a move to the Sunbelt of the West (Arizona, southern California, New Mexico, and Texas), while in Europe it was to the Mediterranean or South Africa. Ocean voyages were also recommended for those with TB because of the slow tempo and the clean air, especially on cruise ships. Living at high altitudes where the air was unpolluted and crisp was considered beneficial, so health seekers founded communities in Colorado. Some recovered, but others did not, so clearly a change in scene did not cure one of TB. Yet, despite the fact that consumption was not an environmental disease that could be cured by sunshine or fresh air, this view remained pervasive. Indeed, it was a critical factor in establishing the political map of Africa.

Cecil Rhodes (1853–1902), who did more than any man of his time to enlarge the British Empire, was an ambitious and ruthless colonialist. He created the state of Rhodesia (now Zimbabwe and Zambia) by forcing the annexation of Bechuanaland (now Malawi and Botswana) and acquiring the lands of the Matebele tribe by devious means. As the leading official of the British South Africa Company, he consolidated all of the diamond mines in the Kimberley area into the DeBeers mines. This created a monopoly over the supply of diamonds, and in the process Rhodes became extremely wealthy. After his death, part of his fortune was used to establish the Rhodes scholarship at Oxford University. All this was due to the fact that, at age 16, Rhodes was stricken with TB while attending Oxford University. To recuperate, he joined his brother, a cotton grower, in Natal, South Africa. Later, the two became involved in the diamond mining business in an inland area now known as Kimberley. Upon regaining his strength, he returned to England and received his degree from Oxford University in 1873. But back at Oxford he suffered a relapse and was told he had only 6 months to live. Rhodes was unable to resist the lure of Africa's riches and climate, so he returned. In 1888, he tricked the Matebele ruler (Lobengula) into an agreement whereby he secured additional diamond mining concessions and soon was able to get complete control of the territory. Through the British South African Company, he secured a monopoly of the Kimberley diamond production and created the De Beers Consolidated Mines (with the slogan, "diamonds are forever").

In 1881, Rhodes became a member of the Parliament of the Cape Colony, a seat he held for the remainder of his life. In Parliament he favored a policy of containing the northward expansion of the Dutch Boers (Afrikaners) in the Transvaal Republic. In 1885, largely owing to his persuasion, Great Britain established a protectorate over Bechuanaland. By 1890, Rhodes had become the prime minister and a virtual dictator of the British Cape Colony. He planned and promoted a Cape-to-Cairo railroad—from the Cape of Good Hope to Cairo, Egypt—dreaming that one day the map of Africa would be "painted red," which meant controlled by Great Britain, since at the time mapmakers used the color red to show British territories. At first, the Transvaal was of little interest to anyone, but when gold was discovered in 1887, the prospect of wealth lured tens of thousands of non-Boers, many of them English, into the Transvaal to seek their fortunes. Transvaal's president, Paul Kruger (after whom the Krugerrand, the gold bullion coin, is named), refused to grant these foreigners political rights. Rhodes' personal and business sympathies were with the British who were living in and being discriminated against in the Transvaal, and in 1895 he saw this as an opportunity to overthrow the Boer-dominated government. He organized a raid on the Transvaal by Rhodesian troops with the aim of triggering an insurrection against the Kruger government, but the raid (called the Jameson Raid, after its leader Leander Jameson) was poorly planned and executed and failed. As a result, Rhodes was censured by the British government and forced to resign as prime minister in 1896. Relations between the British and Dutch, however, continued to deteriorate, and by 1899 this resulted in the outbreak of the Boer War. Rhodes was in Kimberley at the time and was trapped there during a 4-month siege of the town by 5,000 Boers. Here he played an important role in directing the war and in building morale for those defending Kimberley, but 2 months before the war ended, he had a heart attack and died at age 49. At the time of his death, Rhodes had been instrumental in bringing almost 1 million square miles of Africa under British domination. The country, called Rhodesia, declared its independence from Britain in 1965; northern Rhodesia is present-day Zambia, and southern Rhodesia is now Zimbabwe.

Finding the Germ of TB

In 1865, a French military physician, Jean-Antoine Villemin (1827–1892), succeeded in transmitting tuberculosis to rabbits. Villemin recovered pus from the lung cavity of a tuberculous patient who had died 33 hours earlier and injected this under the skin of two healthy rabbits. Two other rabbits served as controls and were injected with tissue fluid from a burn blister. Some months later, when the rabbits were examined, TB was found in the lungs and the lymph nodes of only those that had received the pus. Although it would seem that Villemin's demonstration of transmission would be proof enough of the contagious nature of TB, his announcement before the prestigious French Academy of Medicine was greeted with derision. His severest critic was Hermann Pidoux, who said that consumption in the poor was due to conditions of poverty, including overwork, malnutrition, unsanitary housing, and other deprivations; when consumption developed among the rich, Pidoux claimed it was brought about by their overindulgence in their wealth, laziness, flabbiness, overeating, excessive ambition, and habits of luxury. In both cases the result was, in Pidoux's words, "organic depletion." Such blind rejection of the infectious nature of TB remained until 24 March 1882, when the German physician Robert Koch made a presentation before a skeptical audience—Germany's most prestigious scientific group, the Berlin Physiological Society—in which he claimed to have discovered the microbe of tuberculosis. Koch's announcement not only resolved the contagion argument but also resulted in a shift of scientific pre-eminence from France (and Louis Pasteur) to Germany. This shift took place for several reasons, including differences in the style of science. French scientists relied on intellectual deduction and tended to proclaim general laws from a limited set of experiments, whereas German scientists were more methodical, their approach was more realistic, and they relied on repetitive experiments. In addition, after the defeat of the French in the Franco-Prussian War (1870 to 1872), the German economy and its science flourished while post-Napoleonic France declined both in wealth and in scientific glory.

Robert Koch was born in Clausthal in the Harz Mountains of Germany in 1843, received his medical education at the University of Göttingen, and then trained with the eminent pathologist Rudolph Virchow in Berlin. After the Franco-Prussian War, he settled in the small farming town of Wollstein, where he had a medical practice and was the sanitary officer. Anthrax was a problem among the sheep (and sometimes humans) in and around Wollstein. Outfitted with a new microscope his wife had given him for his 37th birthday, Koch began to examine the carcasses of sheep, and in obsessive fashion he carried out experiments in his home laboratory. Soon (in 1881) he was able to isolate and characterize the germ of anthrax, a rod-shaped bacterium he named *Bacillus anthracis*. Koch presented his findings at the International Medical Congress in London in 1881. After returning from the meeting in London, he was determined to find the microbe that caused TB. Within a year he had succeeded. He wrote,

> The aim of the study had to be directed first towards the demonstration of some kind of parasitic forms, which are foreign to the body and which might possibly be interpreted as the cause of the disease. . . . The objects for study are prepared in the usual fashion for the examination of pathogenic bacteria . . . the coverslips are then covered with a concentrated solution of methylene blue

. . . and when the coverslips are removed . . . the smear looks dark blue and is much overstained, but upon treatment with Bismarck brown stain the blue color disappears and the specimen appears faintly brown. Under the microscope all constituents of animal tissue, particularly the nuclei and their disintegration products, appear brown, with the tubercle bacilli, however, beautifully blue.

Koch also carefully and precisely set the criteria (called Koch's postulates) necessary for a microbe to be the causative agent of a disease:

in all tuberculous affections of man and animals, there occurs constantly those bacilli which I have designated tubercle bacilli and which are distinguishable from all other microorganisms by characteristic properties. However, from the mere coincidental relation of these tuberculous affections and bacilli it may not be concluded that these two phenomena have a causal relation, notwithstanding the not inconsiderable degree of likelihood for this assumption that is derivable from the fact that the bacilli occur by preference where tuberculous processes are incipient or progressing, and that they disappear where the disease comes to a standstill. To prove that tuberculosis is a parasitic disease, that it is caused by invasion of bacilli and that it is conditioned primarily by the growth and multiplication of the bacilli, it was necessary to isolate the bacilli from the body, to grow them in pure culture until they were freed from any disease-product of the animal organism which might adhere to them; and by administering the isolated bacilli to animals, to reproduce the same morbid condition which, as known, is obtained by inoculation with spontaneously developed tuberculous material.

Koch's identification of *M. tuberculosis* was not so simple: the microscopic bacillus is colorless and unusually difficult to stain because of its waxy cell wall, and therefore it cannot be easily seen with the light microscope. To make it visible required heating and a special aniline dye (methylene blue). Later, the staining technique was improved upon by Paul Ehrlich, who had heard Koch speak in 1882; Ehrlich found that the bacilli would retain the dye color if they were first stained with red fuchsin followed by an acid wash; they were acid-fast. Koch also devised a method for growing the bacillus outside the body (by culturing them in test tubes containing coagulated serum as a nutrient source) and was able to use these pure cultures of tubercle bacilli to infect guinea pigs. It was methodical Koch, the German, who, though he validated the imaginative approach of Villemin, the Frenchman, ignored his findings and thus relegated him to obscurity. In this case, scientific credit was given to the person who convinced the world, not the one to whom the idea first occurred.

In 1890, under pressure from the Prussian government and its leader, Otto von Bismarck, Koch forsook careful experimentation and prematurely announced that he had discovered a protective substance made from an extract of the bacillus called tuberculin (today known as PPD, *p*urified *p*rotein *d*erivative). When injected into animals, tuberculin produced fever, malaise, and signs of illness. Koch believed that tuberculin sensitized the animal and effected cure. Subsequently, it was found that it did not. Indeed, instead of being a cure, tuberculin turned out to be dangerous and sometimes lethal. Tubercle bacilli sensitize the body to tuberculin so that, when injected in sufficient quantities to a tuberculous individual, tuberculin can kill the patient via a delayed hypersensitivity reaction. Yet Koch's tuberculin did serve a practical purpose as a diagnostic test. Even today, it remains one of the most useful methods for determining previous exposure to TB. The tuberculin skin test is positive in 15% of adults living in the United States, but a positive test does not indicate whether the disease is active. Indeed, only 10% of people exposed to TB develop active disease.

Koch was also able to show, as did Villemin a decade earlier, that bovine TB (caused by *M. bovis*) could be infectious for humans, principally via the milk. At this time, when infected milk was recognized as a source of human disease, the tuberculin skin test became useful as a means for screening infected cows. Cows that had a positive test were killed. This meant that valuable animals would be taken out of production, and the losses could be devastating to farmers who had more than a few tuberculin-positive animals. However, slaughter was unnecessary, since pasteurization of milk rendered it noninfective and protected those who drank the milk. Opposition to this slaughter-based control was vehement; sometimes, militia protection was needed to ensure the safety of the veterinarians who were involved in the testing in rural dairies where pasteurization was not employed. In 1932, there was a "cow war" in Iowa, as 400 angry farmers destroyed the cars of the veterinarians; ultimately, martial law had to be declared. By the end of the 1930s, more than 95% of the counties in the United States were declared to be free of infected cattle, and the threat of milk-borne TB faded away. However, the problem has not disappeared completely, since bovine TB exists in other countries, including Mexico.

The Disease of Tuberculosis

Charles Dickens, in his 1879 novel *Nicholas Nickleby*, described tuberculosis this way:

> There is a dread disease, which so prepares its victims, as it were, for death . . . a dread disease. In which the struggle between soul and body is so gradual, quiet, solemn, and the results so sure, that day by day, and grain by grain, the mortal part wastes and withers away, so that the spirit grows light . . . a disease in which death and life are so strangely blended that death takes the glow and hue of life, and life the gaunt and grisly form of death—a disease which medicine never cured, wealth warded off, or poverty could boast exemption from—which sometimes moves in giant strides, or sometimes at a tardy sluggish pace, but slow or quick, is ever sure and certain.

Dickens could characterize TB, but he did not know that the principal risk behavior for contracting it was breathing. Persons with pulmonary TB may infect others through airborne transmission: coughing, sneezing, and speaking. A sneeze might contain an aerosol with a million microscopic (~10 μm) drops that evaporate slowly, producing floating droplet nuclei. TB bacilli, slightly bent microscopic rods ~2 to 4 μm in length and 0.3 μm in width (Fig. 13. 5), and enclosed within droplet nuclei, move from place to place by riding on the gentle air currents. The bacteria-laden droplets can easily be inhaled. Tubercle bacilli are rather robust and can survive in moist sputum for 6 to 8 months. Inhalation is the major route of infection. Oral infection, through eating or drinking, is less efficient because the bacteria rarely survive passage through the acid in the stomach. Infection may result if as few as 5 bacteria reach the grape-like clusters of the thin-walled air sacs (alveoli) of the lung.

Once in the alveolus, macrophages engulf the bacteria to initiate an infection. From here, the bacilli within the macrophage can be transported to other parts of the body by the lymph channels. For the first few weeks within the tissues of a susceptible individual (one who has not been tuberculous before), the bacilli multiply slowly, dividing once every 15 to 24 hours. At first, there is little damage or reaction, but after several more weeks of microbial multiplication, either at the initial site or at a more distant site, there is an inflammatory response; this becomes more intense, and fluid

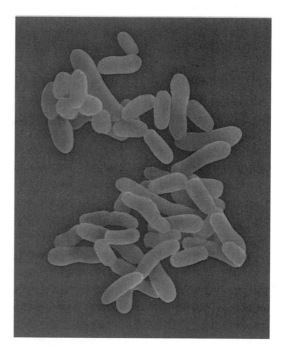

Figure 13.5 The acid-fast rod-shaped bacillus that causes tuberculosis, *Mycobacterium tuberculosis*, as seen with the scanning electron microscope. (Courtesy of Dennis Kunkel Microscopy, Inc.)

(lymph) leaks into the region. The site becomes infiltrated with fiber-secreting cells called fibroblasts, which surround the free and macrophage-enclosed bacilli in an attempt to wall them off. A microscopic tubercle results. The tubercle grows in size, pushing aside normal tissue and producing the characteristic TB lesion—larger visible tubercles in the lung—and the tuberculin test is positive. Primary tuberculosis is a self-limiting infection that often goes unnoticed: it appears to be a cold, and usually there is little impairment of lung function. If there is a protective immune response, which occurs in 85 to 90% of cases, the disease may progress no further, calcification of the tubercles may take place, and the tuberculin test remains positive. However, in ~5% of cases, a latent infection may become active within 2 years of the primary infection, and in another 5%, active disease will return at a later time. This reactivation of infection occurs because some of the tubercle bacilli, even those within macrophages, are not killed.

The alveoli, surrounded by a network of blood vessels, the capillaries, serve for the exchange of oxygen and carbon dioxide. Small alveolar blood vessels are eroded and rupture, hemorrhages occur, and if the tubercles break open, the released bacilli can be disseminated via the bloodstream. The bronchi can become irritated, there is coughing, fluid fills the lung, and breathing becomes more difficult. Now acid-fast *M. tuberculosis* can be found in the blood-streaked sputum. Reactivation of the infection results in the destruction of more lung tissue, producing a cheese-like consistency in which the bacteria survive but cannot multiply because of low levels of oxygen and acidity. Coughing, pallor, spitting of blood, night sweats, and painful breathing signify the spread of the disease. Chest X rays show tubercles and fluid in the lung, and sounds of gurgling and slush can be heard through a stethoscope placed on the chest. Unfortunately, continued inflammation results in liquefaction of the lung tissue, and this oxygen-rich environment provides a rich growth medium; in some cases, there can be more than a billion bacilli in each milliliter. (It is a bitter irony that the tubercle bacillus flourishes in the lung, where oxygen is plentiful, and fresh air rich in oxygen was once believed to be a curative for consumptives.)

In time, the softened and liquefied lung contents are forced out of the lesion into the blood vessels to spread throughout the body; this is how the infection moves into other parts of the body. Untreated, the fatality rate of TB can be between 40 and 60%.

In tuberculosis, it is important to distinguish between infection and disease. In the vast majority of cases, when the bacteria are inhaled, the bacilli are killed by macrophages or they are localized and grow slowly within tubercles. Although *M. tuberculosis* can grow and divide outside the cell, it survives predominantly within macrophages; it can live within the macrophage, because its waxy-lipid cell wall makes it impervious to the killing mechanisms of the macrophage. Despite the formation of antibodies during a TB infection, antibodies are unable to limit the disease, because the bacilli remain hidden within the macrophage. Cell-mediated immunity, especially that involving CD4 T-helper cells (lymphocytes), is critical for disease arrest. The CD4 T cells produce cytokines, especially gamma interferon and tumor necrosis factor, that activate macrophages to kill or limit *M. tuberculosis* growth. The lesion becomes infiltrated with lymphocytes and macrophages, and a delayed hypersensitivity reaction occurs similar to that experienced with a bee sting. However, in tuberculosis, cell-mediated immunity is a two-edged sword—it is required for protection, but it is also involved in tissue damage, i.e., disease. In addition to the CD4 T-helper lymphocytes, there are CD8 T cells (killer lymphocytes) that participate in the immune response; these cells also produce gamma interferon and tumor necrosis factor to activate macrophages; in addition, they punch holes in the membrane of the macrophage by releasing a molecule called perforin. Many macrophages are also killed by tuberculin-like products, causing them to release reactive oxygen species (hydrogen peroxide, hydroxyl radicals, superoxide) and protein-degrading enzymes (proteases) that are detrimental to the host tissues. In addition, the products of the surviving tubercle bacilli activate suppressor T cells that depress both the delayed hypersensitivity reaction and cell-mediated macrophage killing. Thus, the tubercle bacilli are able to spread to other tissues, where the inflammatory cycle and tissue destruction can begin anew. It is ironic that the macrophages that are able to deal with some of the tubercle bacilli by using their cell-killing mechanisms damage the very lung tissues they are supposed to protect.

Today's Diagnosis of TB

TB is a corrosive disease, and much of what has been described as its classic symptoms—pallor, coughing and spitting up of blood, weakness, and emaciation—were indicative of an advanced state of the disease. Much of the improvement in our prevention and treatment of TB has been due to accurate diagnosis and the development of methods to limit the spread of the organism from both within and without the body.

Tuberculin, discovered by Koch in 1890, turned out not to be a cure for TB; however, in the control of TB it serves as the gold standard for diagnosis. In the tuberculin test, a small amount of tuberculin (or PPD) is injected under the skin of the forearm; within 48 to 72 hours after injection, an inflamed area develops if the person has been exposed to TB. The redness persists for up to 7 days. A positive skin test does not indicate an active infectious case of TB but merely that the person has been exposed.

Although *M. tuberculosis* is gram positive, it is weakly stained by the Gram method because of its waxy cell wall. The method of staining—red fuchsin followed by treatment with nitric acid—devised in 1882 by Paul

Ehrlich is still used today and makes identification much easier than the method originally described by Koch. Parenthetically, in 1887, when Paul Ehrlich showed the clinical symptoms of TB, he stained his own sputum and demonstrated that he was in fact infected. He had a mild infection, went to Egypt to rest, and recovered.

In 1895, Wilhelm Roentgen (1845–1923) discovered X rays. X-ray photography, or radiography, made the tubercular lesions caused by the disease visible long before its symptoms became noticeable; this allowed for treatment of the disease at a much earlier stage. Although the X-ray photograph did not become a reliable diagnostic tool until the 1920s, even at that time treatments for TB remained less than satisfactory.

Another tool, the stethoscope, was developed in 1816 by the French physician Rene Laennac. Laennac was asked to examine a rather overweight woman. Realizing he could not directly tap on her chest to determine whether there was fluid in her lungs, he recalled from his boyhood that one could hear the scratch of a pin if the ear was placed in contact with one end of wooden cylinder and the other end was scratched. In an imaginative stroke, he took a paper notebook, rolled it up tight, applied it to the woman's chest, and listened. Much to his surprise, he clearly heard the sounds of the woman's heart and lungs. Further experimentation eventually led him to develop the first stethoscope, a hollow wooden tube about a foot long. Laennec's stethoscope has been improved upon, but it continues to provide physicians with an acoustic "picture" of the conditions within the lung. However, the stethoscope cannot identify the specific cause of the abnormal chest sounds. On the basis of his clinical experience, Laennac theorized that phthisis and scrofula, as well as miliary TB, were different forms of the same disease; this idea was opposed by the German scientists, who were convinced that a specific clinical picture had to be the result of a single infectious agent. However, when Villemin demonstrated transmission of TB to rabbits and guinea pigs, it became clear that Laennac was correct. Laennac, who devoted his entire life to the study of TB, died from it in 1826.

"Catching" TB

By the 1940s—before antibiotics were introduced and strict public health measures were instituted—the lethal incidence of epidemic TB was declining (Fig. 13.6). Why? Some suggested that it was due to the emergence of less virulent strains of the mycobacterium, whereas others proposed that it was due to an increase in the resistance of the human population. Still others contended that public health measures, such as segregating the infected in sanatoria and forbidding the practice of spitting in public, led to the decline in TB. Higher standards of hygiene, including cleaner cities, destruction of tubercular cows, pasteurization of milk, and hand washing, all led to reduced

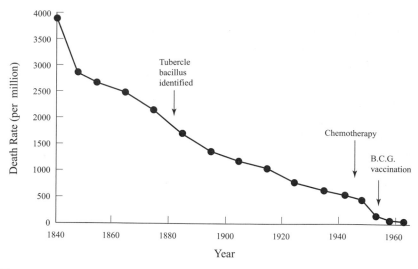

Figure 13.6   The death rate from tuberculosis in England and Wales. (From T. McKeowan, *The Role of Medicine*, Princeton University Press, 1976, p. 9.)

transmission, and better nutrition and a higher standard of living also helped. A recent mathematical model suggests that all these factors were contributory, and the decline was simply due to the natural behavior of this "white plague." Blower and colleagues have calculated that the great epidemics of TB did not occur in Europe until the 17th and 18th centuries (and somewhat later in the Americas) because before this time the population size was too small for R_0 (the value that indicates the number of secondary cases produced in a naive population from the introduction of a single infected source) to be greater than 1. However, with urbanization and crowding, as well as industrialization and abundant poverty and malnutrition, the population size exceeded the threshold, and the value of R_0 became significantly greater than 1. Indeed, Blower and colleagues estimate that R_0 might have been as great as 10! When the course of the epidemic was fast paced, the disease was found primarily in younger individuals, but as the epidemic matured there was a shift toward infections in older individuals, and these were the result of reactivation.

According to Blower's model, TB epidemics operate on an extremely long time scale—100 to 200 years—and the declining phase may take at least 30 years. These slow dynamics are "the result of the gradual development of a large pool of latently infected non-infectious individuals. Some of these individuals, however, gradually develop disease (and become infectious) over their lifetime." Before 1985 the highest rates of TB in the United States were found in individuals 45 to 64 years of age. This is characteristic of a mature epidemic in which infection is the result of reactivation. Such a natural decline produced endemic equilibrium levels that gave the impression that TB was disappearing as a result of public health interventions. However, since 1985 there has been an increase in the incidence of TB, and the infections occur in a younger age group (as was seen in the heroines of the *Lady of the Camellias*, *La Boheme*, and *La Traviata*). This age distribution is characteristic of an early epidemic with fast developing cases predominating and is due to HIV infections, increased poverty, dismantling of control programs, and increased immigration from countries where TB is present. This young epidemic—due to recent transmission—now is superimposed on the mature one. This model of TB transmission makes two predictions: an increased population of newly developed cases should occur in younger individuals, and the proportion of primary infections (due to recent transmission) should also increase. Surveillance data validate these predictions. However, it is important to note that the model provides a cautionary note: if control strategies focus only on preventing new infections (and neglect reactivation or slow developing cases), then it may take many decades to eliminate TB.

Controlling Consumption

Prior to 1940 and as long as tuberculosis was diagnosed in its late phase, relief of symptoms was the primary method of treatment. There were many fads, and many of these did no good. Early treatments for TB included creosote, carbolic acid, gold, iodoform, arsenic, and menthol oil, administered either orally or as a nasal spray. Some physicians prescribed enemas of sulfur gases and drinking of papaya juice. None worked to cure TB. During the late 19th century, surgical techniques were used to "rest" the lung; these included pneumothorax or collapsed lung treatments that required removal of several ribs and reduction in the size of the thoracic cavity. In some cases, lung resections, i.e., removal of infected lung tissue, were performed. Despite a lack of success in curing the disease, these measures were continued well into the 1940s.

More than a thousand years ago, Hippocrates recommended a change in climate for phthisis, and this notion of the benefits of "clean air and sunshine" lasted well into the 20th century. Particularly in the 1800s, there developed a movement in both Europe and North America to treat TB patients in open-air hospitals—sanitaria (from the Latin *sana* meaning "cure"). Life in the sanitarium was a reaction to the stuffy, overheated rooms of the patient's home and workplace. Patients were made to take outdoor treatment whatever the weather. Wide-open windows and outside balconies were the places where the patients took the cure, and this was in the hot summer as well as freezing winter. The art of wrapping oneself in a blanket became an essential ritual. Fresh air was taken with a vengeance. Sanitaria were considered to be indispensable for cure, and patients were provided with brilliant sunshine, fresh air, quiet, rest, and good nutrition—but no anti-TB medicines, since none were available. Life in a luxury sanitarium (in Davos, Switzerland) is immortalized in Thomas Mann's novel *The Magic Mountain*, and one of the most famous sanitaria in the United States was the Trudeau Institute in the Adirondack Mountains at Saranac Lake, N.Y., established by Dr. Edward Livingstone Trudeau.

In 1873, Dr. Trudeau came down with TB after taking care of his consumptive brother, who had died 8 years earlier. Trudeau suffered from fever and weakness and began to cough up blood. Believing the end to be near, he decided to retire to the Adirondack Mountains, where he hiked, hunted, fished, swam, and anticipated death. But within a few months of this outdoor regimen, he gained weight and recovered his energy, and his fever abated. In 1882, after Dr. Trudeau read an account of the benefits of a sanitarium in Europe, he raised money through donations and used his own funds to establish the first sanitarium in the United States at Saranac Lake. The sole industry of the town became the sanitarium.

The sanitarium movement soon went into high gear. In 1900, the United States had 34 sanitaria with 445 beds; 25 years later, there were 356 sanitaria with 73,338 beds. However, sanitaria did not cure TB. In the prechemotherapy era (1938 to 1945) among patients in sanitaria with advanced TB, the death rate was 69%, whereas for those with minimal disease it was 13%—about the same as in the general population. However, the one significant benefit of the sanitaria was that they were able to isolate (quarantine) the contagious individuals, and physicians could maintain complete control of their patients.

Soon after Koch's discovery of the tubercle bacillus, anti-tuberculosis campaigns began, first in Europe and then in North America. In 1889 to 1890, Hermann Biggs of the New York Department of Public Health issued an education leaflet that contained information on how to prevent the spread of consumption. Included were

1. an education campaign using newspapers and circulars to call attention to the dangers of TB, possibilities for treatment, and precautions to prevent it;

2. compulsory reporting of TB by all public institutions;

3. assignment of inspectors to visit the homes of patients in order to enforce sanitary regulations regarding the disposal of sputum and to arrange for disinfection;

4. provision of separate wards in hospitals for TB patients with pulmonary disease;

5. provision of special consumptive hospitals to be used exclusively for the treatment of TB, and

6. provision of laboratory facilities for the bacterial examination of sputum.

Biggs' Riverside Hospital was established to confine, voluntarily or not, tubercular individuals whose "dissipated and vicious habits" endangered the health of the community. It was more a prison than a hospital, and in many respects this approach to TB was reminiscent of the establishment of colonies and hospitals for those suffering from leprosy. Biggs' idea was that protection of the public health was more important than individual freedom. He was the chief of the health police. Much of this attitude toward public health was based on a misguided meaning of Darwinian evolution: TB was no longer a romantic disease; it was a sign of corruption, and pruning out the unfit would benefit mankind. Public health authorities impulsively attributed illness to the environment—how and where you lived—which was under your control. The reasoning went something like this: if you choose to live in filth, lack ambition, or are too lazy to work yourself out of the tenements or are indolent and doze by the fire instead of taking brisk walks in the fresh air and sunshine, then you brought TB on yourself. A consequence of this line of thinking was that, at all costs, the infected should be prevented from infecting the others in society. TB, which once had been regarded as beautiful and a source of creative inspiration, was now marked as a hereditary weakness. Some even went so far as to suggest that succumbing to TB was in a person's genes.

The United States declared a "war on consumption" beginning in the early 1900s. This consisted of vigorous anti-TB campaigns, the most famous being Christmas Seals of the National Tuberculosis Foundation, which later became the American Lung Association. Early anti-TB campaigns blamed capitalism for the disease because it was a found in the crowded cities full of impoverished factory workers. At other times, TB victims were stigmatized by their "choice" of living as paupers; some claimed that those with TB were lazy, lacked ambition, and were inferior. It is ironic that the whole TB eradication movement rose, flourished, and vanished without ever establishing its effectiveness against the disease. The greatest benefits of the "war on consumption," which was in full swing by 1915, were education, loss of stigma, abandonment of TB as a spiritualizing and romantic force, a greater understanding of contagion, the initiation of attempts to clean up tenements, and improvements in medical diagnosis and care—but not quarantine.

Treating TB

Prior to the 1940s, there were no drugs to cure TB. In 1939, Selman Waksman, a soil microbiologist and a Russian immigrant to the United States working at the Rutgers Agricultural College in New Jersey, began a systematic effort to identify soil microbes that could produce substances that might be useful in the control of infectious disease. Waksman based his program of antibiotic research on the earlier (1928) observations of Alexander Fleming. By 1940,

Waksman and his graduate students had developed methods for growing soil microbes and screening them for their antibiotic properties. Soon it was discovered by Waksman's graduate student, Albert Schatz, that a mold named *Streptomyces griseus* (identified by Waksman in 1919) obtained from the throat of a sick chicken that had been eating soil was an antagonist and limited the survival of tubercle bacilli in both soil and sewage. Schatz developed methods to grow large amounts of *Streptomyces*, and in 1943 the inhibitory substance, streptomycin, was isolated from the cultures. Schatz's work was a part of his doctoral dissertation, and he and Waksman published their findings in 1944. Streptomycin inhibited the growth of tubercle bacilli both in the body and in test tubes. By 1945, it was in clinical use for TB, and by 1947, streptomycin was available in large quantities. In 1952, Waksman (but not Schatz) received the Nobel Prize for the discovery. Later, the mode of action of streptomycin was discovered: it inhibits the synthesis of the waxy cell wall of the tubercle bacillus, thus leaving the bacilli naked and unprotected from the onslaught of the killing machinery of the macrophage.

An antibiotic victory over TB was short-lived, however. Soon there were signs of streptomycin-resistant TB bacilli. In 1949, streptomycin treatment was supplemented with *p*-aminosalicylic acid, and in 1952 isoniazid, a drug first synthesized from coal tar in 1912, became the mainstay in the treatment of drug-resistant TB. Isoniazid also blocks the synthesis of the mycolic acids that are a main constituent of the waxy cell wall of *M. tuberculosis*. Streptomycin and other drugs did not eradicate TB, but they did effectively eliminate sanitaria. In 1954, the Trudeau Institute was closed, and by the 1960s almost all sanitaria had vanished. In 1963, rifampin (derived from another *Streptomyces* sp., *S. mediterranei*), an inhibitor of the synthesis of tubercle bacillus RNA, was introduced for treatment. To minimize the emergence of drug resistance, patients are now treated with a drug cocktail called MDT, for *m*ultiple *d*rug *t*herapy; the first of these combinations was isoniazid plus rifampin.

How does drug resistance come about, and why is it prevented by a drug cocktail? Drug resistance is the result of natural selection; that is, those genotypes that are the most able to survive and reproduce their kind pass on their genes to future generations and increase in frequency over time. In tuberculosis, drug resistance develops this way: an average TB patient may have a billion tubercle bacilli, and 1 in 10,000 of these carries a mutation, a genetic change, that allows that bacillus to evade the lethal effects of the anti-TB drug. Once the patient is treated with that drug, only the mutant survives: these are the drug-resistant bacilli. The result is that there are now ~10,000 tubercle bacilli that can reproduce in the presence of the drug, increasing their numbers until almost the entire population is resistant. But let us assume that a second, equally effective drug is added along with the first drug to the billion tubercle bacilli. Again, only one in 10,000 is resistant to this drug. When the drugs are added together, there is less than 1 resistant bacillus. By adding a third drug, the possibility that a bacterium will survive is reduced even further. Thus, multiple drug therapy makes it possible to delay or prevent the emergence of drug resistance.

In the absence of drug resistance, the combination of isoniazid and rifampin administered for 9 months is curative for TB. If rifampin is not used, 18 months is the minimum duration of therapy for cure. By the addition of pyrazinamide for the first 2 months, the treatment period can be shortened to 6 months. Currently, the most commonly used regimen consists of isoniazid, rifampin, and pyrazinamide administered daily for 8 weeks, followed by isoniazid and rifampin given daily, twice a week, or three times a week for 16 weeks. More than 85% of TB patients who receive both isoniazid and rifampin have negative sputum cultures within 2 months after treatment has begun.

The emergence of multidrug-resistant strains of *M. tuberculosis* has provoked great concern, because the disease caused by these strains is often fatal. Multidrug-resistant strains usually develop in people who begin treatment but lapse after a few weeks, allowing larger numbers of mutant bacteria to survive; these bacteria resist the drugs and grow and multiply. In the late 1980s, it became clear that a substantial number of patients with TB were not completing treatment. To address this problem, the Centers for Disease Control recommended that "direct observation of therapy" by a trained health care worker be considered for all patients to ensure that they swallowed each and every pill. Maintaining this vigilance is not easy, however, since it requires trained medical staff, infrastructure, and money. In practice, one-quarter of TB cases are not diagnosed before they become infectious, and many who begin treatment do not stay the course.

There is evidence from studies of twins that susceptibility to TB has a genetic basis, but no major TB susceptibility gene has ever been identified. Susceptibility to TB is dramatically enhanced in HIV-infected persons: the spread of AIDS has been paralleled by a resurgence in TB cases. HIV-infected individuals who adhere to the standard regimen for TB do not have an increased risk of treatment failure or relapse. However, the use of protease inhibitors and reverse transcriptase inhibitors for HIV treatment has complicated the treatment of TB in HIV-infected individuals. Administration of these drugs with rifampin can result in lower levels of antiviral drugs and toxic levels of rifampin. Rifabutin may be substituted for rifampin, with fewer side effects, with some antiviral drugs but not with others.

Vaccination against TB

Why isn't there a vaccine against TB? There is. In the 1920s, two French bacteriologists named Albert Calmette (1863–1933) and Camille Guérin (1872–1961) used a technique first used in 1882 by Louis Pasteur—they attenuated bovine TB (*M. bovis*). They grew the bacteria on a nutritive medium containing beef bile for 231 generations over a period of 13 years, during which time the strain became weakened. This non-disease-producing bacillus, which could elicit immunity, became the vaccine called bacillus Calmette-Guérin, or *M. bovis* BCG, which was cross-reactive with human TB. When first introduced in a trial in Lubeck, Germany, in 1930, of 249 babies inoculated, 76 died. This was not due to inoculation with BCG itself but to live, virulent *M. tuberculosis* stored in the same incubator; however, the adverse publicity so poisoned the public's confidence that it was difficult to achieve popular support for BCG vaccination for decades. This incident, however, may have had one beneficial effect: it encouraged the search for anti-TB drugs.

Injection of BCG produces a mild infection, induces immunity, and has never resulted in a virulent infection. It does, however, produce some adverse reactions and is not 100% effective. (Estimates of protective efficacy range from 30 to 80%.) The mortality from TB in countries that have used BCG since 1950 has declined as markedly as it has in countries not using BCG. Despite this, most public health workers believe BCG has contributed to the decline of TB. BCG, however, is not used in the United States because of fear that the benign bacilli in a live vaccine might become virulent. Further, vaccination with BCG renders the tuberculin test negative, and its use as a vaccine would be of little benefit to those already infected.

Where TB Is Found

Today, one person is infected with TB every second, and someone in the world dies of TB every 18 seconds. Though the death rate is low (<0.01%), in the next decade it is expected that 30 million people will die of TB. Yet in the United States, TB is largely a forgotten disease, having been replaced in the public's consciousness by cancer, AIDS, and cardiovascular diseases. TB still kills, and the new strains are more dangerous because they are drug resistant.

It is estimated that one-third of the world's population is infected at present. In 2003, there were 8 million new cases and 3 million deaths—with all but 400,000 occurring in developing countries. Incidence is associated with poor socioeconomic conditions. In the United States since 1984, there has been a large increase in cases in men aged 25 to 44—the group at greatest risk for AIDS. The rate of TB in the United States reached a peak in 1992 and has decreased steadily. In 2000, there were 16,377 cases in the United States, and in 2004 there were 14,511 cases, or 4.9 cases per 100,000 people—a record low. Statistics from the Centers for Disease Control and Prevention indicate that 62% of the TB cases occur in African Americans, Hispanics, and Asians—40% of whom are under age 35. In Caucasians, TB is primarily a disease of the elderly, especially those in nursing homes. More than 20% of the U.S. cases occur in immigrants who come from resource-poor countries.

When humans were hunter-gatherers and lived in small roving bands, TB was not a serious threat, but with the development of agriculture and animal husbandry the numbers of people increased, as did exposure to new pathogens. A soil bacterium infected cattle or game, and it became *M. bovis*; perhaps by crossing the species barrier, it evolved into *M. tuberculosis*. Coincident with urbanization were epidemics of TB. The poor—undernourished, crowded, and living under unhygienic conditions—became the breeding ground for the "white plague." Today it is recognized that tuberculosis is not only an infectious disease but a societal one as well. Understanding this contagious illness demands the recognition that social and economic factors are as important as the way in which the tubercle bacillus causes damage to the human body, how it manages to evade the immune system, and how it is able to overcome the most potent anti-TB drugs. In the past, fear of the disease, fueled by prejudice, led to public and institutional reactions that included mandatory testing and isolation and stigmatized immigrants and those with different lifestyles. Tuberculosis will continue to affect the emotional and intellectual climate of human populations throughout the world. Now, almost 125 years after the discovery of the cause of TB, it must be realized that for "the people's plague" to be eradicated, the subtle interplay between disease and society must be fully appreciated. Until that time, TB remains a disease that could re-emerge to threaten us once again.

Figure 14.1 Job stricken with a plague, as depicted in a 17th century woodcut. Job lies on a pile of dung dressed in a loincloth and covered with boils. (Courtesy of the Wellcome Library of Medicine.)

Chapter 14

Leprosy, the Striking Hand of God

There is a morbid interest in and dread of leprosy, largely stemming from frequent references in the Bible. To the readers of the Old Testament, leprosy was an abomination. Job says, "Pity me, pity me, and pity me, you, my friends for the hand of God has struck me." Although the Bible does not specify that Job suffered from leprosy, it was assumed that he did, and for a time he was considered the patron of lepers. Indeed, in medieval art, Job is always depicted as being covered with black spots (Fig. 14. 1); body spots are also seen in illustrations of lepers. In Leviticus, this account of the disease is given: "When a man shall have the skin of his flesh a rising, a scab or bright spot, and it be in the skin of his flesh like the plague of leprosy; then he shall be brought unto Aaron the priest or unto one of his sons the priests."

The mention of leprosy in both the Old Testament (Exodus, Numbers, Deuteronomy, Samuel II, Kings II) and the New Testament (Matthew, Mark, Luke) has surely contributed to the fear of the sores associated with leprosy, as well as the notion that what is blemished is unclean and is also displeasing to God because it is defiled. Christ's curing of lepers became a metaphor for divine salvation. The fictional account of Ben Hur, as well as the parables of Christ and his encounter with the 10 lepers on the road to Galilee, gave rise to a massive literature on the imagery of leprosy as a disease of the soul and one that was highly contagious.

Job 19:13–21 (Jerusalem version) reads,

> My brothers stand aloof from me, and my relations take care to avoid me, my kindred and my friends have all gone away, and the guests in my house have forgotten me. The serving maids look on me as a foreigner, a stranger, never seen before. My servant does not answer when I call him; I am reduced to entreating him. To my wife my breath is unbearable, for my brothers I am a thing corrupt. Even the children look down on me ever ready with jibe when I appear. All my dearest friends recoil from me in horror; those I loved best have turned against me.

Often we think of ourselves as being more enlightened and above such attitudes toward the diseased, but considering the hysteria associated with the outbreak of AIDS and the draconian measures that were initially suggested for controlling its spread, we may not be so very different in our approach to disease from the way leprosy was viewed in biblical times and the Middle Ages.

A Look Back

Leprosy probably arose in the Far East, about 1400 BC, for there are accurate descriptions in the sacred Hindu writings of the Veda, and there are also descriptions in Chinese literature. The earliest accounts of leprosy occur in an Indian text, Charaka Samhit, written between 600 and 400 BC. The *Nei Chang*, a textbook of medicine, describes a disease wherein the patient has stiff joints, the eyebrows and beard fall off, the flesh becomes nodular and ulcerates, numbness results, and finally the bridge of the nose changes color and rots. The *Nei Chang* was written between 250 and 230 BC.

It has been claimed that leprosy was brought from India to Greece in the fourth century BC by the soldiers of Alexander the Great, but it is also possible that leprosy spread from the Far East to the West along the trade routes, arriving in the Mediterranean about the time of Christ. Certainly, Hippocrates (ca. 470–380 BC) did not describe leprosy, because he never saw it. Rather, the best description of leprosy in Europe comes from Aractus, a contemporary of Galen, in AD 150. Leprosy then spread further west; in the graveyards of Britain and France there is evidence of the disease in bones dating from AD 500 to 700. Emperor Constantine (AD 274–337) suffered from leprosy, and pagan priests believed that bathing in the blood of sacrificed children would cure him. It did not!

The Hebrew word *saraath* was used to describe many skin conditions; it has been variously translated as "defiled" or "accursed" or simply "scaly." When the Bible was translated into Greek some 300 years before Christ, *saraath* became *lepros*. St. Jerome Latinized *lepros* into *lepra*, and in the first English translation of the Bible in 1384, John Wycliff (1324–1384) translated *lepra* as "leprosy." In the Middle Ages, and even before, leprosy was considered a disease of the soul. The association of the disease with sinfulness stigmatized the leper. Medieval medical authors refer to lepers as being crafty, irascible, and suspicious, and above all having a burning desire for lustful sex. Sometimes the disease was known as satyriasis—an insatiable sexual appetite (from the Satyr of Greek mythology, a forest deity with the ears and tail of a horse given to merriment). The belief that venereal disease was due to immorality was "logically" extended to leprosy; in this way, leprosy became accepted as divine punishment for the sins of the flesh. By the 16th century, there was also confusion between syphilis and leprosy.

The disfigurement of the face and hands (Fig. 14.2) contributed to the alienation of the leper, and the sores on the body led to the belief that leprosy was contagious. Lepers were not considered nice people, and they were cast out of society. In 1179, the Third Lateran Council issued a decree urging the segregation of lepers from society: the leper was told not to mix with the crowds, to use his own container in drawing water, and not to touch anything unless he paid for it first. He was to wear a distinctive garment and announce his presence with a bell or clappers (Fig. 14.3). There was a ritual burial.

At times the lepers were treated with kindness, but more frequently, as in the time of Henry II (1132–1189) and Phillip V of France (1293–1332), the leper was strapped to a post and burned alive. Edward I (1239–1307) was considered kinder—he buried them alive. Some good Christian people did sympathize with the plight of the leper, though, and almshouses or refuges were established for them. Many were called Lazar Houses or lazarets, after

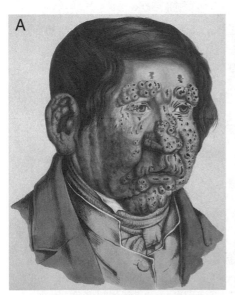
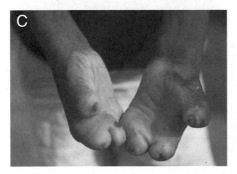
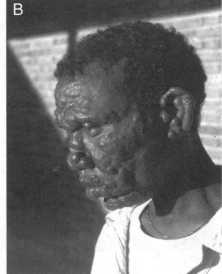

Figure 14.2 The clinical appearance of infections with *Mycobacterium leprae*, the cause of leprosy. (A) Drawing of a leprosy patient in *On Leprosy* by Danielsen and Boeck (1847). (B and C) Facial and hand deformities associated with leprosy. (Panel A is courtesy of the Wellcome Library of Medicine and panels B and C are courtesy of Wallace Peters.)

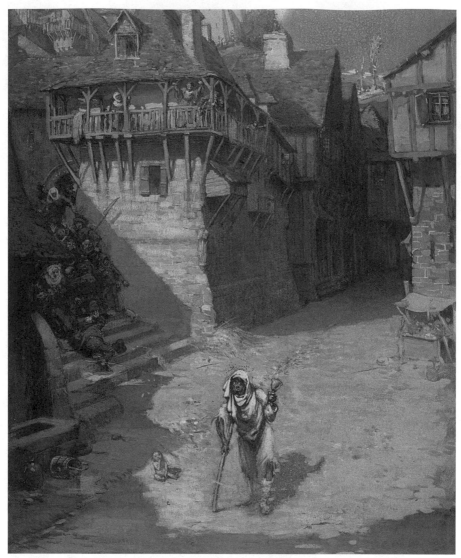

Figure 14.3   Leper announcing his arrival in the city with the ringing of a bell. Water color by R. Cooper. (Courtesy Wellcome Library of Medicine.)

Lazarus of Bethany, who was believed to suffer from leprosy and who replaced Job as the patron saint of the leper.

The pandemic of leprosy reached epidemic proportions in the 12th century and had its peak in Europe in the 13th and 14th centuries; records showing the construction of 19,000 lazarets document this increase in leprosy in Europe. But by the 16th century most of the lazarets had been closed, and by the 18th century they had all but disappeared. The historian William H.

McNeill has suggested that leprosy may have retreated in the 1300s with the rising incidence of pulmonary tuberculosis, which provoked greater resistance to leprosy. However, scientific evidence for this notion is scant. In addition, it has been hypothesized that the increased supplies of woolen textiles for clothing may have broken the chain of skin-to-skin contact needed for the spread of the disease, and at the same time contributed to the spread of typhus. This too, remains conjecture.

Knights participating in the Crusades contracted leprosy, and in 1048, they formed their own spiritual order, the Order of Lazarus. The return of these knights to Europe was probably a contributing factor to the spread of leprosy throughout Europe. Leprosy spread from Spain and Africa to the Americas in the 16th and 17th centuries. It has been claimed that the last leprosy patient in Great Britain died in Scotland in 1798. But as late as the mid-nineteenth century there were thousands of cases in Scandinavia, and with the importation of Chinese laborers into the Pacific islands by the colonizing Europeans, the disease was spread further. The first reference to leprosy in Hawaii was in 1823; two generations later, 5% of Hawaiians were infected. In the early 1900s, it was estimated that about 1% of the deaths in Fiji were due to leprosy.

In the early 1860s there was a "rediscovery" of leprosy as a disease of those of foreign birth. Indeed leprosy, by this time, had reached epidemic proportions on the Hawaiian Islands. In an effort to halt its spread, the Hawaiian Kingdom followed a practice of quarantine using isolation and segregation. The Americans and Europeans who lived in Hawaii feared the disease, and the Western imperialistic attitude served to strengthen the negative image of those who were diseased. Officials rounded up the lepers and loaded them onto ships bound for a settlement at Kalawao, on an isolated peninsula of Molokai's north shore. Many of those transported were shoved off the ships hundreds of yards from the land and forced to swim to shore. Some of the weaker ones never made it. In 1865, a leper colony was formally established on Molokai that housed a total of 142 individuals. However, some of those banished to the island probably had other diseases. Although the Chinese were singled out as the source of leprosy, 97% of the lepers on Molokai were Hawaiian. This was because the prevailing view was that leprosy was a very contagious disease, and the Hawaiians were characterized as lacking in social utility. This made their isolation on Molokai seem totally justified.

In 1873, Father Damien, a Roman Catholic priest from Belgium, joined the colony as its resident priest. His assignment was to be for 3 months, but he stayed for 16 years. The determined Father Damien cleaned and bandaged sores and built homes, a hospital, a reservoir, and a plumbing system, and for the next decade and a half he buried many hundreds of leprosy victims. He provided comfort and compassion to those who had been shunned by their friends and families because they were considered to be unclean and sinners. Father Damien did not avoid contact with the lepers of Molokai, and when he died in 1889 at the age of 49, the story was spread that he had contracted leprosy during his residency. He came to be called the "Martyr of Molokai" and was beatified in 1995. (However, since it was known earlier—perhaps in 1876—that he already showed signs of the disease, he may have contracted the disease before coming to Molokai.) Father Damien's leprosy caused much fear and reinforced the belief that the disease was highly infectious. Although Hawaii became a territory of the United States in 1900, the involuntary isolation of those with leprosy on Molokai continued until 1974.

In the United States proper, there was also a fear of the Chinese—a condition that increased after 1850. Exclusionary laws were enacted to prevent them from bringing their diseases to the United States. By 1870, leprosy was considered one of the diseases specifically associated with the Chinese. Calls for action arose, and in 1894 the state of Louisiana established a Louisiana Home for Lepers, located 85 miles from New Orleans. Run by the Sisters of Charity, it was a neglected asylum. The quarantine of lepers at the Louisiana Home served to reinforce the stigma of the disease.

By 1917, at the height of U.S. racial discrimination and xenophobia, a leprosy bill was passed to establish a leprosarium. The old Louisiana Home for Lepers became the American National Leprosarium at Carville, and in 1921 the U.S. flag was raised over the institution. It was surrounded by a barbed wire fence with a sign reading "No Trespassing." The rules and regulations were more like those of a prison than a hospital. Patients were sent there by special leper trains, some arrived in shackles accompanied by armed guards, and their mail both ingoing and outgoing was disinfected. Babies born to the patients were given up for adoption. Yet the only crime of these children was that they might have contracted leprosy. Patients were not permitted to marry until 1952, despite the fact that drugs to control the disease, such as the sulfone dapsone, were introduced in the late 1940s. The American National Leprosarium at Carville, La., was purposely placed between the men's and women's penitentiaries as a means for preventing mingling of the sexes, since it was hoped that the fear of contracting leprosy by passing through this "zone of contagion" would scare the inmates from considering escape for conjugal visits. By 1956, the Carville Leprosarium became a voluntary hospital under the Public Health Service and was renamed the Gillis W. Long Hansen's Disease Center. Up until the 1960s, strict public health laws forbade those with leprosy to use public transport, to fly over certain states, to use public restrooms, or to live freely in society. The basic freedoms guaranteed to most Americans were denied them. Although over the years some deregulation did take place, Carville continued to be a shameful reminder of the stigma and the stereotype of leprosy in the United States. In 1997, the last 135 people hospitalized (of the maximum number of 450) were "set free" as President Bill Clinton transferred the facility back to the state of Louisiana, where it has become a school and training center for at-risk youth. Carville patients will receive $33,000 annually, have lifelong medical care for their disease, and be able to reside anywhere they desire.

The Disease of Leprosy

What is leprosy and what causes it? The first objective and scientific appraisal of leprosy came in 1873 with the discovery of the leprosy bacillus, *Mycobacterium leprae*, by Gerhard Armauer Hansen (1841–1912) in Norway. (Indeed, the preferred name for the disease today is Hansen's disease.) *M. leprae*, like its cousin *M. tuberculosis*, is an acid-fast bacillus, but it differs in several significant ways—it cannot be grown in tissue culture and it grows very slowly and only in humans, mice, and nine-banded armadillos, taking about 2 weeks to divide. (Tuberculosis bacilli divide every 12 hours.) In the mouse, the bacteria grow best in the footpad, where the temperature is 30°C, and as few as 1 to 10 bacilli can initiate infection, but to get a million bacilli it takes 5 to 6 months. In the armadillo, the disease becomes disseminated, and one can get a billion bacilli (~200 g) in 15 months. *M. leprae* has been found in 10% of the armadillos in certain regions of Louisiana.

There are no known vectors or reservoir hosts, and there is no satisfactory way of detecting past infections or preventing inapparent ones. Therefore, what we know of the disease and its spread is derived from clinical cases. Leprosy is a disease of low infectivity, and it is also a spectral disease, that is, one showing different manifestations: tuberculoid leprosy, localized in skin nodules where the bacteria may be abundant, and disseminated or lepromatous leprosy, with bacteria in macrophages and the skin. Most patients show the tuberculoid type, which develops 1 to 2 years after exposure, whereas the lepromatous condition requires longer periods. The manifestations seem to depend on the immune status of the host. Tuberculoid leprosy is associated with severe nerve damage. The disease involves cell-mediated immunity with T-helper cells and interleukin-2 secretion in tuberculoid leprosy, but since the bacteria multiply within the Schwann cells that insulate the nerve, there is damage to the nerves, and anesthesia results. In lepromatous leprosy, the T-helper cells do not respond to the bacilli—gamma interferon is not produced, macrophages are not activated, and as a consequence the bacteria multiply within the macrophages and the disease spreads with multiple organ involvement, leading to facial deformity and blindness. There is osteoporosis and shortening of the digits. The testes are damaged, and there is no natural remission. Death results most commonly from renal failure, pneumonia, and tuberculosis. (Lepromatous leprosy patients do have antibodies to *M. leprae* antigens, but they are not protective.) Natural resistance to leprosy is highest in African blacks and lowest among Caucasians.

It is believed that after the bacteria enter the body (via the nose or through open wounds), they somehow find their way to the Schwann cells that surround the nerve cells. The affinity for Schwann cells by *M. leprae* is due to the high affinity of the bacterium for a specific region of the molecule laminin found on the outer surface of these cells; once they adhere to the laminin using a bacterial surface protein called H1p, they invade. The mechanism of invasion is unknown, but once inside the Schwann cell they are temporarily protected from the host's immune system. Over time, however, the immune system attacks the infected Schwann cells, destroying nerves in the process. Accompanying the loss in sensation are secondary infections that "whip up the immune system into a frenzy . . . and under siege, soft tissues, even bones become degraded, particularly in the limbs and digits" (Young and Robertson).

Where Leprosy Is

Leprosy continues to infect millions of people; it is estimated that there are 15 to 20 million cases worldwide, found mostly in the tropics. Even today, leprosy leads to a lifetime of social ostracism and misery. The fear of contagion, coupled with the possibility of disfigurement, has contributed to an acceptance of terrible cruelty toward lepers. The higher the standard of living, the less the incidence of leprosy. Leprosy is often an endemic disease. It is usually acquired in childhood, but infection can occur even at age 70. Many who are exposed develop no disease. At the beginning of 1999, of the 122 countries in which leprosy was considered endemic in 1995, 94 had reached the elimination target, and prevalence had been reduced by 85% since 1985. However, the number of new cases has continued to increase; the reasons for this are not understood. Most of the new cases (~750,000 annually) occur in India, Brazil, Indonesia, and Myanmar, but also in the Philippines, Venezuela, Cameroon, Uganda, Nigeria, Malawi, Burkina Faso, Zambia, Thailand, and Japan. In many parts of the world, males are more frequently infected than females (2:1), but the reasons for this remain unclear as well.

Leprosy Today

Because leprosy bacilli cannot be cultured in the laboratory, it is difficult to assess the onset of the disease. *M. leprae* can be grown in nude mice and nine-banded armadillos, so microbiological analysis and studies on immunology and treatment can be done. The diagnosis of leprosy, however, has not changed substantially in 100 years: it relies on microscopic examination and evaluation of the response of the patient to a pinprick or heat. To date, examination of individuals with tuberculoid lesions stained for acid-fast bacilli is the diagnostic method, and this was the one used by Hansen in 1873. Newer molecular methods, such as the polymerase chain reaction, are being tested to detect the presence of *M. leprae* nucleic acids.

Leprosy can be treated. Dapsone, developed in the 1940s, controls the growth of the bacilli. Its mode of action is by interference with the synthesis of bacterial nucleic acids by blocking the conversion of *p*-aminobenzoic acid to folic acid; since folic acid is obtained in the human diet, dapsone does not affect the cells of the body. In 1991, the World Health Organization recommended multiple drug therapy (MDT) using dapsone, clofazimine, and rifampin, with treatment lasting 6 to 24 months before cure is complete. It has been estimated by the WHO that 8 million patients have been cured by using MDT and that 840,000 are currently receiving MDT.

Mycobacterium bovis BCG vaccination for tuberculosis has had a positive effect on leprosy by reducing the incidence by 20 to 80%. In 1998, India announced approval of a leprosy vaccine called Leprovac. It contains a heat-killed, fast-growing, nonpathogenic mycobacterium called *Mycobacterium w* (*Mw*), which shows the closest antigenicity of all mycobacteria to *M. leprae*. The vaccine stimulates the immune system of leprosy patients by disrupting immune tolerance and provoking an immune response that kills and clears *M. leprae* from the body. In this manner *Mw* resembles the action of the tuberculosis vaccine, BCG, which also stimulates clearance of leprosy bacilli. Because it has been shown that the cytokines gamma interferon and interleukin-2 induce clearance of bacilli in leprosy patients, both vaccines may act by triggering the release of cytokines that then serve to enhance the killing capacity of macrophages. Leprosy patients must be given 8 doses of Leprovac at 3-month intervals to effect clearance; however, for ethical reasons, the patients are also given MDT.

"Catching" Leprosy

How is leprosy spread? The bacilli are shed with the skin, but most bacilli are found in the nasal secretions. Some leprosy patients shed 100 viable *M. leprae* each day from the nose. There remains the question of whether leprosy can be spread by mosquitoes. The route of entry into the body is still uncertain, but it may be through inhalation since the bacteria cannot penetrate the skin directly. Leprosy seems to occur in clusters of families, suggesting that susceptibility depends on genetic factors, but since few of the members of a family ever contract the disease, leprosy is considered a disease of low virulence. Leprosy patients have no greater susceptibility to tuberculosis than do those without infections. There appears to be no cross protection between tuberculosis and leprosy, although in some cases BCG can protect against both diseases.

Leprosy has been a dreaded and disfiguring disease since time immemorial. Although today we know that for transmission to occur there must be prolonged contact with the infected individual, and that the disease is curable, we still do not know the precise mode of transmission or how nerves are destroyed. But, most important, as with AIDS, we still do not know how to completely remove its stigma from our consciousness so that discrimination of those afflicted is abandoned.

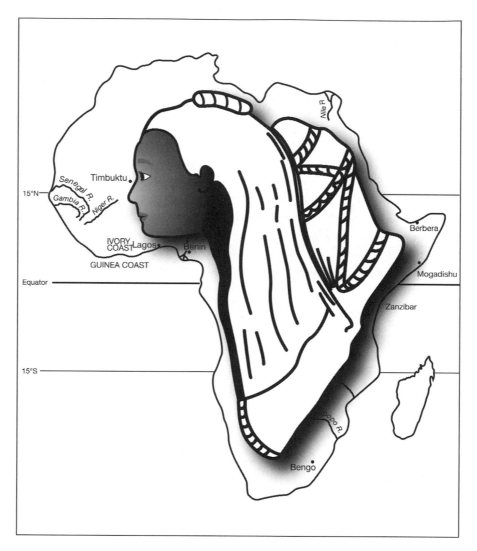

Figure 15.1   Depiction of Africa looking west with a burden on her back.

Chapter 15

Six Plagues of Africa

A writer once described the profile of Africa as resembling that of "an old woman carrying a heavy burden on her back [Fig. 15.1] with her head turned to the west." Surely, one such burden she carried was disease. Disease has shaped the African continent and determined its history, and with the modern plague of acquired immunodeficiency syndrome (AIDS), it will undoubtedly affect its future.

Africa, with 12 million square miles, contains nearly one-fifth of the world's land surface and is three times the size of the United States. The real Africa, what some historians have called "Black Africa," stretches from about 15° N to the Limpopo River Valley in the south. Before the 19th century, Black Africa had 80 million people. It was virtually unknown to the rest of the world, and to those who knew about it, the knowledge was scant or ill informed. The reason for this was that there was little communication with the rest of the world except for brief incursions by the Phoenicians, the Romans on the Nile, and a few Arab caravans in search of slaves. Black Africa also appeared to lack the climatic attractions and plunder of the New World, and it was difficult to penetrate for several reasons. First, there were few deep bays or gulfs for the shelter of ships. Second, most of the rivers were not navigable because of sandbars at the mouth or rapids a short distance upstream. Third, beyond the shore lay miles of impassable mangrove swamps, and further inland was a tropical forest that obliterated the paths of traders, explorers, and natives. Fourth, for the Europeans the climate was oppressive, and there were savage animals, hostile natives, and parasitic diseases.

The early humans in Africa initially acquired their parasites from the forest animals. Over the millennia, these parasites produced the endemic tropical infections such as malaria, yellow fever, sleeping sickness, river blindness, Guinea worm, and blood fluke disease (see chapter 3, page 44). The African people acquired some immunity to these diseases, and though debilitated they survived. This was the pattern for tens of thousands of years. The Africans of Black Africa were not subject to "crowd diseases" such as smallpox, tuberculosis, or measles; however, all that changed as more and more Europeans began to intrude.

Slavery and European Exploration

Prince Henry the Navigator of Portugal (1394–1460) inspired the passion for exploration of Africa. However, until the mid-1400s, Europeans could not sail down the west coast of Africa and return home because of the strong winds from the north all year long. All that changed by 1440, when better sailing ships had been developed, and ships could return to Europe from Africa's west coast. This advance in shipbuilding technology and navigational tools made possible the further exploration of Africa. In 1442, two of Prince Henry's ships brought back with them a dozen Africans whom they had captured on the west coast in the course of an unprovoked attack on an African village. From this time onward, contact with the remainder of the world increased, and the health of the Africans was disrupted more and more.

In 1460, Captain Alviso da Cadamosto, sailing in the service of Prince Henry, described the Gambian coast of West Africa in the following way: "there are tree-fringed estuaries, inlets and animals such as elephants, hippopotami, monkeys, and in the villages I saw Muslim traders." Initially, the Portuguese had little inclination to venture further into the interior, because there seemed to be little of commercial value, but after 1482 Portugal sent an expedition to obtain gold that was available in the villages that lined the Bight of Benin. By the 16th century, 10% of the world's gold supply was being sent to Lisbon. In the early 17th century, the English attempted to obtain the riches of Africa's interior, but their efforts failed. However, with the settling of the Americas, a demand was created for another African commodity, one more valuable than gold—slaves.

In the mid-1400s, the Portuguese began buying small numbers of slaves for sale in Europe; however, this trade in human cargo accelerated once America was discovered. Slaves became an important part of the American economies, which were heavily dependent on crops such as cotton, sugar cane, and tobacco. Slaves were also needed because the local labor force, the Amerindians, had died from European diseases, and the Europeans themselves were highly susceptible to the tropical diseases endemic to the Caribbean. Africans, on the other hand, were relatively immune to tropical diseases such as yellow fever and malaria, so they could work in the Americas. As a consequence, Europeans bought slaves for work in the American colonies; this trade increased in the 1500s when sugar plantations in Brazil and the Caribbean were established. The Spanish kings initially employed the Portuguese as intermediaries for obtaining their slaves, but as their conquests in the West Indies and mainland America increased, so too did the demand for slaves. They inaugurated a system of special contracts under which they bestowed a monopoly upon chosen foreign nations, corporations, and individuals to supply slaves for their American possessions. Soon, other European countries, including Britain, Holland, and France, were in competition with one another for these sources of human flesh. In 1562, the first British contractor, John Hawkins, with Queen Elizabeth as a "silent partner," embarked on a career of murder and brigandage using his ship *Jesus*. Ten years later, Elizabeth knighted Hawkins and commended him for "going every day on shore to take the inhabitants with burning and spoiling their towns." Toward the middle of the 17th century, the British became direct exporters of slaves from the west coast through the African Company and from the Mediterranean coast of Morocco through the Barbary Merchants, among whose directors were the earls of Warwick and Leicester. Though the French were also one of the players in the slave trade, when the Napoleonic Wars began (in 1792) and Britain blockaded France and other trading nations, it was primarily the British ships that moved as many as 50,000 slaves per year. The Atlantic slave trade reached its peak between 1740 and 1790; it is estimated that, from 1515 to 1870, Europeans brought 11 million people from Africa to the Americas on slave ships.

The slave trade also encouraged the local African chiefs to sell their prisoners for European products such as cloth, guns, and iron—materials that the Africans found easier to buy than to develop on their own. By 1780, Arabs and Africans began a slave trade on the east coast of Africa. East coast African slaves were shipped to Zanzibar and other parts of eastern Africa and then to countries around the Red Sea and the Persian Gulf. Slavery uprooted Africans from their homes, and disease spread among them with the greatly increased population movement. The incursions of slavers led to dispersal of rural clans, resulted in tribal warfare, and also contributed to the ever-wider distribution of disease.

The River Niger has fascinated historians and explorers since the time of Herodotus in 450 BC. The prevalent hypothesis in the 18th century was that the Niger rose near Bambuk and flowed eastward to the sea. In 1788, the Association for Promoting the Discovery of the Interior Parts of Africa (the African Association) was formed in London by 12 English gentlemen. Their aim was to open up trade with the gold-rich lands of the interior, particularly Tellem, a city on the Niger that was said to be built entirely of gold. They picked Mungo Park (1771–1805), a medically trained Scot, who was a surgeon with the East India Company, to lead the first expedition to establish communication between the rivers Niger and Gambia. In 1795, Park and his party of 30 sailed down the west coast of Africa to the mouth of the Gambia River, where the English had already established a fort. After traveling up the Gambia, they moved into the dense jungle of the interior, and a year later they reached their destination (Sego) on the Niger some 750 miles from the coast. The city of Tellem was not found, and by that time Park's team had run out of money, so they returned to England (1797) via the coast. On his return, Park published *Travels in the Interior Districts of Africa* (1799) and raised sufficient funds for a second expedition to find Tellem and the legendary city of Timbuktu. In April 1805, Park departed with 44 men to penetrate and navigate the Niger; during the 16-week trek to reach the Niger River, they suffered from attacks of dysentery, black vomit (yellow fever), and ague (malaria). Three-quarters of the men had died by the time they reached the river, and after 8 months only five were left to follow the Niger's course. Park and the remainder of his party disappeared into the jungle on 19 November 1805. Of the men who started with him, not a single one returned to England, yet only six were killed by something other than malaria or dysentery. In *Travels*, Park recorded some of the diseases he'd encountered. He observed that although malaria was severe in his party, it was mild among the African adults. He also described elephantiasis, yaws, and leprosy. He saw many cases of gonorrhea but not a single case of syphilis; he also saw cases of smallpox, but no measles.

When news of Park's disappearance reached England, the director of the African Society, Joseph Langley, mounted an expedition to find him. The team followed Park's route up the Gambia, crossed through the jungle to reach the Niger River, and then paddled their canoes toward Tellem. Langley wrote in his 1808 account of the trip *Dark River* that Tellem was not a city of gold but instead consisted of a small village of mud huts populated by cannibals. Returning to the river, the Langley team and their canoes were carried downstream; eventually, they reached a friendly village, were picked up by a British ship, and returned to England to tell their tale.

The lower course and termination of the Niger River remained unknown in the early 1800s. Richard and John Lander were commissioned by the British government to fill in this gap. Leaving England in January 1830, the two brothers arrived on the Guinea Coast (at present, the region between Senegal and Angola) in March, reached the banks of the Niger in June, and then, using a compass as their only scientific instrument, began their river journey by canoe. After 18 months of hardship and adventure, they brought back definitive news that the Niger emptied into the Bight of Benin. During their expedition, they found evidence of Mungo Park near Bussa Falls. Richard and John Lander informed the government of their discoveries in their *Journal of an Expedition to Explore the Course and Termination of the Niger* (1832). In late 1832, following the report of the Lander brothers, two British ships set out northward on a tributary of the Niger under the command of Major McGregor Laird, with the purpose of founding commercial settlements on the river. After a month on the river, only one member of the party was fit for duty. When the ships returned to the sea in August 1883, only 20 Europeans among the 48 who made the journey, survived the fever.

In 1841, the British sent a second, larger British Niger expedition under Captain H. D. Trotter, which moved from the Niger delta with three steamboats, 145 Britons, and 158 locally recruited Africans with the mission to preach the end of slavery and cannibalism. When the steamers reached a point about 100 miles from the sea, fever broke out. As they pushed on further upstream, the number of sick was so great that the expedition had to be terminated, and they returned to the coast. After 9 weeks on the river, 130 Britons had come down with fever, and of these, 50 died. Among the Africans, only 11 developed fever, and none died.

The Congo River is the second largest river in Africa. It drains an area about as large as Europe without Russia and flows for 2,000 miles through dark forests. Although its mouth had been discovered by a Portuguese explorer (Diego Cao) in 1482, its interior course was uncharted. In February 1816, Commander James Tuckey of the British Royal Navy pioneered a river route up the Congo River; however, the ship could only proceed as far as the rapids, so he landed his party and they moved further along its banks on foot. By September, he and his 44 men were attacked by "remittent fever" and black vomit; 35 died.

The human losses on the Niger and the Congo provoked horror as well as an intense interest in malaria on the part of the British. The disease agent responsible for malaria was unknown and would not be discovered until 1880. Even so, a breakthrough in treatment—a triumph of empiricism—occurred in 1640 with the discovery of the fever bark, called by the Indians of Peru *kina-kina*. The drug was brought to Europe by the Jesuits. However, despite its successes in curing malaria, there were failures because samples of bark varied greatly in potency. Finally, in 1820, when the cinchona alkaloids (which were the active principle) were isolated and crystallized by two French chemists, Pierre Pelletier (1788–1842) and Joseph Caventou (1795–1877), and the drug was named quinine, then it could be prescribed in a powder form of known strength, and its reliability increased. By 1854, quinine powder was generally available. Quinine was not without its disadvantages. It has a bitter taste and causes ringing in the ears, nausea and vomiting, and, in some cases, deafness. It also is eliminated from the body rapidly, so it had to be taken daily.

The dramatic effect of quinine was seen in 1854, on the third Niger Expedition, when not a single man died of malaria. However, its commander, Dr. William Baikie, did not realize that malaria could relapse, so he stopped taking quinine on leaving the African coast. He died from a relapse of malaria on his way home to England.

Livingstone and Stanley

Dr. David Livingstone (1813–1873) was born in Balantyre in the lowlands of Scotland to a poor religious family. At age 10 he was sent to work as a "piecer" in the cotton factory, but even so he continued to read and study. Young David probably read Mungo Park's *Travels*, and from that time forward, Africa became for him a land whose secrets would be revealed only to those who were bold and brave.

Although the 1833 Act of Abolition abolished slavery in the British Empire, the practice did continue in other nations. In response, there came into being an active church-sponsored British missionary movement. The Missionary Society for the Extinction of the Slave Trade and for the Civilization of Africa was founded in London in 1839 to 1840. Unlike the African Association, the missionary movement concerned itself with the salvation of the soul of the African people rather than the accumulation of geographical and scientific information. By age 20, Livingstone had decided to become a doctor and a lay missionary; he attended medical college at Glasgow University and then gained admission to the London Missionary Society. At age 27 he embarked for Africa, and 13 weeks later, after arrival at Capetown, he made his way by ox-wagon to the northernmost mission station at Kuruman. It was from here that he ventured north into the jungle. In 1848, accompanied by two big-game hunters, William Oswell and Mungo Murray, he discovered Lake Ngami. This discovery made his reputation as an explorer and convinced him that his primary inclinations were in exploration rather than as a sedentary preacher. In 1850, Livingstone, now married and with a family of three, set out to revisit Lake Ngami; he hoped to find a water highway capable of being quickly traversed by boats to a large section of the well-peopled interior territory. His goal was to establish missions and drive out the slaver trade that was then being carried out by the Portuguese, who were swapping arms for people at the rate of one gun for one youth. At Ngami, Livingstone and his children had their first experience with malaria, but he was already well aware of the efficacy of quinine; he administered from his kit a large dose that brought relief from the fever. The "Livingstone prescription" consisted of 3 grams of calomel (mercurous chloride), 3 grams of quinine hydrochloride, and 10 grams of rhubarb plus grape or cinnamon flavoring (to reduce the bitterness of the quinine) mixed with a bit of alcohol. In 1851, Livingstone and Oswell ventured forth again and this time discovered the lower Zambezi River. Although this glorious, magnificent, and beautiful river was "discovered" by Livingstone (and others over the next 20 years), it was well known to the Africans living in the interior, and on most expeditions they served as his guides. However, it was Livingstone and other European explorers who provided the observations of latitude, longitude, and altitude that enabled accurate maps to be prepared. It is for this reason that their finds have been called discoveries.

Concern for the medical dangers and the welfare of his family prompted Livingstone to dispatch them to England from Capetown. This was not an act of pure altruism on his part; indeed, it freed him to continue his explorations unencumbered by wife and children. On 8 June 1852, he left Capetown by wagon and began his 4-year epic trans-African journey that covered 6,000 miles. Usually, Livingstone traveled only 6 miles a day, and his equipment for the journey consisted of two guns, three muskets, biscuits, a few pounds of tea and sugar, 20 lb of coffee, a bag of spare clothing, medicines, a box of books, a slide projector, a chronometer, thermometer, and compass, a horse blanket for a bed, a gypsy tent, and beads for bartering with the natives. Throughout the trek—partly on foot, partly by oxen, and partly by canoe— malaria, prolapsed piles, fatigue, attacks by hostile tribes, and torrential rains hampered him. Livingstone was aware that the scourge of Africa was malaria and that "it was destined to preserve inner-tropical Africa for the black races of mankind." He also recognized that eradication of malaria would be critical if missionary and commercial penetration of Africa were to succeed—he remarked that the inventor of the mosquito net was deserving of a statue in Westminster Abbey. Although Livingstone was aware of the correlation between the myriads of mosquitoes and his persistent bouts of malaria, he never pursued the connection, believing the fevers were due to exposure to an east wind arising from the swamps, to drinking milk in the evening, or to drinking water that flowed over granite. Livingstone was the first to recognize the relationship between the tsetse fly and nagana (an animal form of sleeping sickness) in cattle; he also described the river blindness worm (*Onchocerca*) in the anterior chamber of the eye, and when he reported on the bloody urine he saw in his porters, he provided evidence for the presence of blood fluke disease.

Between August 1852 and May 1853, Livingstone traveled from Kuruman to Linyanti, and from November 1853 to May 1854 he traveled down the Zambezi from Linyanti to the west coast city of Luanda. After 4 months of rest, he turned inland for an 11-month trek back to Linyanti. En route to Quilimane (which he reached in May 1856) on the Indian Ocean, he discovered the great waterfall called "the smoke that thunders" and named it Victoria Falls after his queen. Livingstone returned to Britain just before Christmas 1856 and was celebrated as a folk hero and a great explorer. He transcribed his African journals into an 1857 best-selling book, *Missionary Travels*. Anxious to return to Africa, Livingstone convinced the British government and the Royal Geographic Society to sponsor a Zambezi River Expedition (1858 to 1864) to explore the river and its tributaries. He also accepted appointment as consul in the District of Quilimane and the Eastern Coast of Africa. Together with six assistants, Livingstone and his family left England in March 1858 and arrived in Capetown in April; a month later, they departed for the mouth of the Zambezi. Recognizing the perils of malaria, the staff were instructed to take daily doses of quinine in a glass of sherry. The Zambezi proved not to be navigable owing to impassable rapids, but forays on foot led to the discovery of Lake Nyassa (September 1859), now known as Lake Malawi. During the second half of 1860 he joined the exploration party of Bishop Mackenzie's

"Universities Mission to the Shire Highlands." Despite quinine prophylaxis, malaria struck the Mackenzie party within a year. This included Livingstone's wife, and she and several other members died of the fever.

After the Zambezi expedition, Livingstone returned to London (July 1864), but this time he did not receive a hero's welcome, since the expedition was considered an expensive failure. In addition, the discoveries of Richard Burton and Hanning Speke and their search for the source of the Nile diminished his accomplishments. Despite this, he wrote a second book, *The Zambezi and Its Tributaries,* and convinced the Royal Geographic Society that there was still much to be discovered about the lakes and the central African watershed. In March 1866, at age 53, Livingstone embarked on his last journey into the interior of Africa. The plan was to start out in Zanzibar, march northward from Lake Nyassa to Lake Tanganyika, and continue along its east coast to Ujiji where the great caravan route to the interior ended. He reached Ujiji in March 1869, but the travel was exhausting and he was plagued with malaria, bleeding piles, and tropical ulcers. On this venture he became acquainted with "earth eating," or *sufura,* finding it prevalent among the slaves. This condition, known today as geophagy, is a sign of hookworm disease, but Livingstone did not make the connection. For the next 2 years, he traced the lakes and the rivers with which they communicated, but little of this was known in England. Not hearing of Livingstone's whereabouts and fearing for his safety, the *New York Herald* sent its star reporter, Henry Morton Stanley, to seek news of his health. Stanley arrived in Zanzibar in January 1871. Starting out with 192 men and 6 tons of equipment, Stanley met Livingstone at Ujiji on 10 November 1871 with the now famous line, "Dr. Livingstone, I presume." The two remained together for 4 months, and although Stanley encouraged Livingstone to return to England, he refused, believing he was on the verge of a great discovery. Stanley retraced his path to the coast and arrived in England to a hero's welcome. Livingstone remained in Africa, where he died on 4 May 1873. His heart was buried in Africa, but his mummified body was returned to England, where it is interred in Westminster Abbey.

Stanley and the Scramble for Africa

David Livingstone was broad-shouldered, of medium height, and had a determined but gentle demeanor; he was also a manic-depressive. Courageous and enthusiastic, he marched across Africa, Bible in hand, condemning slavery, speaking the native languages, dispensing medicine, and loving the African continent and its people. When he met Stanley at Ujiji, though emaciated and in ill health, he was dressed for the occasion: he wore a gold-braided consular cap, a woolen jersey, grey tweed trousers, and tight shoes. Stanley, on the other hand, rode on horseback, wearing white flannel trousers, a white-powdered pith helmet with a neatly folded headband, and shiny Wellington boots.

Henry Morton Stanley (1841–1904), born John Rowlands in Denbigh, Wales, of unmarried parents, was brought up in crushing poverty. As a child he was placed in a workhouse, where he was exposed to a sadistic master. He escaped, came to the United States as a cabin boy, jumped ship, was adopted by a New Orleans merchant named Henry Hope Stanley, and shortly thereafter took the name of his adoptive father and changed his middle name from Hope to Morton. Stanley fought in the American Civil War, worked as a deckhand and a clerk, and, after discovering he had a way with words, became a newspaper reporter. Stanley was described by Queen Victoria as "a determined, ugly little man with a strong American twang." His face was red and round, with high cheekbones, a wispy moustache, and penetrating gray eyes. In contrast to Livingstone, Stanley was tough and intent on gaining fame, not caring who got hurt in the process; for his entire life, he was overly sensitive to criticism, perhaps because of the stigma of his illegitimate birth. Stanley was not above stretching the truth. He was also a harsh leader who drove his African porters, and he was arrogant in his belief in his own racial superiority. Whereas Livingstone loved Africa and its people, Stanley disliked both, saying, "I do not think I was made for an African explorer for I detest the land most heartily." But Stanley's efficient and powerful manner contributed to several successful explorations, and in the minds of many he remains one of Africa's greatest explorers.

When Stanley heard of Livingstone's death, he promised to return to Africa to carry on his work. Three years later, Stanley was commissioned by the *New York Herald* and London's *Daily Telegraph* to continue the explorations begun by Livingstone. With three British companions and 300 Africans, he left Zanzibar in 1874, circumnavigated Lake Tanganyika and Lake Victoria (1878), navigated down the Lualaba and Congo rivers, and made the trip to the Atlantic (1879) via the Congo River. All the other whites and 150 Africans died during the expedition.

Stanley tried to interest the British government in further exploration and development of the Congo basin but was rebuffed. His expeditions did, however, attract Belgium's King Léopold II, who was both energetic and ambitious. King Léopold, a devoted supporter of the Roman Catholic church, was very rich, and he promoted scientific research, as well as missionary efforts. In 1878 Léopold II founded the International Association of the Congo, financed by a consortium of bankers, and he recruited Stanley to help him explore the Congo under the auspices of this association. In 1879, Stanley arrived at the mouth of the Congo River and traveled upstream. He founded Vivi, the first capital, across the river from present-day Matadi, and moving farther upriver reached a widening he named Stanley Pool (now Pool de Malebo). At Stanley Pool he founded a trading station and the settlement of Léopoldville (now

Kinshasa) on the south bank. The north bank of the river had already been claimed in 1880 by de Brazza for France, leading ultimately to the colony of French Congo. In 1881 Stanley claimed for Belgium and King Léopold II that part of the Congo that had not been claimed for France.

Stanley returned to Europe at the end of 1881; however, he was back in Africa by December 1882, signing more than 450 treaties on behalf of Léopold II with persons described as local chieftains who had agreed to cede their rights of sovereignty over much of the Congo Basin. Stanley's series of journeys and treaties throughout the Congo led to the establishment of the Congo Free State (1885–1908) with its capital Léopoldville (now Kinshasa) and its other major city, Elizabethville (now Lumumbashi). From 1908 to 1960 the country was called the Belgian Congo, from 1960 to 1994 it was called Zaire, and since 1994 it has been called the Democratic Republic of the Congo.

The period of European colonial expansion, termed by some writers as "the scramble for Africa" began in the 1880s. European powers were able to pick up huge parcels of land cheaply and with little risk from armed resistance since they now possessed two critical resources: quinine and firearms (repeating rifles and machine guns). The power differential between Europe and Black Africa was enormous and would remain unchanged for the next 70 years. In 1884, Germany laid claim to Togo, Cameroon, Tanganyika, and Namibia. The British, who already had a presence in west Africa (Gambia, Nigeria), secured land around Mombassa and Berbera to ensure a safe sea route to India, and then Germany proclaimed sovereignty over Dar-es-Salaam. Great Britain formalized occupation of the Guinea Coast (an old geographic designation for the West Coast of Africa that now stretches from Senegal to Angola), and the French established themselves on the right bank of the Congo. In this way, the delicate balance of power in Europe was preserved, but power could be expressed by European nations in the less sensitive area of colonial lands. A gain by one country was countered by another in a kind of overtrumping in a card game.

The Conference of Berlin, held in 1884 to 1885, was convened to settle disputes among the European nations in Africa. By international agreement at the Berlin Conference, all signatories had to be notified of a power's intent to annex. The term "spheres of power" entered into the treaty language, and in 25 years almost all of Africa was partitioned in a manner that had no reference to history, geography, or ethnic considerations. To Europeans, these were simply lines traced on a map that indicated a possession. By 1914 nearly 11 million square miles of Africa was in the hands of European nations. Only 613,000 square miles was independent. At the outbreak of World War II there were only three independent African states: Egypt, South Africa, and Liberia.

Separately, the Berlin Conference recognized Léopold's International Association of the Congo as an independent entity. Later, the association would become the Congo Free State. By the General Act of Berlin, signed at the conclusion of the conference in 1885, it was agreed that activities in the Congo Basin should be governed by certain principles, including freedom of trade and navigation, neutrality in the event of war, suppression of the slave traffic, and improvement of the condition of the indigenous population. The conference recognized Léopold II as sovereign of the new state.

> To meet the conference's legal requirement of "effective occupation," Léopold II proceeded to transform the Congo Free State into an instrument of colonial hegemony. Indigenous conscripts were promptly recruited into his nascent army, the Force Publique, manned by European officers. A corps of European administrators was hastily assembled, which by 1906 numbered 1,500 people; and a skeletal transportation grid was eventually assembled to provide the necessary links between the coast and the interior. The cost of the enterprise proved far higher than had been anticipated, however, as the penetration of the vast hinterland could not be achieved except at the price of numerous military campaigns. Some of these campaigns resulted in the suppression or expulsion of the previously powerful Afro-Arab slave traders and wary merchants. Only through the ruthless and massive suppression of opposition and exploitation of African labor could Léopold II hold and exploit his personal fiefdom.

Though Léopold II was respected as a diplomat and as a philanthropist through his support of research to prevent sleeping sickness and his proclaiming how much he loved the African, his actions suggest that the real motive was greed. The Belgian goal was not to teach Africans to work in the "childish" pursuits of their own culture but to be organized in rational routines of productive wage labor in the European manner for European employers. Such labor was considered a civilizing influence, and a profitable by-product was cheap labor. Agricultural programs required the Africans to raise designated crops—cotton for export and food for the towns and mines within the colony, neither of which threatened European interests nor ensured the health and well-being of the native population. Léopold II, through his European civil servants and officers, established a system of state socialism and issued decrees by which all land, all rubber, and all ivory were the property of the State—himself. He made it illegal for the natives to sell these resources to other Europeans or for other Europeans to buy them from the natives. His decree was to "neglect no means for exploiting the forests." Vast profits were accumulated by Léopold II in a simple scheme: each village was ordered to bring in a specific quota of rubber, and the workers were to neglect all other activities until this quota was met. If they failed, the women were taken away as hostages or became parts of harems, and other native troops were sent into the villages to brutalize, kill, and spread terror. To prove that they had killed the villagers, they had to bring in one right hand for every cartridge used. Because of this practice, and the accelerated spread of sleeping sickness by the movement of hostages from one province to another, in 15 years the native population was reduced from 20 million to 9 million.

The increased population movement and the punitive expeditions to quell uprisings brought European diseases to the virgin territories of Africa. Forced labor, a feature of colonial rule, also spread disease: for example, when rubber workers were recruited from the Congo Free State, thousands left their villages and moved up the tributaries of the Congo; these new settlers brought their diseases far across the continent. The men of Mozambique, Zimbabwe, and Malawi took their diseases to the gold mines of Johannesburg and brought back with them to their homelands new strains of disease organisms. Those who lived in the highlands of Kenya had little experience with malaria, so when they were recruited to labor on harbor works in Mombassa, they died of malaria at an appalling rate.

The imposition of efficient administrations by the Europeans on their African colonies tended to concentrate people in larger villages and also forced them to grow crops that were highly remunerative, such as coffee, tea, and groundnuts, but that removed acreage needed for food production. The movement of people between village and forest spread disease, and the roads and railways provided increased opportunities for disease movement. Different communities with different backgrounds, resistance patterns, and diseases became mixed, and this had disastrous consequences for the Africans. Finally, a different kind of destabilizer entered the picture: firearms were brought to Africa in large numbers. Some were obsolete firearms that had been replaced by breech-loading rifles in the American Civil War, but they were snatched up by the local tribal chiefs. There was tribal warfare, and with an expansion of tribal domains there was more movement and more exchange of diseases. Civil wars, social upheavals, economic disruptions, and endemic diseases, as well as diseases introduced by the European colonists—e.g., tuberculosis, whooping cough, diphtheria, plague, typhoid fever, and cholera— became more and more a factor in Africa's decline in health. When David Livingstone traveled in Africa between 1841 and 1873, he noted that the Africans were not a sickly people, but after 1880 they were described very differently by another writer: they were "a hive of gangrenous germs."

The Europeans were quite aware that the native Africans were comparatively free of the indigenous diseases. The nature of their immunity was unknown, and some of the European whites feared that if it was survival of the fittest, they would eventually be replaced by the better-endowed blacks. Such a thought was not very comforting to the whites, so they concocted fanciful explanations. One was that the Africans had a greater facility for sweating and this threw off the noxious vapors and humors in the blood. But this theory was abandoned when a group of ex-slaves from America were settled in Freetown in Sierra Leone and came down with the same fatal diseases as the Europeans.

The Europeans eventually overcame the "immunity problem" by chemotherapy (and force). With quinine, it was possible for Europeans to enter Black Africa and live without fear of death. Quinine therapy, combined with the introduction of the breech-loaded rifle, gave the Europeans increased confidence, and exploration moved into high gear with a move toward the Rift Valley. East Africa was not only strategically placed, but it also had great plains, immense game, and snow-capped mountains. The British of India were encouraged to spend their leaves there, and a new brand of intrepid explorers was born. These were of the safari type, and their adventures could be recorded by sketches and photographs. Big-game hunting and ivory trading became all the rage. Missionaries came to Africa in increasing numbers and opened up the continent even more. There were handbooks on how to travel and how to avoid or cure disease, and the number of European visitors increased. Soon, the terror of Africa for Europeans disappeared.

Endemic Diseases of Africa

Sleeping sickness

One of the most deadly African diseases is sleeping sickness. It is undoubtedly of African origin and today remains indigenous to that region. The disease, in humans and cattle, occurs in the fly belt of Africa between 20° N and 20° S (Fig. 15.1), extending over 10 million square miles, an area larger than the continental United States. The disease precludes farmers from keeping cattle and small ruminants and accounts for low livestock productivity. Indeed, the production of this region of Africa is 1/70 that of the animal protein produced in Europe on the same amount of land. The losses in milk, meat, hides, manure, and tractive power result in an estimated annual loss of 5 billion dollars, and 30% of the 150 million cattle are at risk.

The story of human sleeping sickness begins with fishes and frogs. In 1841, Professor G. Valentin of Berne, Switzerland, gave a very brief description of a protozoan he had found in the blood of a trout, and he noted that it moved by means of its undulating membrane. Two years later, David Gruby, working in Paris, discovered a similar organism in frog blood; because its swimming motion suggested the action of a corkscrew, he called it *Trypanosoma sanguinis*. The name *Trypanosoma* (see Fig. 1.2C) is derived from two Greek words, *trypano* meaning "auger" or "screw-like" and *soma* meaning "body"; the species name *sanguinis* comes from the Latin word for blood. Trypanosomes were also described from the blood of field mice, voles, and rats, but they were considered mere curiosities and of no economic, medical, or veterinary importance. All that changed in 1880 when Griffith Evans, an English veterinarian serving in Punjab, India, found trypanosomes in the blood of horses, mules, and camels suffering from a fatal wasting disease called surra. Inoculation of blood containing trypanosomes into healthy animals produced surra. Evans was convinced that the trypanosome (later named *Trypanosoma evansi*) was a parasite, but he did not discover how horses, mules, and camels became infected. (Leonard Rogers discovered the vector, a biting stable fly, in 1899.) In 1892, Alphonse Laveran, who had discovered the malaria parasite in 1880 and had then become interested in other blood parasites, summarized all that was known about trypanosomes in 11 pages and 18 references!

In the early 1890s, the British colonial farmers of Zululand were faced with the decimation of their European breeds of cattle by a wasting disease called nagana, a word meaning in Zulu "in low or depressed spirits." At the farmers' urging, in 1894 the governor of Zululand requested that Surgeon-Captain David Bruce (1895–1931) of the British Army carry out an investigation to determine the cause of nagana. Bruce was trained as a bacteriologist and initially believed that the disease was caused by a bacterium, but none could be cultured. However, when he examined the blood of diseased cattle, he discovered the cause: a rapidly vibrating body, lashing about among the red blood corpuscles—a trypanosome. (In 1899 that trypanosome was named by Plimmer and Bradford *Trypanosoma brucei*, after its discoverer.) Bruce then went on to establish Koch's postulates for nagana: if he injected blood from cattle suffering with this disease into dogs, severe wasting symptoms resulted and abundant trypanosomes were found in the dog's blood. He wrote, "the clinical features of nagana are defined by the constant occurrence in the blood of a protozoan parasite." In 1895, Bruce discovered the vector for nagana: the bloodsucking tsetse fly (genus, *Glossina*) (Fig. 15.2). But where did the tsetse fly acquire the trypanosome? Bruce hypothesized, and then proved, that wild game—buffalo, wildebeest, and bushbuck—were the source of infection and that transmission was the result of the bite of the tsetse flies that infested the area in which these game animals lived. Bruce's

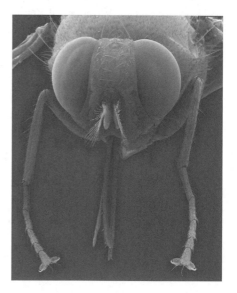

Figure 15.2 The head of a tsetse fly (*Glossina*), the vector of African sleeping sickness and nagana. (Courtesy of Dennis Kunkel Microscopy, Inc.)

work on "the fly disease," nagana, set the stage for a fuller understanding of the human disease known as African sleeping sickness.

Between 1375 and 1406, an Arab historian had reported human sleeping sickness (Fig. 15.3) in Mali. It went largely unnoticed. During the late 1700s, increased numbers of British missionaries and explorers went to Africa, where they established settlements and colonies. In 1702, an English naval surgeon, John Atkins, described a disease in Africans living along the Guinea Coast. Atkins believed its cause was the accumulation of phlegm around the nerves that impaired their function and that a poor diet and a lack of mental exercise led to the symptoms of "the sleepy distemper." But in 1803, Thomas Winterbottom, a physician working in the colony of Sierra Leone, published an account of "the African lethargy." He offered no cause but recognized a telltale clinical characteristic: swelling of the cervical lymph nodes (Fig. 15.4). This swelling, called "Winterbottom's sign," was recognized by slave traders as an indicator that the African would suffer from lethargy; either they would not buy such slaves or, if they had, would rid themselves of them as soon as the sign appeared. In hindsight, it can be recognized that Winterbottom had described a clinical condition of individuals suffering from Gambian sleeping sickness, but before 1901 there was no record of human disease caused by a trypanosome. In May 1901, a 42-year-old Englishman who worked on the steamships plying the Gambia River came down with a fever. He was admitted to the hospital and was treated with quinine for malaria without success. When his blood was examined, no malaria parasites could be found; however, there were trypanosomes. In 1902, this trypanosome of humans was named *Trypanosoma gambiense* by Joseph Everett Dutton of the Liverpool School of Tropical Medicine. This was the first demonstration of trypanosomes in a European patient suffering from a disease resembling sleeping sickness. To carry out further work on African

Figure 15.3   A victim of African sleeping sickness. (Courtesy of Wallace Peters.)

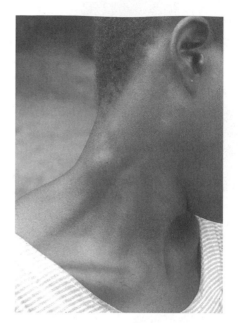

Figure 15.4 Swelling of the lymph node in the neck known as Winterbottom's sign. (Courtesy of Wallace Peters and Peter Janssen, Antwerp, Belgium.)

trypanosomes, Dutton traveled from England to the Gambia. There he found that a very small proportion of the local natives also contained the parasite in their blood but suffered no ill effects. He suggested that they probably served as a reservoir to infect the Europeans. Dutton's death in 1903 (of relapsing fever) cut short his brilliant research career and prevented him from completing his studies on transmission of Gambian sleeping sickness.

In 1901, when a severe epidemic of sleeping sickness broke out in Uganda, the Royal Society of London sponsored a three-person scientific commission headed by David Bruce to investigate its cause. By 1902, the commission had discovered that the distribution of individuals with Winterbottom's sign corresponded with the distribution of sleeping sickness. Further, examination of the cerebrospinal fluid from a patient with sleeping sickness by one of the members of the commission, Dr. Aldo Castellani, showed trypanosomes. However, Castellani, who was supposed to investigate the evidence for a bacterial cause, became convinced that the disease was due to a streptococcus, not the trypanosome, and he did not appreciate his own finding until much later. Castellani refused to tell Bruce of his discovery unless he agreed that the discovery be published under his name alone and that no credit or information be given to the other member of the commission, D. N. Nabarro. Bruce agreed. Then he and Castellani were able to show that in the cerebrospinal fluid from 34 other cases of sleeping sickness, 20 had trypanosomes, whereas in 12 control cases none were found. Castellani published the findings on his own and took full credit for himself. Bruce honored his agreement with Castellani but wrote in a report, "For the sake of a future historian, at the time of the arrival of the commission Dr. Castellani did not consider . . . that this trypanosome had any causal relationship to the disease but thought it was an accidental concomitant like Filaria. As Dr. Castellani has not entered into any detail respecting these matters in his reports, it is thought advisable to supplement his account with the above, as the history of the discovery of the cause of such an important disease must always be of interest." Bruce suspected that the tsetse fly was involved in the transmission of Gambian sleeping sickness. Later, when it was found that monkeys were susceptible to the disease and after a tsetse fly had fed on a human with sleeping sickness it could transmit the disease to monkeys, Bruce concluded, "sleeping sickness is . . . a human tsetse fly disease."

In 1910, J. W. Stephens of the Liverpool School of Tropical Medicine discovered a new species of trypanosomes. It was from a patient with sleeping sickness who had acquired the disease in 1909 in Rhodesia, an area where *T. gambiense* and its vector (*Glossina palpalis*) did not occur. When Stephens inoculated this patient's blood into rats, the trypanosome caused a more acute disease and had a slightly different morphology from the more benign *T. gambiense*. In collaboration with H. B. Fantham, Stephens described the new species, which he named *Trypanosoma rhodesiense*. In 1912, David Bruce headed another sleeping sickness commission in an area near Lake Nyssa. He found *T. rhodesiense* in the blood of one-third of the 180 game animals examined. Bruce compared the morphology of *T. rhodesiense* with that of *T. brucei* from cases of nagana and found them to be identical. He concluded, "*T. rhodesiense* is neither more nor less than *T. brucei*, and the human trypanosome disease in Nyasaland is nagana." The vector was suspected to be the most abundant tsetse, *Glossina morsitans*. Thus, early in the 20th century, the causative agent, the vectors, and the pathogenesis of African sleeping sickness came to be discovered.

T. gambiense gives rise to a mild chronic infection found in western and central Africa and is transmitted by riverine species of *Glossina* (*G. palpalis* and *G. tachinoides*) that are associated with human habitation. Although occasionally pigs can serve as a reservoir host, the tendency is for parasitemias to be of low abundance in humans, so transmission to other animals is infrequent. Asymptomatic individuals may harbor parasites in the blood for long periods of time and could be a source of infection for the vector. The disease is an anthroponosis: fly, to human, to fly, to human. *T. rhodesiense* is the East African form and is the more virulent subspecies, producing an acute infection. It may also be the subspecies that has more recently evolved, i.e., within the past 100 years. *T. rhodesiense* is transmitted by the dry savannah bush flies *G. morsitans*, *G. swynnertoni*, and *G. pallidipes*. In wild game animals (bushbuck, gazelles, wildebeest) the disease is mild, but in humans it is highly pathogenic, killing in 4 to 6 months if untreated. The disease is an anthropozoonosis: fly bites infected ungulates; fly becomes infected; fly transmits disease to human.

The spread of sleeping sickness was increased about 1890 because of another infection brought into Africa by cattle from outside the continent: rinderpest, a virus of cattle and game with high mortality. The killing off by rinderpest upset the balance of the eastern savanna where there was a delicate balance between cattle, game, human, and tsetse. When rinderpest killed the cattle and game, the tsetse sought an alternative host—humans. When the game was gone, the amount of vegetation also increased, because grazing was reduced. Now the flies had new breeding grounds, and sleeping sickness extended its grip. Rinderpest also decimated the cattle herds of the pastoral Africans so that protein deficiency became more severe and human susceptibility to other diseases increased.

Control of sleeping sickness is yet to be achieved. Since the vector of *T. gambiense* thrives in hot, humid regions and along swift-running streams, clearing of vegetation has worked to a limited degree. But *T. rhodesiense* has a reservoir in game animals, where it causes little harm, and it also has a tsetse vector that lives in the dry savannah, which is impossible to clear. Today there are reported to be ~30,000 human cases of sleeping sickness per year, with 50 million people at risk; however, the true number of infections may be 10 to 15 times greater. Treatment of sleeping sickness is of limited value and depends on early use of rather toxic drugs administered intravenously. Pentamidine, melarsopol, and suramin are all used.

What about a protective vaccine against sleeping sickness? All efforts to date have failed, largely because the trypanosomes in the blood are covered with a surface coat that is shed periodically, thus providing a constantly changing target for the immune system. It is estimated that there may be 100,000 variants, expressed sequentially. So every time the immune system mounts an attack against one variant surface coat, the trypanosomes stop making that variant and switch to a new surface coat, which has to be reacted to by the immune system anew. Absent a vaccine and with poor drugs for treatment, sleeping sickness and nagana continue to plague Africa.

River blindness

In the region of Africa called the Upper Volta, and in Sierra Leone, Liberia, Northern Ghana, and Central Africa, it is not uncommon to see blind older men being led by young boys to the fields (Fig. 15.5). The cause of their blindness is the worm named *Onchocerca volvulus*. This parasite is also found in the highlands of Guatemala and southwestern Mexico in the coffee-growing areas. River blindness is probably an African disease, brought to the Americas with the importation of slave laborers from Africa to cultivate sugar cane and coffee. However, it was not described from the Americas until 1915. In some African localities, some 80 to 100% of the people are

Figure 15.5   Line of blind Africans walking with canes. In areas in which river blindness is endemic (Chad, Central Africa), it is not uncommon to see blind adults being led to the fields by children who have not yet lost their sight. (Courtesy of the National Library of Medicine.)

infected with *Onchocerca*. It is estimated that worldwide there are 20 million infections.

River blindness is transmitted by a blackfly called *Simulium damnosum*, which carries the microscopic infective juvenile stages (called microfilaria) of *Onchocerca* (Fig. 15.6). The disease is most common along rich, oxygenated water in rapidly flowing rivers where the blackflies breed—the same areas where people congregate for bathing, fishing, drawing water, and defecating. Thus, the river is an excellent place for the fly to meet its human host. The infected flies bite, and the infective stages migrate out of the mouthparts and enter the wound. The worms become mature at the site of the bite and become enclosed in a subcutaneous connective tissue nodule (Fig. 15.7A). Here they are trapped, but the adult female worms give rise to live microfilaria that migrate throughout the body and are a source of infection for the biting blackflies. Development in the fly may take 1 to 3 months, and an adult female (which can live for up to 2 years) can produce 1,500 microfilaria each day.

Why is it called river blindness? The microfilaria live in association with the bacteria *Wolbachia*, and in concert they cause the pathology. The circulat-

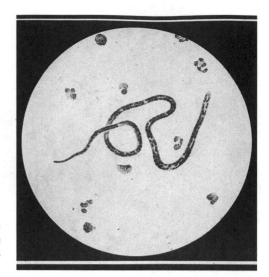

Figure 15.6 The microfilaria of *Onchocerca*. The microscopic worms are about 300 microns in length, or 20 times the width of the white blood cells.

ing microfilaria migrate into the eye, invading the cornea, the retina, and the choroid, creating mechanical damage. Owing to an inflammatory response to the bacterium, there is a partial loss of vision. Eventually, the optic nerve itself is affected, and total blindness results (Fig. 15.8). It is the repeated exposure of humans to the bite of the infected river-dwelling blackfly that ultimately results in blindness.

The search for and identification of the cause of river blindness took 70 years. In 1875, John O'Neill, a British Naval surgeon attached to the HMS Decoy at Cape Coast Castle in the Gold Coast (Ghana), became interested in a scabies-like skin disease known as "craw-craw." After examining material from the skin nodules in a drop of water under a microscope, he observed threadlike undulating worms, the microfilaria, and concluded that these were "the immediate cause of the complaint." In 1890, Rudolph Leuckart, using samples collected and sent by missionaries from this same region, described the morphology of the adult female and male worms. The adult worms were coiled together to form a knotted ball with female worms 6 to 70 mm in length and the males half that size. In 1901, William Prout, working in Sierra Leone, named the worms *Onchocerca* (for "hooked tail") *volvulus* (for "roll about"). Despite these findings, the life cycle of the parasite, as well as its transmission from person to person, remained a mystery. Then in 1915, Rodolfo Robles discovered the presence of onchocerciasis among coffee plantation workers in Guatemala; he removed a nodule from the forehead of a boy and found within it the worms described earlier by Leuckart. Robles ruled out the possibility that transmission took place via water. He noted that the laborers who lived outside the disease zone and entered endemic areas only during the day to work became infected with

A

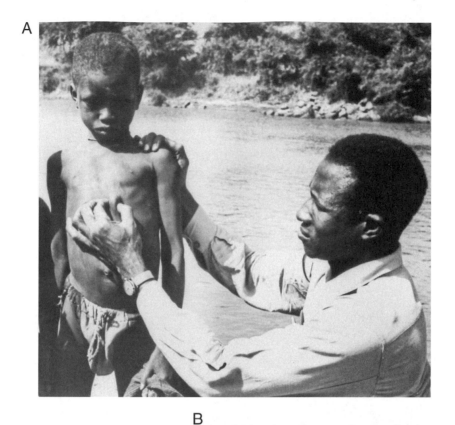

B

Figure 15.7 River blindness. (A) Nodule containing the adult worms of *Onchocerca*. (Courtesy of the National Library of Medicine.) (B) Adult *Onchocerca* removed from the nodule. The adult female releases live offspring (microfilariae). (Courtesy of WHO/TDR/Baldry Irwin.)

onchocerciasis. He hypothesized that the vector of the disease was a day-biting insect, more specifically, blackflies that were found in the endemic areas. Robles, however, never verified his hypothesis. Eleven years later in 1926, Donald Blacklock of the Liverpool School of Tropical Medicine, working in Sierra Leone, picked up where Robles had left off. Blacklock found that the microfilaria were found not in the blood but in the skin, and he postulated that the vector must be an insect that could damage the skin and dislodge the juveniles in order to feed on blood. He discovered a biting, blood-

Figure 15.8   Blindness as a result of *Onchocerca* infection. (Courtesy of WHO.)

feeding blackfly, *Simulium damnosum,* near streams where the disease was endemic. Blacklock collected several wild blackflies and proceeded to infect them by allowing them to bite infected individuals. He then proceeded to trace the development of the parasite in the gut, thorax, head, and proboscis of the flies and in this way provided evidence for the transmission of the disease. This work was later confirmed by other workers in other parts of Africa, and in 1931, Hissette reported that 20% of the patients in the Belgian Congo with onchocerciasis were blind.

The economic and social consequences of river blindness are profound, and surely much of the local history is tied to the progressive depletion of human resources. Some inhabitants in Ghana became so accustomed to blindness that they could not understand why all people did not become blind. In some villages in the Upper Volta River, the number of blind adult men can be up to 30%. Ferrymen plying the river, and coming in close contact with blackflies, could look forward to a working expectancy of no more than a few years before suffering from photophobia and then being condemned to darkness for the remainder of their lives. As a consequence, over 25 million acres of fertile land, capable of feeding 17 million people a year, lies fallow.

How is river blindness diagnosed? Small snips of skin are removed from the shoulders or heads of infected individuals, placed in a drop of salt solution, and observed under the microscope for the wriggling emergence of microfilaria. Other methods use slit lamps to detect microfilaria in the cornea and anterior chamber of the eye. More recently, detection has been accomplished by screening for parasite DNA and parasite antigens.

How can river blindness be prevented or, better yet, eliminated? (i) Destroy the blackfly breeding sites. This is sometimes especially difficult to effect with rapid flowing rivers and streams. (ii) Kill the blackfly vector by spraying a biological control agent such as *Bacillus thuringiensis*. (iii) Surgically remove the nodules to decrease the parasite burden and the source of microfilaria. (iv) Treat humans with the drug ivermectin. Since 1987, ivermectin, which kills the microfilaria, has been the drug of choice; however, in many countries where the disease is endemic and transportation systems are inadequate, there remain difficulties in distribution of the drug.

Guinea worm

The Guinea worm has been known since antiquity, particularly in the Middle East, Africa (Cameroon and Senegal), and India. Numbers 21:4–8 reads, "And they journeyed from Mount Har by Way of the Red Sea to . . . the land of Edom . . . and the Lord sent fiery serpents among the people; and much people of Israel died . . . And the Lord said unto Moses 'Make thee a fiery serpent and set it upon a pole and it shall come to pass that everyone . . . when he looketh upon it shall live.'" It is believed that this biblical reference is to Guinea worm and to the caduceus, the symbol of modern medicine. But some contend that the caduceus, the staff carried by the gods Hermes and Mercury, is not of Greek origin but refers to the Jews in Egypt who were able to remove the worms by wrapping them around a wooden stick (Fig. 15.9). The existence of Guinea worm in Egypt in the time of the Pharaohs has been confirmed by the recent finding of calcified worms in mummies from the period of the New Kingdom in the 15th century BC.

In 1758, Linnaeus recognized the Guinea worm as a roundworm (nematode), naming it *Gordius medinensis* after the city of Medina where the condition was prevalent. In 1868, the Englishman Henry Bastian provided a detailed description of the worm and called it *Dracunculus* after the Latin word *draco* meaning "snake" or "serpent." The adult female Guinea worm is a meter in length and 2 mm in diameter, whereas the male is approximately 2 cm long. Despite the female worms' being so long and so active under the skin—resembling a wriggling varicose vein—the entire life cycle was not demonstrated until 1905 by Robert Leiper. Both adult worms live in the skin, and the female is fertilized by the male, who dies soon thereafter. The female discharges between 1 and 3 million juvenile worms through a blister in the skin; this discharge is facilitated by contact with cool water (Fig. 15.10). After

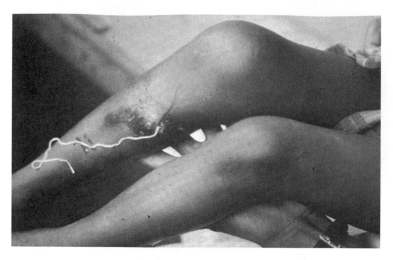

Figure 15.9   An adult female Guinea worm emerging from a painful and enlarged ulcer. In some cases the adult worm may be removed by slowly and carefully pulling it out and winding it around a pencil or stick. (Courtesy of Wallace Peters and Brian Greenwood.)

release of the offspring, the parent worm shrivels and dies, and the tissues absorb the remnants. The blistering of the skin is due to an intense allergic reaction and can be painful, hence the name "fiery serpent." The juvenile worms are fed upon by a small one-eyed shrimp (*Cyclops*), commonly called a "water flea" (Fig. 15.11), and over a period of 1 to 3 weeks they develop

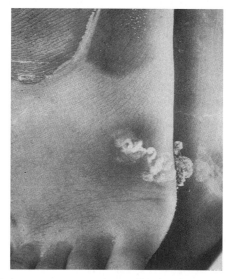

Figure 15.10   Release into water of juveniles of the Guinea worm. (Courtesy of Wallace Peters and Anthony Duggan.)

Figure 15.11 The one-eyed cope-pod *Cyclops*, with a pair of egg sacs. *Cyclops* eats the juvenile Guinea worms to become infected. Photo by James M. Bell/Photo Researchers, Inc. © 2005 Photo Researchers, Inc.

further into infectious juvenile stages. Humans drinking raw contaminated water ingest the infected water fleas. The water flea is digested by gastric juice in the stomach, the juvenile worms are liberated, and within 2 weeks they migrate from the intestinal tract through the abdominal cavity, and then to the skin and joints. Maturation to an adult worm may take up to a year.

Although infection with Guinea worm is not a cause of mortality, it is a burden in terms of morbidity, incapacity, and debilitation. Infected humans suffer from allergic reactions, rash, nausea, and dizziness. The blisters, which can be quite painful, often become infected with bacteria, leading to abscesses, arthritis, buboes, and tissue damage. Persons with Guinea worm may be unable to work during planting and harvesting, may have difficulty in caring for their children, and may suffer permanent crippling.

There are no therapeutic drugs for this infection. Treatment typically involves rolling the emerging adult female around a stick or performing surgery. Under the best of circumstances, after worm extraction the person may be disabled for 2 to 4 weeks, but more often, incapacitation may last for several months owing to the presence of many worms, sensitivity of the area in which the worms reside, and secondary infections of ulcerated areas.

Eradication programs for Guinea worm were begun in the early 1920s and focused on water sanitation as a way to eliminate the *Cyclops*. These efforts have continued with great success in 13 sub-Saharan countries, where the disease remains endemic; control can easily be accomplished by passing the raw drinking water through a fine mesh screen to remove infected water fleas. As a consequence, the number of cases has dropped from 3.2 million in 1986 to less than 75, 000. The Carter Center (Atlanta, Ga.) estimates that Guinea worm disease will be the second disease to be eradicated by human intervention.

Plagues Out of Africa

Yellow fever, the Louisiana Purchase, and the Panama Canal

Two of the most significant African diseases to establish themselves in the New World were malaria and yellow fever. Both were important factors in determining the pattern of human settlement and survival in the Americas. Before Christopher Columbus, it appears that neither malaria nor yellow fever existed in the Western hemisphere, and some contend that yellow fever was successfully transferred from West Africa to the Caribbean for the first time in 1648 when epidemics broke out in Yucatan and Cuba. What delayed the establishment of yellow fever was that this disease could not become epidemic in the New World until a specialized mosquito, *Aedes aegypti*, could find a suitable niche in the Americas.

Aedes is a highly domesticated mosquito that breeds in small bodies of still water—water casks or cisterns are excellent sites for the female to lay her eggs. The *Aedes* mosquito crossed the ocean aboard ships, becoming established in areas where the temperature stayed above 72°F. This peculiarity of mosquito–human association meant that mosquitoes carrying yellow fever could remain on shipboard for weeks and months. Yellow fever is usually a fatal disease, and a voyage lasting several months could decimate a crew—no one knew who would be next. The British feared the "Yellow Jack"—the flag symbolizing an infectious disease on board ship.

Yellow fever was called "Bronze John on his Saffron Steed" and "the Saffron Scourge" because of its symptoms: a yellowing of the skin, fever, nausea, and black vomit due to hemorrhages in the stomach, as well as headaches and muscular pain. Coma and death ordinarily result from kidney and liver damage. Not all who suffer with yellow fever die; recovery may begin in two weeks, and convalescence may require several more weeks. Those who recover are immune for life.

Yellow fever was called the "scourge of the American South" because it was restricted geographically due to climate and slavery. The climate of the American South, which made possible the cultivation of tobacco, cotton and sugar cane, also provided a favorable environment for the breeding of *Aedes aegypti*. Yellow fever came to the Americas by ship during the slave trade. The fear of "Bronze John " drove many people from the coastal communities, and those who could afford it left the South during what was called the summer "sickly season." Since yellow fever can become endemic only where the climate is warm enough to permit year-round mosquito activity, it was restricted to those parts of the American South where there was no cold and frost during the winter. Where there was a cold climate transmission ceased, but each summer it could be reintroduced from Latin America and Africa.

The first recorded yellow fever epidemic in the Western Hemisphere was in 1648 in Yucatan, Cuba, and the Barbados. By the 1690s it was in North America, especially in port cities with dense populations and good opportunities for transmission. There was an outbreak in Boston in 1693 after the arrival of the British fleet from the Barbados. Epidemics broke out in Philadelphia and Charleston in 1699, and in New York in 1702. Between 1763 and

1792 there was no yellow fever in the British colonies, but in 1793 there was an epidemic in Philadelphia after a ship arrived from Tobago which caused 5,000 deaths in a total population of 60,000.

During the 1800s there were 39 epidemics in New Orleans—a city called "Bronze John's port of entry." One of the last outbreaks in the United States occurred in 1878 in Memphis, Tenn. After the 1820s, the disease was confined to areas south of the Mason-Dixon line. The North was no longer troubled with yellow fever about the time that slavery was abolished. However, in New Orleans, the major port of entry from Latin America, almost every summer between 1790 and the Civil War there were epidemics.

In New Orleans yellow fever had an interesting pattern: (i) It was confined to cities since *Aedes* is a cosmopolitan species and in cities population densities were higher. (ii) It affected newly arrived immigrants and northerners because they lacked the immunity of the local residents. (iii) Blacks were rarely affected, demonstrating a natural immunity. As a consequence, yellow fever had its major impact on the urban south, not the rural south.

The peak of yellow fever epidemics in the United States took place in 1850, when there was an intense debate over slavery. Some feared an uprising of the immune blacks; others suggested that yellow fever was God's punishment of the whites who engaged in slavery, whereas others claimed that slavery was good for the southern plantation system since the deaths due to yellow fever were less catastrophic for black laborers.

The epidemic nature of yellow fever with its high mortality and dramatic symptoms attracted more attention than did other diseases, and it had great impact in the Northern press. Consequently, the negative impressions of the American South were reinforced, leading to a lack of migration from North to South, and the level of absenteeism in the South increased during the epidemic season of the summer. Yellow fever was bizarre, exotic, and dreadful. Not understanding its cause or the means of transmission, people in the American South accepted it as a way of life. Little wonder that the southern United States became insular.

Until 1880, when Patrick Manson discovered mosquitoes to be transmitters of elephantiasis, few suspected that yellow fever was not due to a miasma. Indeed, the mosquito vector for yellow fever was not discovered until 1900, and the actual cause of the disease was learned only much later, 1927. Thus, it is not surprising that until 1905 there was no real control of yellow fever, and when there was, it was 250 years after the disease had first reached the shores of the United States.

Yellow fever, caused by the arbovirus, an *ar*thropod-*bo*rne virus, was isolated by A. H. Mehaffey and J. Bauer in 1927 from the blood of a mild human case from Ghana and used to infect rhesus monkeys. For natural transmission to occur, the *Aedes* mosquito must feed on the yellow fever victim within the first 3 to 4 days of illness. The virus must then be incubated for 10 to 12 days in the mosquito before it is infective. The mosquito is then infectious for the remainder of its life.

A New Orleans physician described yellow fever in this way: "Often I have met and shook hands with some blooming handsome young man today and later I have been called to see him in black vomit, with profuse hemorrhages from the mouth, nose, ears, eyes and even the toes; the eyes, prominent, glistening yellow and staring; the face discolored orange. . . ." Three to six days after an infected *Aedes* bites, fever and headache, dizziness, and muscle ache occur. In some cases, the disease goes no further, but in others the fever rises sharply, the gums and stomach bleed, and there can be delirium and coma. The skin turns bright yellow—jaundice—because the yellow fever virus attacks the liver; as a result, the liver cannot function properly, and yellow bile pigments accumulate in the skin. Although only 5% of those infected commonly die of yellow fever, the death rate can be much higher (20 to 50%) during an epidemic.

In 1937, Max Theiler, a South African working in the Rockefeller Foundation laboratories in New York City, came upon an attenuated strain of the yellow fever virus (called 17D). This live attenuated virus conferred immunity without pathogenicity. Today this remains the one and only practical live virus vaccine against yellow fever.

Yellow fever and the Louisiana Purchase. In 1803, the United States purchased Louisiana from the French for about 15 million dollars. The area was 827,987 square miles and included all the land between the Mississippi River and the Rocky Mountains, and from Canada to the Gulf of Mexico. Today it includes 15 states. This is how the United States acquired it. In 1682, the French explorer Robert La Salle led 50 men from the Great Lakes region down the Mississippi River and claimed the entire Mississippi Valley for France. He named the region Louisiana after his king Louis XIV. Louisiana became a French colony, and by 1718 New Orleans was its capital. The French were disappointed in the revenues from the colony, so in 1762 they ceded the trading rights to Spain. After several skirmishes between the French and Spanish settlers, Spain took firm control in 1769. Sugar planting and processing began in 1795, and by 1800 it was the major cash crop. Louisiana prospered with sugar cane.

During the American Revolutionary War, Spain allowed the Continental Congress to use New Orleans as a supply base for moving goods up the Mississippi River to the struggling colonies. In 1800, France secretly persuaded Spain to return Louisiana. This brought concern to the new president, Thomas Jefferson, since Spain had allowed the Americans to deposit their goods duty free in New Orleans and then permitted them to export them to Europe and elsewhere. At about this time, there was a rumor that Spain also planned to give up parts of its colonies to France, including Florida. If this were carried out, the fear of President Jefferson was that the French would cut off American access to the Gulf of Mexico, especially New Orleans. In 1801, Jefferson's Secretary of State, James Madison, learned of the possible deal between Spain and France, but exactly how much land was to be transferred was not known. The United States tried to block the transfer, but Napoleon Bonaparte turned them down. Then in 1802, the Spanish governor, probably under pressure from Napoleon, suspended the right of deposit for American goods in New Orleans. This put the United States on the brink of war with France, and the United States threatened to send 50,000 troops to New Orleans to take the port by force. Instead, a land deal was effected. The reason for averting war: yellow fever.

In 1697, the French established a colony on the western side of Haiti, with the Spanish taking the eastern side. Haiti was a major sugar producer, largely because this labor-intensive crop used slave labor imported from Africa. In 1791 and 1794, there were slave revolts, and one of the leaders, Francois L'Ouverture—himself the son of slaves—declared himself governor of Haiti. It was clear to Napoleon that he had to assert his authority, but Haiti was critical for another reason—it would be a staging area for Napoleon's invasion of Louisiana and ultimately the establishment of Napoleon's North American empire. Were Napoleon to succeed, this would interfere with the British interests in North America.

In late 1801, to subdue the rebellion, Napoleon dispatched 20,000 well-equipped soldiers under the command of his brother-in-law, General Victor Emanuel Le Clerc. The French landed in Haiti in early 1802 and after a few battles were in control. But there were still minor skirmishes with rebel forces, and Le Clerc was concerned by the number of sick—there were already 1,200 in the hospital, and by April the numbers had increased. The cause of the illness was yellow fever, and soon one-third of the French force was ill. By June, the French soldiers were dying at a rate of 30 to 50 per day. In October, Le Clerc himself died of yellow fever. By 1803, the disease had killed 20,000 additional replacements, and the French gave up. Of all the French dispatched to Haiti, only 3,000 were left alive; it is estimated that 50,000 of their comrades died. On 11 April 1803, Napoleon told his finance minister Tallyrand that he could ill afford an American campaign, and he feared the British might seize the territory or form an Anglo-American alliance against him. Napoleon renounced all claim to Louisiana, and the Louisiana Purchase treaty was signed on 2 May 1803.

The colony of Haiti had been one of France's richest, and in 1798 it had 520,000 inhabitants. By 1804, not a single white person was left; the total population consisted of 10,000 mulattos and 230,000 blacks. Why was Napoleon's army decimated by yellow fever? (i) They were a naive population. (ii) The indigenous Haitian population was immune to yellow fever, yet they served as a reservoir for the disease. (iii) The climate was favorable for mosquito transmission—1802 to 1803 was particularly rainy. (iv) Most of the cities in

Haiti were burned to the ground, so hospitals and medical supplies were unavailable. (v) The hot, humid climate stressed Napoleon's soldiers. (vi) The French army bivouacked in the low-lying areas where the mosquitoes were more abundant.

Yellow fever and the Panama Canal. The British Empire existed between the reigns of two great queens: Elizabeth (1558–1603) and Victoria (1837–1901). British domination was based in part on its ability to rule the waves by controlling strategic points around the globe. These not only served for provisioning, coaling, and repair of ships, but for military ventures as well. The first of these strategic bases was Gibraltar, but later came Capetown, Hong Kong, Singapore, Ceylon, the Falklands, Aden, and Suez. Britain held control of the Mediterranean, Red Sea, and Indian Ocean. However, it was forced to let go of the Isthmus of Panama—not because of fortresses or cannons, but because of yellow fever.

The success of Ferdinand de Lesseps in building the Suez Canal (1850 to 1869) encouraged the French to attempt to build a Panama canal. (The Suez Canal—connecting the Mediterranean and the Red Sea—shortened the route between England and India by 6,000 miles, yet the canal itself is only 100 miles long.) The French acquired the land rights to build the Panama Canal from Colombia and also gained title to the Panama Railroad for $20 million. (It had been built in 1855 by a group of New York executives at a cost of $5 million.) The French began digging in 1882, but by 1889, after removing 76 million cubic yards of earth, they gave up because of mismanagement, poor skills, theft, and disease. More than 22,000 workers died during that period. The DeLesseps Panama Canal Co. went bankrupt in 1889.

The California gold rush stirred U.S. interests in a canal to connect the Atlantic and the Pacific Oceans as early as 1849. A canal through the Isthmus of Panama became even more enticing during the Spanish American War of 1898, because of the difficulty of sending ships from San Francisco to Cuba to reinforce the Atlantic fleet. (The distance from New York to San Francisco with the canal is 5,200 miles; without it, it is 13,000 miles, because ships have to go around South America.) In 1899, Congress authorized a commission to negotiate a land deal, and the French sold their rights and the railroad for $40 million. However, Colombia refused to sign a treaty involving a payment of $10 million outright and $250,000 annually, because the figure was considered to be too low. A group of Panamanians feared that Panama would lose the commercial benefits of a canal, and the French worried about losing its sale to the United States. So with the backing of the French, the Panamanians, encouraged by the United States, began a revolution against Colombia. The United States, in accord with its treaty of 1864 with Colombia to protect the Panama Railroad, sent in troops. The Marines landed at

Colon and prevented the Colombian troops from marching to Panama City—the center of the revolution. On 6 November 1903, the United States recognized the Republic of Panama, and 2 weeks later the Hay-Bunau-Varilla Treaty was signed, giving the United States permanent and exclusive use of the Canal Zone (10 miles wide, 50 miles long). Dealing directly with Panama, the United States paid $10 million outright plus $250,000 annually, beginning in 1913. The United States also guaranteed Panama's independence.

Construction began in 1907, and the canal was completed in 1914. More than 43,400 persons worked on the canal, most of them blacks from the British West Indies. The cost of construction was $380 million. In 1936, the annual U.S. payments to Panama increased to $430,000; by 1955, they had increased to $1,930,000; and in the 1970s, they increased to $2,328,000. In the 1960s, there were riots, and other concessions were granted. Panama continued to exercise greater and greater control over the canal and took complete control on December 31, 1999.

The successful construction of the Panama Canal is related to the effective control of disease. In 1880, Carlos Finlay, a Cuban physician, showed that *Aedes* mosquitoes could transmit yellow fever; however, not everyone believed his results with human volunteers. Of course, few at that time believed in mosquito transmission, most claiming that yellow fever was carried by filth. In 1900, U.S. Army Major Walter Reed carried out the definitive transmission experiments and reported his findings in 1901. Reed used human volunteers and screens to keep out mosquitoes. The volunteers were paid $100 for participating and a $100 bonus for getting yellow fever. Several volunteers died. Reed's experiments clearly showed that *Aedes* mosquitoes were capable of transmitting the disease.

In March 1901, Major William C. Gorgas, the sanitary engineer on the Canal project, initiated measures to eliminate mosquitoes and their breeding sites. During construction of the canal, all patients with yellow fever were kept in screened rooms. Workers lived in screened houses. Drainage and kerosene spraying were the mainstays of mosquito control. The magnitude of the accomplishment can be best appreciated by the number of deaths. During de Lesseps' canal days, the death rate was 176/1,000. When the canal was completed, the death rate from all causes was 6/1,000. The last fatal case of yellow fever in Panama occurred in 1906.

Annually, about 1,000 cases of yellow fever are reported worldwide, but the actual number may be 200 times greater. It can be maintained in a forest transmission cycle between monkeys and mosquitoes, and humans can acquire this jungle yellow fever. It can also be transmitted human to human even in an urban setting, since *Aedes aegypti*, the domestic vector, can be present at high density owing to its capacity for breeding in artificial containers used to store water.

Malaria and the American South

There is no convincing evidence that malaria existed in the Western Hemisphere before the arrival of Columbus and the conquistadors. Malaria was probably first introduced into Mexico and Central America by the Spanish explorers, as evidenced by the writings of a physician with Cortez, who in 1542 clearly described the disease in the soldiers. Although malaria did not exist in pre-Columbian North America, its mosquito vector, *Anopheles*, did. In short, before the colonists and explorers arrived in the Americas, there were anopheles mosquitoes but no malaria.

During the 16th and 17th centuries, malaria was introduced into the United States by early settlers coming from England, especially from the fen country as well as Kent, where endemic malaria was called the Kentish disorder or the lowlands disease. In the American colonies, both the landed gentry and indentured laborers provided the seedbed for infection of *Anopheles* in Maryland, Virginia, and the Carolinas. The early French and Spanish explorers and settlers along the coast of the Gulf of Mexico were also a source of malaria for those living in the American South and the lower Mississippi Valley. But it was only after the introduction of slaves from Africa—to work in the rice fields and on the sugar plantations—that malaria became a prominent feature of the American South, and this condition persisted until the 20th century.

By the time of the Revolutionary War, malaria was endemic from Georgia to Pennsylvania; it was then carried up the coastal rivers and across the Appalachians into western New York and Pennsylvania, and then down the Ohio River and its tributaries into Kentucky and Ohio. As the pioneers moved westward, they carried with them their earthly possessions and their diseases, and so it was that malaria moved into Indiana and Illinois and crossed the Mississippi River into Missouri. Moving relentlessly, the fever plague of malaria reached New England, lower Michigan, and Wisconsin and as far north as Minneapolis. French settlers sailed down the Mississippi River, bringing malaria as far south as the Mississippi delta and into the Tennessee Valley. As people migrated west in search of new places to settle and chose the fertile areas of river bottoms and alongside streams and creeks, malaria became established in Iowa, Nebraska, eastern Texas, the Dakotas, the San Joaquin and Sacramento valleys of California, and even the Willamette Valley of Oregon because these were also the areas where anopheline mosquitoes flourished. As a consequence, by 1850, with increased numbers of settlers and the opening up of more and more land for agriculture, malaria became endemic in much of the United States and was hyperendemic in the Ohio and Illinois River valleys, but especially the Mississippi Valley from St. Louis to New Orleans.

In the American South, malaria was transmitted principally by *Anopheles quadrimaculatus*, a mosquito well adapted to warm standing pools of water, flooded rice fields and tree holes, and with a distinct preference for human blood. In the valleys of California, *Anopheles freeborni*, a mosquito associated with irrigation agriculture, preferring animal blood, and restricted by the dry, hot summers, became the transmitter of "miner's fever." Mosquitoes have a flight range of about 1 mile and live for several weeks. Blood is required for egg laying. Depending on the temperature and humidity, a new generation of mosquitoes can develop every 10 to 14 days, but mosquito populations do rise and fall seasonally. Incidence usually increases in the late spring and early summer, then declines slightly, and rises again in late August and early September. By the end of the 19th century, with few exceptions, malaria was no longer found in the North. This was because the land was cleared and drained, reducing mosquito breeding sites; the introduction of cattle provided a source of blood other than that from humans, and the cold winters killed off infected mosquitoes. In the South, however, with its warmth, humidity, and breeding sites (ponds, the bottom lands, and the cultivated fields of rice, sugar, tobacco, and cotton), mosquitoes were present year-round and so was malaria, which became endemic. Because of the slave trade, there was a continuous introduction of infections from Africa. In the American South, malaria struck down newcomers, contributed to high infant mortality, and reduced the vitality of the inhabitants. It also made the southerner more susceptible to other disorders. Endemic malaria gave rise to the classic picture of a thin, sallow-skinned, dull, lazy southerner, an image that remained until the 20th century.

Individuals who survive an initial attack of malaria continue to harbor the parasites in the blood for months to years, but because of immunity, they may show no signs of clinical disease. However, such individuals can serve as a reservoir for infecting mosquitoes and thus contribute to the spread of disease. Most adult southerners—black and white—had some measure of acquired immunity, but blacks had a distinct advantage. Ninety percent of those from West Africa were naturally resistant to vivax malaria, and 10 to 20% were endowed with sickle cell trait and, as a result, were resistant to falciparum malaria (see pages 151–152). However, outbreaks of malaria did occur among the slaves and their descendants, because resistance was not absolute: protection from one type of malaria by a specific trait did not confer resistance to the other type of malaria. When slaves were transferred to the South and West from older plantations, they could also develop malaria because their acquired immunity was specific, and thus they were susceptible to the newer strains of malaria.

When federal troops returned from the Mexican War (1848 to 1850), they brought malaria to Texas, New Mexico, Arkansas, and the Atlantic states. During the Civil War, malaria was present, especially in the Union troops from the North who were garrisoned in the South. In the period 1861 to 1866, malaria represented 25% of all reported diseases among the Union troops. There were

1,213,814 cases of malaria and 12,199 deaths among the Union forces. Though malaria was more prevalent among the Confederate troops, it was less fatal, probably due to a degree of acquired immunity. When the Civil War ended and the Union troops returned home, there was an increase in malaria in the northeastern United States, but it was still considerably less prevalent than it was in the South. In the South during the Reconstruction Period, mosquitoes flourished in the fallow agricultural lands, because there were fewer individuals to work in the fields: the labor force had been depleted by death and disease during the Civil War. Although the northern border of the malaria belt began to retreat between 1890 and 1920, it did not fall back as significantly in the South. Indeed, the mortality (per 100,000 people) in Missouri in 1860 was 112, and by 1910 it was 17. However, in Illinois and Wisconsin, where it was 57 and 37, respectively, it had declined to less than 1 by 1920.

As early as 1900, after the discovery of the malaria parasite by Laveran and the discovery of its *Anopheles* vector by Ross and Grassi, it was recognized that to reduce the incidence of malaria one would have to block transmission. This required a reduction in the numbers of both mosquitoes and human infections. Killing mosquitoes, however, was found to be labor-intensive and expensive. In addition, it was soon realized that there needed to be an integrated program involving the screening of beds and houses in malarious areas, the confinement of malaria patients under screens, the treatment of humans with quinine, the elimination of breeding sites by drainage and filling, the use of mosquito larvicides such as oil and Paris green (arsenic trioxide), and education. In 1917 to 1918, the U.S. Army pursued mosquito control in the South where troops were stationed and trained, but it was of limited success. It was clear that ridding the entire United States of malaria was too costly and too difficult to be handled by local governments. From 1933 to 1935 (and during the height of the Depression), federal resources were committed to an antimalaria campaign using the Civil Works Administration, the Emergency Relief Agency, and the Works Project Administration. The Office of Malaria Control (the forerunner of the Centers for Disease Control and Prevention) was established in 1942. Despite this, malaria remained entrenched in the South because southern leaders would not accept the hypothesis of mosquito transmission, and endemic poverty limited local funding for mosquito control. Even the meager efforts at control, such as they were, were interrupted by World War II. After the war, the attempt to eradicate malaria became a national program. By 1946 there was widespread use of the insecticide DDT and "flit guns" loaded with pyrethrum to kill adult mosquitoes. The Tennessee Valley Authority set in place mosquito abatement programs to avoid the problems associated with damming streams and rivers for flood control and hydroelectric power as well as recreational and navigational purposes. In a 10-year period (1942 to 1952), malaria in the United States was reduced to negligible proportions at a cost of $100 million.

Of the three chronic disorders associated with the American South—hookworm, pellagra (a dietary deficiency disease) and malaria—malaria had the most profound effect, since it struck young and old, rich and poor. Although not invariably fatal, it was always debilitating. Malaria was eliminated from the American South by a combination of factors: public health measures, diversified farming, shifts in population (for example, cotton growing moved to the Southwest), mechanization of farming, and an improved standard of living. As malaria receded from the South, so too did the distinctive image of a fever-ridden area populated by poverty-stricken, lazy sharecroppers growing cotton and tobacco.

George Washington's ally. The American Revolution had its battle heroes: Washington, Lafayette, Rochambeau, de Grasse, Pulaski, John Paul Jones, and Nathan Hale. But one hero hardly ever mentioned was a small fickle female named *Anopheles*. Here is how she became Washington's ally and Britain's foe. The Revolutionary War began on 19 April 1775, when a group of colonists fought British soldiers at Lexington, Mass. Ill feelings between the British government and its colonies culminated in Britain's attempt to force the colonies to depend on it for manufactured goods, as well as their restriction of the colonists' rich trade in rum, slaves, gold, and molasses in Africa and the West Indies. From 1776 to 1778, both sides gained victories. New York fell to the invading British, but after the British general John Burgoyne surrendered at Saratoga, the French entered on the side of the United States. In April 1778, Sir Henry Clinton was in command of the British army in the American colonies. He was under orders to take Philadelphia, evacuate the city, and then concentrate his forces in New York City in order to meet the Continental Army under General George Washington. By summer, Clinton began to move his army to New York City by land instead of by sea. This was a serious blunder on his part, for in the march northward to their strategic destination, Clinton's men had moved through the mosquito-infested lowlands of New Jersey. By the time the men were garrisoned in New York, in early August, seven out of 100 soldiers had died of malaria, and many more were ill with fever. One of Clinton's lieutenants wrote, "the troops more sickly than usual and the same for the inhabitants."

On 29 August, six British warships were anchored off the coast of New York to reinforce Clinton's troops, and 2,000 men disembarked—all sick with scurvy (see page 370). To add further to Clinton's problems, 10,000 of his men were relocated to the West Indies, Georgia, and Florida. As a consolation, he was promised reinforcements at a later date. Clinton's strategy was to attack and destroy Washington's Continental army at West Point. However, because the available troops in the West Indies, Florida, and Georgia were now ill with malaria and typhus (see chapter 6), Clinton was forced to redraw his battle plans: he would subjugate the South by landing at Charleston, S.C., and then move inland to capture Virginia. Clinton embarked on 26 December 1778, with 8,500 men. Major General Charles Cornwallis was his second in command, and by May 1779, Charleston and most of South Carolina were under British control. The situation appeared to be so well in hand that Clinton returned to New York City and left Cornwallis in command of this southern contingent. Cornwallis moved his forces northward to Virginia, but by July 1779, two-thirds of his officers and men were unable to march owing to malaria fever. Over the next few months, they continued to suffer from relapses. Still they pushed on. Fever plagued the men and Cornwallis himself as they marched toward Virginia. Although battles, maneuvers, retreats, and advances characterized the next 2 years of the Revolutionary War, it was the particularly severe malaria among the British troops that would make the difference in the future course of the war. Between January and April 1781, Cornwallis' army had dwindled from 3,224 to 1,723, and by May he had only 1,435 men fit and ready for battle. Because Clinton expected Washington to attack New York City, the question was whether Cornwallis should adopt a defensive posture and remain in Charleston or move north to assist him. If he took the latter course, then a port in Virginia would be critical to resupplying his depleted army by ship. Cornwallis chose to remain in Virginia at the port of Yorktown on the River York, but he did so unhappily. He complained of being surrounded by acres of unhealthy swamps and felt that he would soon become prey for the Continental army. The plan was for Washington's army to march south, surround Yorktown, and with a French blockade of the harbor, Cornwallis' escape would be prevented. On 14 August 1781, Washington received word that the French fleet (which had arrived in the Caribbean) was heading for the Chesapeake Bay. Six days later, Washington's army crossed the Hudson River. Clinton expected Washington to attack him at Staten Island, but instead, Washington and his troops headed south for Yorktown where they were to be met by eight French ships sailing from Rhode Island. By 5 September, the French ships had arrived in Chesapeake Bay, defeated the British fleet, and controlled the mouth of the York River. Cornwallis was bottled up. Clinton advised Cornwallis that Washington was not going to attack New York City but instead was moving troops against Yorktown. Clinton also told Cornwallis he was dispatching 4,000 troops from New York City to assist him. Such help never came, because the men that had been garrisoned in New York were decimated by malaria acquired during the previous two autumns. On 17 September and again on the 29th, Cornwallis wrote to Clinton, "This place is in no state of defense . . . if relief doesn't come soon you must prepare to hear the worst . . . medicines are wanted." By medicines, he meant pulverized fever bark (quinine). Cornwallis, with one-third of his besieged forces too sick from malaria and no relief in sight, surrendered at Yorktown on 19 October 1781. After 7 years, the Revolutionary War was over. Great Britain gave up all hope of conquering the colonies, and on 3 September 1783, it recognized the new republic in the Treaty of Paris. It may have been the superior tactical strength of the combined American and French forces that brought victory to the Continental army and independence for the United States, but Washington's army had a secret ally: the malaria-carrying *Anopheles* mosquito.

Hookworm and the American South

Yellow fever and malaria are microparasites, whereas hookworm is a macroparasite (see Fig. 1.3B). Both the adult male and female hookworms live in the small intestine, where they attach to the wall, abrade it, and suck blood (Fig. 15.12). After the eggs of the threadlike female worms (~10 mm in length and 0.4 mm in diameter) are fertilized by the smaller male (~7 mm in length and 0.3 mm in diameter) they pass out with the feces; for the juvenile worms to hatch out of the egg requires wet, warm, and sandy soil. When the juvenile worm comes in contact with the skin, it burrows into it, enters the bloodstream, and finally gets into the digestive tract, where it matures, feeds, and reproduces. The adult worms can live in the small intestine for up to 10 years, and a female can produce about 9,000 eggs each day.

In Europe, hookworm infections were considered an occupational disease of miners. In Hungary in 1786, 1,200 coal miners were attacked by it, and in 1802 and again in 1820, there was an outbreak in the French coal mines.

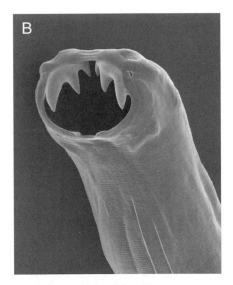

Figure 15.12   Hookworms (white and threadlike) attached to the wall of the small intestine (A), where they suck blood using their razor-like teeth (B). (Panel A courtesy of Wallace Peters. Panel B courtesy of Dennis Kunkel Microscopy, Inc.)

Similar outbreaks occurred in Germany and in the mines of Cornwall, England. In Cornwall it was called "Miners' bunches" due to a telltale sign of infection: inflammation at the site where the juvenile worms had penetrated the skin. In 1880, during the construction of the St. Gotthard tunnel between Switzerland and Italy, 10,000 workers developed severe hookworm anemia, and there was an exceedingly high fatality rate, especially among the Italians. Disease was the result of infected miners' passing eggs in their feces and then having skin contact with the rungs of ladders soiled with mud and feces.

In the 18th century, the clinical picture of hookworm infection was described among African slaves in the West Indies. It was unknown among the early European settlers in the Americas but was commonly seen in their African slaves. By 1808, it was observed that anemia and geophagy (literally "dirt eating") were common in both slaves and poor whites living in the cultivated lowlands of the American South where the soil was sandy and the climate was humid and warm. Because the blood-feeding hookworm produces anemia, it was referred to as the "vampire of the South."

The ancient Chinese medical literature described hookworm disease as the "able to eat but lazy to work yellow disease." The Ebers papyrus of 1550 BC refers to the characteristics of infections in ancient Egypt, and the Greek physician Hippocrates in 450 BC described it thus: "the skin is yellow, the intestine disturbed and the person has an appetite for eating clay." Depending on the particular species, an adult hookworm can cause a daily blood loss of 0.03 ml to 0.15 ml. In a typical infection, blood loss can be as large as a cupful. Victims of hookworm disease show a characteristic set of symptoms: anemia, diarrhea, fever, and sometimes an insatiable appetite to eat clay or dirt. (It has been claimed that clay eaters are attempting to replace their iron stores since red clays are high in iron content.) The individual is lethargic, the complexion is sallow, and physical development is slowed, as is mental acuity. It is particularly prevalent in children, where it may result in grotesque deformities.

Humans can be infected by two kinds of hookworms: *Ancylostoma duodenale* and *Necator americanus*. Hookworm infections probably existed among the ancient Egyptians, and before 1838, the disease had been described from Italy, Arabia, and Brazil. In 1838, the Italian physician Angelo Dubini discovered and named the minute whitish worm he found in the small intestine of a woman dying in the hospital, *A. duodenale*; however, he did not associate the presence of the worm with a specific disease. In 1843, when he found many worms in the intestine of a man dying with pneumonia, as well as in 20% of the cadavers he autopsied, he suspected that this worm was parasitic. Dubini named the worm *Ancylostoma* (*ancylo* [hooked], and *stoma* [mouth]) because the worm had teeth and *duodenale* because of its location in the duodenum of the small intestine. In 1847, Franz Pruner, a German working as a personal physician to the Abbas Pasha of Egypt, described the same worm. However, it was not until 1853 that Wilhelm Griesinger and Theodor Bilharz associated the presence of these minute roundworms with anemia. In 1878, Battista Grassi and his associates in Italy identified hookworm eggs in the feces of infected patients. Up until the 19th century, it was believed that the disease was spread by ingestion of eggs through fecal contamination, since human volunteers could be infected by feeding them fully developed eggs but not hatched-out juvenile worms. Indeed, it was found that the infected volunteers were passing eggs in a month's time. But in 1893, when Arthur Looss accidentally dropped some juvenile worms onto his hand and then developed itching and redness, he began to suspect that the juvenile hookworm could directly penetrate the skin. This hypothesis was proven experimentally in 1902 by C. A. Bentley, a medical officer working on a tea plantation in India. Bentley took two samples of fecal-contaminated soil: one was associated with "ground itch" due to the presence of hookworm juveniles, and the other was soil contaminated with feces but without worms. A portion of each soil sample was heated and dried for 8 hours, and the remainder was left unheated. Each soil sample was applied to the wrists and held in place with a bandage. "Ground itch" developed only in the unheated soil sample containing living hookworm juveniles.

For hookworm to be an important parasite requires several conditions: lack of sanitary facilities for disposal of feces, bare feet, and the presence of a moist, warm climate and sandy soil. All were present in the American South, and not in the North. Initially, it was assumed that the hookworm in the United States was *A. duodenale*; however, in 1902, Charles W. Stiles identified differences in the parasite and named it *Necator americanus*. By 1902, one health worker had boldly claimed that hookworm was an endemic disease of the American South. Indeed, this was substantiated by further work. Although Stiles was ridiculed as the discoverer of the "germ of laziness," he persisted in calling for a public health program to control the disease. By 1910, the Rockefeller Sanitary Commission for the Eradication of Hookworm, with a grant of $1 million, began its work. It was dedicated to the proposition that sanitary privies, the wearing of shoes, and health education would eliminate the disease, and by 1914, together with chemotherapy (mebendazole), it succeeded in significantly reducing the incidence. However, the greatest reduction in the number of infections in the American South required persistent and multifaceted control programs, which did not occur until 1950.

Hookworm remains a worldwide problem, with an estimated prevalence of 630 million people, largely due to social customs, inertia, and economic constraints. Hookworm is most prevalent in rural, poorly nourished individuals with low resistance to infection. In the well-nourished individual, there is resistance, sometimes called "self-cure," evidenced by lower numbers of worms upon reinfection and lower egg production by the worms.

Where did hookworm come from? It is suspected that humans first acquired *A. duodenale* at the time of the domestication of the dog (3000 BC) in the Fertile Crescent. From there it spread with the migration of the human population across the continents. *Ancylostoma* originally infected people in the warmer, non-arid areas north of the Tropic of Cancer. Mediterranean, Middle Eastern, Indian, and Asian immigrants brought the parasite to the Americas. *N. americanus* had as its original range the region south of the Tropic of Cancer in the South Pacific Islands, Indonesia, southern Asia, and Africa. It was probably introduced into the Western Hemisphere with the importation of slaves from Africa. Today, *Ancylostoma* is found in southern Europe, North Africa, and northern Asia, whereas *Necator* predominates in the Western Hemisphere and equatorial Africa.

If clinical descriptions of hookworm disease stem from 1742 in the West Indies, and from the American South 100 years later, why did it take nearly 200 years to eliminate the disease from the southern United States? Hookworm eggs are microscopic, so the cause of the infection may be missed. But this cannot explain the situation completely, since the adult worms, minute though they are, were seen in an autopsy as early as 1840. A more plausible reason is that the disease was not epidemic, and hookworm anemia was masked by the abundance of malaria as the primary cause of anemia in humans living in the American South. More important, however, there was no motive for eradication until 1902. After that time, the United States developed a distinct national pride and a vision for itself. This resulted in two kinds of proposals: physical exclusion of the deviants or their integration into the society. It was the latter approach that sought to bring the poor "white crackers" and the dirt eaters as well as blacks—the so-called aberrant and distinctive groups—into the American nationality, and this probably provided the medical impetus to look for the cause of the group's distinctiveness, as well as the discovery of the cause and cure for hookworm. In addition, social reform played a role in the public health measures that eventually made the South truly American.

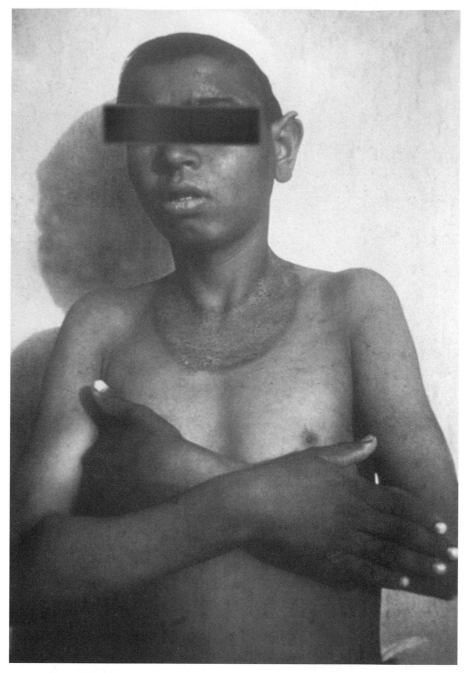

Figure 16.1 The sign of the butterfly (pellagra) in a young girl (Courtesy CDC).

Chapter 16

Plagues without Germs

Paul de Kruif (author of *Microbe Hunters*) caught the essence of late 19th century thinking about plagues:

> The vast exciting Battle of the Germ Theory was on! Every medical man and Professor of Diseases who knew—or thought he knew—the top end from the bottom of a microscope set out to become a microbe hunter. Every week brought glad news of the supposed discovery of some new deadly microbe. . . . One enthusiast would shout across the continents that he had discovered a kind of pan-germ that caused all diseases. . . . But in the midst of the danger that foolish enthusiasm would kill the new science of microbe hunting, Koch kept his head. . . . "One germ, one kind of germ only, causes one definite kind of disease— every disease has its own specific microbe, I know that."

This confidence in one germ, one disease led to the identification and control of many infectious diseases; however, such hubris also fueled the unscientific bias that hampered the uncovering of the causes of deficiency diseases—plagues without germs.

The Red Plague

The red plague, pellagra, is a disease almost unheard of today but, in its time, it was a killer. It was first described in Spain, in 1735, as Mal de la Rosa, literally "the red disease," because it generally appeared first as a skin rash that covered the hands and feet and sketched an ugly butterfly pattern across the neck (Fig. 16.1). This mark of the butterfly became a stigma that set those who were afflicted in a caste apart, reminiscent of the treatment accorded lepers. In the early stages, the reddening of the skin might resemble a sunburn, or it might be confused with the inflammation due to contact with poison oak, but when the skin crusted and peeled away, showing the smooth and shiny skin below, then a diagnosis of pellagra was certain. One physician described it this way: "If you ever shake hands with one of these people you never forget the sensation. The skin of the hand is dead, harsh and rough." The flaming of the skin was the signature of the disease, but a loss of balance, a staggering gait, and senseless mutterings also characterized the disease. The sufferer was weak and melancholy. There were other changes: the tongue became reddened, and the afflicted person had a burning sensation in the mouth, as well as diarrhea. Some went insane, and others died. The clinical symptoms of pellagra were frequently referred to as the 4 D's: dermatitis, diarrhea, dementia, and death.

Pellagra was a horrible and obstinate disease that usually afflicted the poor. The disease was not confined to Spain; indeed, the word "pellagra" is derived from the Italian words *pelle* meaning "skin" and *agra* meaning "rough," because that's what the peasants of Italy called it. Pellagra was associated with people who had a diet consisting mainly of corn. In 1863, the French physician I. Salas came to the conclusion that the cause of pellagra was not the corn itself but "spoiled" corn infested with the parasite verdet. Salas and other European physicians sought a cure, but none was found. Pellagra had all the epidemiological hallmarks of an infectious disease. Indeed, using this line of reasoning, and the emergence of the germ theory of disease in the 1880s, physicians sought pellagra's cause in microbes and searched for its vector. Louis Sambon at the London School of Hygiene attributed pellagra to "unsound" corn and claimed that its vectors were bloodsucking insects, blackflies, and buffalo gnats. But neither microbe nor vector could be found.

In the early 1900s, just at the time that malaria, yellow fever, and hookworm were on the wane, pellagra struck the American South with a vengeance. It was a strange kind of "spring fever" that appeared just before mid-February, gradually grew worse until May or June, and then began to decline. It was first recognized among the insane hospitalized in Alabama and among tenant farmers and mill workers. Those in the butterfly caste were the pariahs of the American South, and the disease was considered to be peculiar to that region, since its distribution was identical to that of the Confederacy, with a corner of Missouri and Kentucky tacked on. The image of pellagra and the American South did not fit with attempts by political and business leaders to portray their southern states as places of wealth, power, and prosperity. Many southerners, embarrassed and ashamed, denied the existence of pellagra. At the end of 1909, pellagra was an epidemic reported from 26 states in the United States, although it had probably existed at a fairly low level in the Deep South much earlier. An expatriate from the South, Walter Page, who had made his mark in the newspaper business in New York and Boston, urged support for research, and the U.S. Public Health Service entered the fight against pellagra in 1909. Three years later, the number of pellagra cases had not decreased but, rather, increased. In the American South, the fatality rate from pellagra could be as high as 40%, and in 1916, it was the second leading cause of death in South Carolina. The continuing spread of pellagra became a national concern. Despite all efforts, no cause for the disease was found, there was no treatment for it, and the disease did not develop in animals injected with tissue or urine extracts from pellagrins (people suffering from pellagra). From 1730 to 1930, it is estimated that pellagra caused half a million deaths worldwide. Between 1900 and 1947, there were roughly 3 million cases in the United States, with 100,000 deaths. By 1950, however, pellagra had been eradicated.

Searching for the "Germ" of Pellagra

The uncovering of the cause of pellagra is much like a detective story. The principal detective was Joseph Goldberger, whose studies remain a model of epidemiological research. Goldberger was born in Hungary and came with his parents to the United States when he was 6 years old. At age 16, when he finished grammar school in New York City, he began to study civil engineering at City College of New York. After hearing a lecture by the eminent physician Dr. Austin Flint, he dropped engineering for medicine and entered Bellevue Medical College in 1892; he graduated with honors in 1895. In 1899, Goldberger joined the U.S. Public Health Service. His first assignment was at Ellis Island, where he was a junior medical officer looking for disease among the immigrants. His next post was at Reedy Island, a quarantine station for ships bound for Philadelphia. The vessels had to be examined, and their holds were fumigated to kill rats, lice, and mosquitoes in order to prevent diseases from entering the country. Subsequent assignments familiarized him with yellow fever, malaria, typhoid fever, dengue, parasitic worms, typhus, measles, and diphtheria. In the process of working with these infectious diseases, he contracted but recovered from both yellow fever and typhus. Goldberger had become a skillful microbe hunter.

In his fifteenth year as a Public Health Service physician (1914), Goldberger received a letter from the Surgeon General. "I would like to have you take charge of pellagra investigation as soon as possible and am writing you in order that you may know of my plans. It is needless to state that I shall assist you in the way of furnishing assistants." At that time a staff of very able men in the Public Health Service had devoted 5 years to the pellagra problem. In their report of 1912 to 1914, the Pellagra Commission had concluded, "pellagra is, in all probability, a specific infectious disease communicable from person to person by means at present unknown." Despite this conclusion, however, they found no infectious agent.

Goldberger, the microbe hunter, did not set up a laboratory or unpack his microscope. Instead, for a whole year (1914 to 1915) he traveled throughout the "pellagra belt" of the American South, visiting mainly state insane asylums, where the number of pellagra cases was highest. At South Carolina State Hospital, he discovered a paradox: the nurses, orderlies, and doctors who treated the patients slept in the same rooms and had direct skin contact with those with pellagra, yet not a single one of them came down with the disease. It appeared that pellagra was not a contagious disease spread by direct contact.

In Milledgeville, Ga., at the Georgia State Sanitarium for the Insane, Goldberger suggested to the physicians that pellagra occurred only among the poorest and that the difference between the poor and the not-so-poor was the food they ate. The physicians at the insane asylum countered this supposedly simple notion, stating that at the Sanitarium, both nurses and orderlies received food prepared in the same kitchen as those with pellagra. Goldberger investigated the dining habits: the orderlies, nurses, and doctors did receive their food from the same kitchen as did the patients, a diet of corn meal mush, hominy grits, and cane syrup, but they had it supplemented with nice cuts of meat, vegetables, and milk, while the patients did not.

In the summer of 1914, Goldberger went to Jackson, Mississippi, where, at the Methodist Orphan Asylum, 168 out of 211 children had pellagra, and practically all the cases were in children between the ages of 6 and 12. This peculiar age distribution was puzzling until the distinct differences in living conditions among the children were investigated. The little ones, under 6 years of age, received milk, but the older ones did not. The older children, over 12 years old, were rewarded for doing small chores with a higher ration of meat; these older children, always hungry, as are all teenagers, also grabbed and stole food. In short, the older children supplemented their basic diet of biscuits, hominy grits, corn mush, molasses, and sowbelly with fresh animal components such as milk, eggs, butter, and lean meat.

Goldberger suggested a simple experiment: supplement the diets of those who were most likely to get pellagra. At the orphanage, every child under 12 years of age got 14 ounces of milk a day, and those under 6 years of age got 20 ounces. Eggs were added to the daily menu, and fresh meat was now to be served 4 days a week instead of once a week. Remarkably, pellagra faded away from the orphanage, and no new cases appeared.

In 1915, Goldberger proposed a more convincing experiment, using the convicts at the Mississippi Rankin Prison Farm, especially those serving life terms. This group seemed ideal: they were used to being watched, they endured restrictions, time was of little consequence to them since they were imprisoned for life, and they could be rewarded in a substantial way by receiving a full pardon. The Governor of Mississippi approved the use of a dozen convicts, and the experiment began in April 1915 at the Rankin Prison Farm, where pellagra never had previously occurred. The food for the experimental group was a monotonous diet, called 3M, because it consisted of meat (salt pork), meal (cornmeal), and molasses. It contained no red meat, no fresh vegetables, and no milk, The daily menu was:

Breakfast: Biscuits, fried mush, corn syrup, grits, brown gravy, coffee with sugar
Lunch: Corn bread, cabbage, sweet potatoes, grits, molasses
Dinner: Fried mush, biscuits, rice, gravy, corn syrup, coffee and sugar

The men were placed in clean, screened quarters and were given light physical work. The controls for the experimental group were 80 convicts who were kept under similar conditions except for their diet, which included an ample supply of fresh vegetables from the prison farm. After the first month, the convicts on the unsupplemented diet, the experimental group, had headaches, dizziness, and a burning sensation in their mouths, but no rash. Indeed, it took 6 months before the telltale butterfly rash of pellagra appeared on the necks and hands of five convicts. None of the controls developed pellagra.

Then, in collaboration with an expert statistician of the Public Health Service, surveys on the geography, economics, diets, and habits of those living in the Carolinas and other southern states were carried out. It was found that the poorer classes were dependent on foods that did not include red meat because of the depression of 1917, when the price of animal protein food rose 40% over that of all other foods. This situation was most acute in the Southern cotton mill towns.

An epidemiological study conducted by Goldberger and his associates from 1917 to 1921 showed that the lower the income, the greater the incidence of pellagra. Towns in the Cotton Belt were ravaged by pellagra, whereas in towns where cotton was less profitable and the people had greater access to red meat, milk, and fresh vegetables, there was no pellagra. Pellagra was clearly tied to economics: if the price of cotton fell, then wages also fell, and the health and nutrition of the workers, especially those working in the mills, suffered too. Federal protests about the poor economic conditions in the American South incensed southerners; they denied the problem existed and refused federal aid. Southerners would not accept the harsh reality of poverty during the 1920s and claimed their region was being maligned. The mill workers and tenant farmers of the South came to be excluded from the remainder of the community, and the lower classes became more segregated from the mainstream society. The South was depicted as "a region full of little else but lynchings, shootings, chain gangs, poor whites, Ku Kluxers, hookworm, pellagra and a few decayed patricians whose chief intent was to deprive the poor and the spiritual singing Negro of his life and liberty" (E. W. Etheridge, *Disease and Distinctiveness in the American South*). This harsh criticism further hardened the southerners' position, fostered isolationism, and contributed to the perception that the American South and its people were distinctly different from the rest of the country.

In his report to the Surgeon General, Goldberger remained cautious about ascribing pellagra to a dietary deficiency. Indeed, in 1918, a distinguished and influential nutritionist, Elmer V. McCollum of Johns Hopkins University, held the view that pellagra was a contagious disease. Little wonder that as late as 1928 there were 120,000 people in the United States who suffered an attack of pellagra. Why? The answer was that despite indirect evidence for pellagra as a dietary-deficiency disease—a "hidden hunger"—

few were prepared to accept such a simple explanation, and there was little incentive to combat its causes. Goldberger wrote, "the problem of pellagra is in the main a problem of poverty. Education of people will help; but improvement in basic economic conditions alone can be expected to heal this festering ulcer of our people."

To some, including McCollum, the correlation of diet with pellagra did not provide convincing evidence of a cause-and-effect relationship. McCollum claimed that "poor diet predisposes to infection as do unsanitary conditions." The contagionists (see page 165) held to the belief that pellagra was an infectious disease, and they were skeptical, critical, and scathing in their comments on the results of the Goldberger experiments with diets. Therefore, beginning in 1916, Goldberger decided to carry out an experiment that would be both daring and convincing and that would prove that pellagra was not an infectious disease. He began by having "filth parties." Twenty friends, his wife, and he volunteered to be the susceptible animals to be injected with the blood from a pellagra patient. Gelatin capsules were filled with the skin scrapings, urine and feces from pellagra patients, and each volunteer swallowed a capsule. None became ill with pellagra. In 1918, Goldberger wrote, "The hypothesis that pellagra is of dietary origin, not contagious, is greatly strengthened."

Preventing and Curing Pellagra

Today, it is recognized that pellagra can be prevented by adding fresh meat, milk, eggs, and vegetables to the diet. These foods contain what Goldberger called the P-P—pellagra-preventive—factor. The red plague of pellagra was a plague of corn: hominy grits, corn mush, and corn bread all lack the P-P factor. Once the dietary cause was appreciated, the key to understanding the disease of pellagra had been found. But though the means of prevention and cure was clear, identification of the specific factors involved required chemistry and the use of laboratory animals. Goldberger then got lucky: he discovered that rats and dogs would show pellagra-like symptoms—black tongue and an inflammation of the skin (dermatitis)—if kept on a diet deficient in the P-P factor. The diet (called Basic Diet 123) consisted of corn meal, cowpeas, casein, sucrose, cottonseed oil, cod liver oil, sodium chloride, and sodium carbonate. It required only 60 days on this diet to produce pellagra in dogs. The P-P factor was heat resistant and was contained in dried brewer's yeast. Yeast would both prevent and cure pellagra. Simply sprinkling dried brewer's yeast on Basic Diet 123 prevented black tongue and death in dogs. Goldberger demonstrated that pellagra-preventive foods were both inexpensive and available in the American South and that brewer's yeast worked best. This discovery saved thousands of lives in 1927, when the Red Cross distributed brewer's yeast to pellagra victims after the Mississippi River flooded, driving people from their homes and destroying crops.

Nine years after Goldberger died (in 1929) at the age of 55 from cancer, the P-P factor was isolated and chemically characterized. In 1937, Conrad Elvejhem and his colleagues at the University of Wisconsin attempted to identify the "anti-black tongue factor" in liver extracts. Previously, they had found that when liver extract was added to the Basic Diet 123, dogs did not suffer from severe weight loss and did not develop black tongue. In 1938, they isolated the factor from liver extracts and determined its structure to be nicotinic acid. They were also able to show that nicotinic acid (or its amide) would completely substitute for the liver extract. Yet Elvejhem and his colleagues cautiously concluded, "That a deficiency of this material may be the cause of black tongue is most interesting. Clinical trials are needed to show the same effectiveness in human pellagra." These human experiments were carried out in 1939 by T. D. Spies and associates. Nicotinic acid and its amide were renamed niacin. Niacin, a water-soluble B vitamin, was the specific preventative of black tongue and pellagra, and it was Goldberger's P-P factor.

Oddly enough, it was the Great Depression that changed dietary habits and eliminated pellagra, since people with time on their hands grew their own food, primarily in vegetable gardens. World War II reinvigorated the economy of the American South, and bread and cornmeal were enriched with synthetic vitamins, including niacin. Pellagra vanished. With the waning of pellagra, Southerners appeared to be no lazier than the rest of the U.S. population, and their distinctiveness and insularity decreased.

Pellagra clearly shows that the proposal that microbes cause all diseases is misleading. Some diseases are related to the food eaten and to economic conditions. Understanding the cause of a disease might eliminate it more easily; simply identifying the cause of a disease, however, does not always result in an immediate cure. The cure must be endorsed by the community at large—health workers, as well as those who are susceptible—and the solution must be applied in a cost-effective and socially acceptable manner. A disease cure without acceptance by the public may never realize its full potential.

Why was pellagra associated with a corn-based diet in the United States but not in Mexico? In ripe corn, niacin is a constituent of the coenzymes NAD (nicotinamide adenine dinucleotide) and NADP (nicotinamide adenine dinucleotide phosphate), which are in turn stored in a complex with starch. In this form they are not water soluble and cannot be absorbed. These bound forms release niacin only on treatment of the mature corn with alkali, that is, by cooking with limewater. Prolonged boiling in water releases only 20% of the niacin from corn meal; in 1% limewater, at 80°C, however, all the niacin is released. The Mexican practice of soaking the corn meal in hot limewater before making tortillas releases the bound niacin and prevents pellagra.

Why does a deficiency of niacin lead to the disease symptoms of pellagra? Niacin, as a constituent of the coenzymes NAD and NADP, is needed for normal oxidative metabolism of sugars. (A coenzyme is an accessory factor that is itself not a catalyst; however, it is responsible for activating the catalyst or enzyme by binding to it. Without the coenzyme, the enzyme remains in a functionless state.) The hidden hunger of pellagra is that, without niacin in the diet, food cannot be utilized as a source of energy because sugars are not metabolized. The cells in the body die or do not function properly, and illness results.

Uncovering the mystery of the red plague had a significant additional benefit: it led to the discovery of antimetabolites, that is, drugs designed to inhibit bacterial growth by interfering with their coenzymes or enzymes. Antimetabolites (sometimes erroneously called antibiotics [see page 252]) "trick" the bacteria into accepting a nonfunctional analog of the coenzyme, and this "poisons" their vital functions. Isoniazid and pyrazinamide, drugs used in the treatment of tuberculosis (see page 299), are both structural analogs of niacin (Fig. 16.2) and work by preventing the natural substance

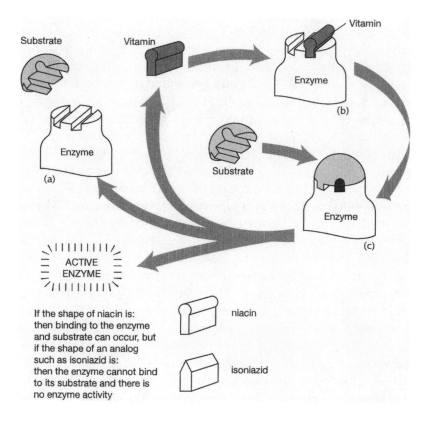

Figure 16.2   Enzyme activity depends on the shape of molecules. The enzyme must fit the substrate (a) in order for there to be enzyme activity. Vitamins (b and c) act as adaptor molecules and are necessary for some enzymes to be active, i.e., to bind to the substrate. Since vitamin analogs are structurally similar to the natural vitamin, they can bind to the enzyme, but once bound the enzyme cannot bind to its natural substrate.

from combining with a bacterial enzyme that would bind NAD; the result is that these drugs inhibit the growth of *Mycobacterium tuberculosis*.

The White Rice Plague

In 1611, the governor general of the Dutch East India Company in Indonesia described a neurological disorder characterized by paralysis of the hands and feet. It was called by the native name *beriberi*, meaning weakness, or "I cannot," because the afflicted person is too sick to do anything (Fig. 16.3). In 1853, trade with Japan and the West opened up when Commodore Mathew Perry landed in Yokohama harbor with a fleet of four gunboats and negotiated a trading treaty. By 1858, Japan had signed similar treaties, under "gunboat diplomacy," with Britain and other European nations. During this early period of Japan's Westernization, doctors from Europe encountered a mysterious illness that some called the "national disease of Japan." The disease manifested itself as a progressive loss of strength, loss of appetite, swelling of the ankles and thighs with great pain in the leg joints, loss of the ability to speak, emaciation, and death from asphyxia and convulsions. Because the loss of sensation in the feet produced a characteristic ankle drop, the disease was called *kakke*, meaning "leg disease." Treatment for kakke was acupuncture

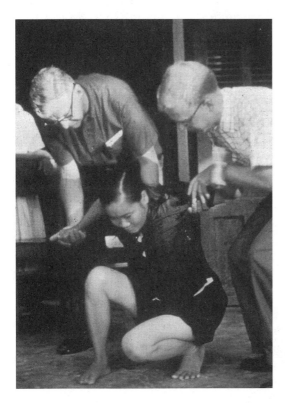

Figure 16.3 This young woman is unable to stand on her own feet because she suffers from beriberi. (Courtesy of the Armed Forces Institute of Pathology.)

or scorching and blistering of the skin in order to draw out the toxic gases in the body, a practice reminiscent of bloodletting, used in Western medicine for centuries. The most common belief was that the disease was a result of a miasma that rose up from the damp soil. The Japanese government recognized that if the country was to fend off military interventions from the West, such as those that had forced Japan to trade with the West, it would need a modern navy with well-trained physicians. Kanhiro Takaki was one of these naval physicians, and in 1870 he observed that 75% of all the patients in the Tokyo Naval Hospital had kakke and were unfit for duty. Hospital treatments for these sick sailors included bloodletting, purgatives, digitalis, and strychnine, yet none effected a cure.

Takaki was sent to Britain for further medical training, and upon his return to Japan (1882) he set about to determine the cause of kakke, which he now recognized to be identical to beriberi. Since Takaki could find no correlation between this disease and the crowding on the ships, he quickly discarded the miasmatic theory of infection, but he did note that the food the Japanese sailors ate was distinctly different from that of the British, especially in its protein content. The high incidence (>55%) of "foot drop" among Japanese sailors during a training cruise between Japan and New Zealand led Takaki to modify the diet of the naval cadets to include more meat, milk, bread, and vegetables and less rice. With this new diet, the foot drop of beriberi did not occur, and Takaki "logically" concluded that the disease was a result of a protein-deficient diet. Few Western medical scientists in the 1880s and 1890s, however, subscribed to Takaki's claims. Instead, following the then-prevalent germ theory of disease, they assumed that beriberi was due to a toxin produced by a microbe.

In 1902, the British began a study of beriberi in Malaysia. The highest incidence of the disease occurred in immigrants from China who worked in the tin mines. Indeed, in 1909 there were 16,716 cases of beriberi in Malaysian hospitals, and of these, 97.6% were Chinese. At first, it was suspected that the cause was the presence of arsenic in the tin ore. When it became apparent that beriberi occurred in Chinese prisoners who'd had no contact with tin ore, the arsenic-poisoning theory of beriberi had to be discarded. The Tamils, immigrants from India, were another ethnic group living in Malaysia; in this group, the incidence of beriberi was 0.5%, yet both the Chinese and the Tamils had rice as the main component of their diet. Could the difference in incidence of beriberi in the Tamils and the Chinese be due to a genetic difference in resistance to a toxin in rice? Two British physicians, Hamilton Wright and Leonard Braddon, carried out a series of dietary experiments in Malaysia and found that the differences were not genetic but involved the manner in which the rice was prepared. Although both ate white rice, the Tamils ate rice that was parboiled. (Parboiling rice consists of soaking and drying rice before it is milled to produce white rice.) They postulated that white rice contains a water-soluble toxin produced by a fungus growing on it, and that parboiling destroys the toxin. Despite the strong implication that a rice diet was associated with beriberi, no toxin could be found.

In 1886, the alarming number of cases of beriberi among the troops sent to Sumatra to subdue pirates who would not accept Dutch rule in the Dutch East Indies prompted the government of Holland to send a scientific team to Indonesia (then Java) to find the germ causing beriberi. Cornelius Pekelharing, Professor of Pathology at Utrecht, headed the team. A member of the team was 28-year-old Christiaan Eijkman, who had studied medicine at the University of Amsterdam on a military scholarship and then worked with Koch in Berlin, as had Pekelharing. After 8 months in the Dutch East Indies, the Pekelharing team wrote its report: "Beriberi has been attributed to insufficient nourishment and misery but the destruction of the peripheral nervous system . . . is not caused by hunger or grief. The true cause must be something coming from the outside, but is it a poison or an infection?"

The search for an infectious agent was a torturous one. Sometimes bacteria could be found in the blood of those with beriberi, and at other times it could not. Convinced that there must be a microbe, they cultured a spherical-shaped bacterium (a micrococcus) from beriberi patients and then repeatedly injected these bacteria into dogs and rabbits before there were any signs of nerve damage. They concluded that beriberi was an infectious disease of an unusual character, since a single infection was insufficient to produce disease, and many injections were needed to overcome resistance. They speculated that beriberi was an airborne infection caused by a micrococcus that survived only under hot and humid conditions. Satisfied, Pekelharing returned to Holland but persuaded the Dutch government to allow Eijkman to remain to study transmission in animals. Working in a small lab in Java, Eijkman tried many times during 1888 to produce beriberi in rabbits, dogs, and monkeys by injection of cultures of the micrococcus; each time, save for one instance, he failed. Then, in 1889, in what turned out to be a stroke of luck, he switched from rabbits and dogs to chickens as his experimental subjects. Why he made the switch to chickens has never been revealed, but it is clear that these feathered friends were pivotal to understanding the cause of beriberi.

Eijkman's chickens, after 3 to 4 weeks in the animal facility, showed the polyneuritis characteristic of beriberi. Because the chickens that had been injected with the bacteria were kept in close proximity to chickens that were not inoculated with bacteria, Eijkman thought the entire facility was infected. But then for some reason (still unexplained) he began to suspect there must be another explanation. He discovered that the animal caretaker had persuaded the cook at the military hospital, without Eijkman's knowledge, to provide him with leftover cooked white rice for feeding the chickens. These were the birds that developed beriberi, and inoculations with bacteria had nothing whatsoever to do with the appearance of disease. When the chicken diet consisted of brown rice, however (a new cook who was employed at the military hospital refused to give the caretaker leftover white rice), there was no beriberi. But Eijkman the microbe hunter did not recognize the significance of the differences in diet. He wrote, "cooked rice favored conditions for the development of microorganisms of a still unknown nature . . . for the formation of a poison causing nerve degeneration."

What is the difference between brown and white rice? In 1870, steel rollers were invented for milling rice, and tons of rice could be milled using steam power. The milling process removed the outermost covering—the pericarp, or husk. Polished rice, produced by milling, yielded an attractive white product that did not spoil as readily as rice prepared by other means, and as a consequence the usable yield from each harvest was increased owing to less spoilage; larger numbers of people could be fed, and the cost declined. The polished or white rice was quickly adopted in the Far East, especially in crowded urban areas and on naval vessels. It was also used in institutional diets such as those in prisons.

Eijkman could find no microbe associated with white rice, so he turned his attention away from germs and focused on the toxin theory, believing that polished rice contained a toxin and that the husk contained an antitoxin. After many years of study, he found neither toxin nor antitoxin. In 1895, Eijkman had an important conversation with Adolphe Vorderman, a government physician in Java. Vorderman told him that beriberi was almost always confined to prisons where the diet consisted of cooked white rice. By 1897, Vorderman was able to show that the incidence of beriberi in Javanese prisons was related to the amount of husk retained in the rice: more husk, less beriberi; less husk, more beriberi. Meanwhile, Eijkman had returned to Holland, where he continued to use chickens as experimental subjects. Over the next 5 years, he carried out extensive trials with rice—both white and brown—using various methods of preparation, and fed it to his chickens. With cooked white or table rice, the chickens developed polyneuritis, but with brown rice they did not. These results convinced Eijkman, now a Professor of Medicine at Utrecht University, and his colleague Pekelharing, that germs played no role in beriberi.

In 1900, a Dutch military physician, Gerrit Grijns, working in Java, confirmed Eijkmans's work, i.e., that rice bran cured polyneuritis (beriberi) in chickens; he then made a great intellectual leap in suggesting that beriberi was a nutritional deficiency disease. He also noted that the disease affected peripheral nerves only, not muscle, and that what was required to prevent it was a substance in small amounts—a micronutrient. From bran (husk) he isolated a factor that was water soluble and lost its potency when heated to 120 to 130°F—it was a heat-labile substance. (In rice that is parboiled and dried, most of the factor diffuses inward and is not lost in the subsequent process of mechanical milling and cooking.)

In 1911, Casimir Funk, a Polish scientist working at the Lister Institute in London, isolated what he believed was the antineuritis substance from rice bran. Since it was an amine and was active (i.e., vital) he coined the term "vital amine" or "vitamine." The material Funk isolated was actually nicotinic acid amide, the antipellagra factor, not the antineuritis factor, but the term he coined, "vitamine," remains in use today (without the "e"). Although Casimir Funk coined the term "vitamin"—based on his belief that these substances were "vital amines"—we know today that vitamins are not necessarily amines but substances vital to the proper function of the body's cells. The current definition of a vitamin is an organic substance that is required in very small amounts for proper cell function and that cannot be manufactured in the body. Consequently, vitamins must be supplied to the body from outside sources, most commonly in our food. Vitamins function as accessory substances to activate specific enzymes.

In 1926, two Dutch chemists, B. Jansen and W. Donath, working in Eijkman's old laboratory in Java, isolated crystals of the actual antineuritis factor. This was truly a tour de force since there is only a teaspoonful of the vitamin in a ton of rice bran. They named the vitamin vitamin B_1, or thiamine.

Preventing and Curing Beriberi

Extensive studies by Eijkman showed that the husk alone contained the substance that was the "antineuritis factor." In 1929, Eijkman received the Nobel Prize for his discovery of the cause of beriberi and the means for its prevention. Beriberi has been virtually eliminated by retaining the husk on the rice grain, or by adding vitamin supplements, but it does reappear occasionally. Beriberi was the main cause of death among prisoners of war of the Japanese during World War II, especially those who surrendered at Corregidor after the Battle of Bataan and who survived the death march. In the Philippine prison camps, the disease appeared within 3 months—exactly the interval seen by Eijkman and Vorderman in Java half a century earlier. Beriberi is also frequently seen among alcoholics who lack solid food in the diet and who also require more thiamine to metabolize the large amounts of alcohol consumed.

What is the role of thiamine? Thiamine (in the form of thiamine pyrophosphate) is a cofactor for three enzymes (pyruvate decarboxylase, alpha ketoglutarate decarboxylase, and transketolase) necessary for the normal cell metabolism of sugars and alcohol. A deficiency of thiamine leads to malfunction in cell activity and cell death, and the result is disease. The required daily allowance for thiamine is 1.2 mg for men and 1.1 mg for women.

The many-feathered story of identification of the cause of beriberi once again demonstrates how, on occasion, even the best prepared scientific sleuth can be handicapped by adherence to prevailing dogma—germs as the cause of disease. If the germ theory had not been so pervasive in the minds of medical scientists during the mid-19th century, then the "usual suspects" would have been more quickly discarded. It was the lack of preparedness—in those who might have been able to analyze the experimental evidence without bias in favor of toxins and germs—that delayed an appreciation of the import of vitamin deficiencies and the emergence of plagues associated with white rice and corn. This "scientific blindness" condemned millions of people to a life of suffering and contributed to countless painful deaths.

Plague of the Seafarer

Next to famine, scurvy has caused more suffering than any other nutritional disease in history. Yet the disease itself was recorded only after the 1500s. This can be explained by the fact that at about that time the technical developments in the design of ships, navigation devices, and more reliable compasses allowed Europeans to sail further and further away and to explore new territories. There was also a strong commercial inducement to find a sea route to the Far East and thus to eliminate the middlemen in the overland trade of silk and spices. Portuguese sailors from 1440 onward explored the west coast of Africa (see page 316), getting farther and farther south until, in 1497, they finally rounded the southernmost tip, the Cape. In 1498, Vasco da Gama sailed from Portugal to Mombassa on the east coast of Africa, a trip that took 200 days. During the voyage he wrote in the ship's log, "Many of our men fell ill here, their hands and feet swelling, and their gums growing over their teeth so that they could not eat." In 1519, Ferdinand Magellan commanded five ships that left Seville, Spain, and after 15 continuous weeks at sea reached the Pacific Ocean. He described the voyage this way: "above all other calamities this was the worst; in some men the gums grew over the teeth, both lowers and uppers, so that they could not eat in any way and thus they died of the sickness; nineteen men died and 25–30 became sick, some in the arms, in the legs or other places, so that few remained healthy." Later voyages by Spaniards to the south of Mexico and Baja California produced similar disease symptoms. The French, not to be outdone by the Portuguese and Spanish, also sought to find a route to the Far East by going north across the Atlantic Ocean. In 1536, Jacques Cartier reached Canada and sailed up the St. Lawrence River as far as Quebec; some of his crew even ventured to present-day Montreal, but then he and his men were trapped by winter. From November until April, the ships were frozen in. Cartier wrote, "some did lose all their strength and could not stand on their feet, then did their legs swell, their sinews shrank as black as coal. Others also had their skin spotted with spots of blood of a purple color: then did it ascend up their ankles, knees, thighs, shoulders, arms and necks. Their mouths became stinking, their gums so rotten, that all flesh did fall off, even to the roots of their teeth, which did also almost fall out." Cartier lost 26 men from this *mal de terre*, or "sickness of the land," before the Indians came to his rescue by offering a cure: extracts of pine needles. Those who survived until the spring recovered.

The British, not to be outdone by the French, also set sail for and explored North America. From 1537 to 1539, Sir Francis Drake circumnavigated the globe and reached California. Only his flagship, the *Golden Hind*, completed the journey, largely because the crew suffered badly from scurvy. A similar fate befell a Dutch expedition to the East Indies in 1591 and again in 1604. Sea voyages over the next century were similarly afflicted, and the disease came to be called "explorers' disease."

A description of scurvy appears in literature in the *Rime of the Ancient Mariner* by Samuel Taylor Coleridge (1772–1834):

It is an ancient mariner
And he stoppeth one of three
By thy long grey beard and glittering eye
Now wherefore stoppst thou me?
He holds him with his skinny hand
There was a ship quoth he

. . .

'Hold off! unhand me, graybeard loon!'
Eftsoons his hand dropt he.
[He recounts his tale.]
One after one, by the star dogged Moon
To quick for groan or sigh,
Each turned his face with a ghastly pang,
And cursed me with his eye.

Four times fifty living men,
(And I heard nor sigh nor groan)
With a heavy thump, a lifeless lump,
They dropped down one by one. . . .

Explorers' disease persisted for more than 200 years. Christopher Columbus and his crews, however, were spared. Columbus' success in avoiding scurvy had nothing to do with his skills as an explorer or leader of men, but simply that the expeditions to the New World were of shorter duration, lasting less than 12 weeks. Scurvy appeared to be characteristic only of long ocean voyages. However, when the British Navy began to employ a strategy of naval blockades, they too suffered from scurvy, although they were never very far from shore. Scurvy killed more men and destroyed more naval and military operations than did any enemy action.

Search for the "Germ" of Scurvy

In 1590, the English sea captain Richard Hawkins said that in his 20 years at sea he had seen 10,000 men consumed with scurvy and "wished some learned man would write of it, for it is the plague of the sea." In 1541, the Dutch physician John Echth wrote, "scurvy is a disease of the spleen." According to his theory of melancholic humor (which was based on and can be dated back to Hippocrates and Galen), sufferers from scurvy had a swollen and hard spleen that blocked normal function and did not allow the black bile formed by the liver to be cleaned up. As a consequence, the black bile was dispersed and then rejected in the form of black spots on the skin and produced a swarthy complexion. By diffusing generally throughout the body it resulted in a lack of energy, and when the swollen spleen was pressing on the diaphragm, breathing became painful. Echth's theory covered the symptoms of scurvy; he then went on to list some of its principal causes among sailors: corrupted water, poor-quality food, lack of sleep, hard physical labor, and cramped quarters with fetid, warm air. No practical cure was suggested.

Two hundred years later, the English naval surgeon James Lind performed the first controlled nutritional experiment aboard the ship *HMS Salisbury*. He kept a group of 12 sailors with scurvy "as similar as I could have them in the same quarters . . . and they all in general had putrid gums, the spots and lassitude, with weakness of their knees." He saw to it that they all had the same diet: for breakfast, gruel with sugar; for lunch, broth made with fresh mutton on some days and pudding on others; and for supper, barley and raisins, rice, and currants. Two men were allocated to each of six different daily treatments: (i) cider; (ii) some dilute sulfuric acid; (iii) two spoonfuls of vinegar; (iv) a half pint of sea water; (v) a paste made of mustard, garlic, radish root, and gum; and (vi) two oranges and one lemon for 6 days, after which the supply ran out.

Without exception, those sailors who received the supplement of oranges and lemons recovered quickly from scurvy, whereas none of the others did. Today, the idea of using a sample of 12 patients would be ridiculed, but in Lind's time this was revolutionary. Indeed, Lind said, "I shall propose nothing dictated merely from theory, but shall confirm all by experience and facts, the surest and most unerring guides." In short, his was the first documented example of a controlled clinical trial. One of the unique features of Lind's approach was his insistence that what mattered above all was prevention; curative measures, though important, were of secondary concern. This remains the prevailing philosophy of human nutrition in present medical practice.

After spending 8 years at sea, Lind left the navy in 1748. In 1753, he published his findings in *A Treatise of the Scurvy, in Three Parts, Containing an Inquiry into the Nature, Causes and Cure, of that Disease. Together with a Critical and Chronological View*. He recommended that ships lay in stores of lemons. The Admiralty, however, did not make the daily issue of lemon juice on all naval and merchant vessels compulsory for more than 50 years. From that day forward, British sailors were called "limeys" because at the time lemons were called limes.

Acceptance of Lind's work was slow in coming for many reasons. First, Lind was a naval surgeon, and naval surgeons were not held in high regard either by the medical profession or by the Lords of the Admiralty. This was because a naval surgeon was a relatively untrained man, often apprenticed to a university-trained physician on shore, who had enlisted in the navy as a surgeon's mate; aboard ship, he acquired the necessary skills (surgery, cautery, setting of broken bones) by observation and imitation. Second, the Lords of the Admiralty had little incentive to seek a cure for scurvy, since it was easier and cheaper to replace a seaman than to replace stale stores with fresh fruits and vegetables. Third, unlike the bubonic plague, scurvy did not produce panic in the streets; it was not a sudden and devastating epidemic but a constant, corrosive, and deadly companion. Finally, the Admiralty, full of snobs and patronage, supported the illustrious Captain James Cook in his claim that malt wort, not citrus juice, was responsible for there not being a single death due to scurvy aboard ship during his 2-year voyage to the South Seas (1769 to 1771) and his 3-year circumnavigation of the globe (1772 to 1776). Indeed, upon Cook's return from his second expedition, he was lauded both as an intrepid sea captain and a medical hero and was given the Royal Society's Copley Gold Medal for his care and attention to life at sea and for keeping his expeditions free from scurvy. Cook was celebrated as "the man who defeated scurvy."

In 1769, 16 years after Lind published his *Treatise*, William Stark, a brilliant young physician in London, was determined to prove the details of the course of scurvy. He planned 24 different diets to be tested over several months to determine which ones were the healthiest. The subject for the dietary experiments was to be Stark himself. Dr. Stark, well-built and 6 feet tall, was in excellent health when the study began, and he kept meticulous records. He lived on a diet of bread and water for 10 weeks, and then he supplemented the diet with sugar and olive oil for another 10 weeks. For another period of time, he ate flour and oils, bread and honey, and lean or fatty meats, such as roast goose or boiled beef. At about this time, Stark realized that he was beginning to develop early signs of scurvy. Although familiar with Lind's work, Stark consulted John Pringle, a well-respected London physician, concerning his condition and asked how the diet should be modified. Dr. Pringle advised Dr. Stark to abstain from salt. This he did, but instead of investigating greens and fresh fruit as a dietary supplement, Stark chose to test the value of honey pudding and Cheshire cheese. This was a fatal choice. On 23 February 1770, less than 9 months after he began his studies of nutrition by self-experimentation, Dr. Stark died. The cause of death was scurvy.

Despite Lind's work, scurvy continued to surface, and in some cases the condition worsened. During the British siege of Gibraltar in 1780, there was a serious outbreak of scurvy, and extract of malt wort (Cook's remedy) was provided; it proved useless. Then in 1783, Thomas Trotter, a naval surgeon on a merchant ship taking slaves from West Africa to the West Indies, made a significant observation: when scurvy broke out among the slaves, they seemed to prefer unripe, acid-tasting guavas rather than those that were ripe and sweet, and they soon recovered. In a controlled experiment upon nine slaves with the disease, Trotter observed that three who received unripe guava recovered as well as the three who received limes; those who received ripe fruit showed little change. In 1786 he published *Observations on the Scurvy with a Review of Theories Lately Advanced on that Disease*, in which he suggested that fruits and berries that were able to cure scurvy were all acidic, but he was unable to explain why vinegar and sulfuric acid were inactive. In 1780, during a serious outbreak of scurvy in the British fleet in the West Indies, the Admiralty appointed Gilbert Blane as physician to the fleet. After only 8 months, he wrote to the Admiralty that "scurvy . . . may be infallibly prevented or cured by vegetables and fruit, particularly oranges, lemons and limes." Yet the Admiralty did nothing. In 1793, when scurvy appeared in the Mediterranean fleet, the chief physician wrote that "no ship . . . should leave port without . . . an ample supply of lemons; and . . . the use of lemon juice, [is] the effectual means of subduing scurvy." But the Lords of the Admiralty continued to dither. Finally, in 1795, Blane, who was now one of the commissioners of the Sick and Hurt Board, commanded that each sailor aboard the ship have two-thirds of an ounce of lemon juice each day. On the 23-week sail to India, without contact with land, not a man died from scurvy. The success of this "experiment" led the Admiralty to authorize lemon juice as a regular issue to the navy. Now Lind could take comfort that he, not Cook, was "the man who defeated scurvy."

From 1810 to 1840, there was little interest in scurvy. The problem had been solved at sea, but during this period there were outbreaks on land, principally in prisons and in pauper institutions. It was recommended that the inmates receive a ration of potatoes. Indeed, scurvy declined among the peasants once the potato became a staple in the diet. This was because its yield was greater than any grain, and one potato contains the human daily requirement of the antiscurvy factor (later identified as vitamin C). In 1846, when the potato crop failed in Ireland and a potato famine resulted, scurvy broke out. In 1848, scurvy appeared among those going to the gold fields of California by ship and even among those crossing the continent by land. Of the 12,000 captured Union soldiers who died in the Andersonville (Camp Sumter) prison in Georgia during the Civil War, 25% died of scurvy. In the 1890s, explorers to the Arctic and children in New York suffered from scurvy. In the case of the children, sterilized or condensed milk was blamed; the cure was feeding the children orange juice. In 1913 to 1914, the Jewish Orphanage in New York stopped giving the children orange juice, since pasteurized milk was being supplied; again, numerous cases of scurvy developed. The recommendation was that freshly cooked potato be added to the meals served in the orphanage. This reduced the scurvy outbreak.

In the 1900s, neglect of Lind's prescription for preventing scurvy led to the return of explorers' disease. In 1901, Captain Robert Scott led an expedition to the South Pole. Despite the availability of lime juice, vegetables, and fruit, the party was racked with scurvy. There were several reasons: lime juice was looked upon with disfavor, since the prevailing theory at the time was that scurvy resulted from ptomaine poisoning; the preserved vegetables and fruits were overcooked and therefore vitamin C deficient. The expedition team survived because the naval surgeon increased the amounts of fruit and lime juice as well as fresh seal and penguin meat. Between 1910 and 1913, Captain Scott led another expedition to the Antarctic. Again, he disregarded the value of lime juice. Though the team was provided with tinned foods, especially vitamin C-rich tomatoes, the sledding parties frequently suffered with scurvy. Scott and his entire team perished in the South Pole from a combination of lethal factors, none of which was ptomaine: unusually harsh and cold weather, a lack of food, frostbite, and scurvy.

Lind correctly identified lemons and oranges as scurvy preventatives, but he could only theorize as to the mechanism. Lind postulated that a cold, wet climate caused the pores in the skin to clog, which blocked perspiration, leading to an accumulation of undesirable humors that produced abraded and degenerated tissues. He ascribed the curative powers of lemon juice to its acidity and its ability to combine with the oily substances in the lymph; by dispersing the globules, as a detergent would, the much smaller particles could be excreted from the skin by perspiration. Lind's theory, however, did not satisfactorily explain why other acids did not cure or prevent the disease. Why lemons and oranges prevented and cured scurvy remained a mystery for nearly a century. Then in 1907, quite by accident, an animal model—the guinea pig—was developed to study the disease. This model allowed for the identification of antiscurvy factors in citrus fruits. Axel Holst, professor of hygiene at Oslo University in Norway, who had worked previously on beriberi, began a collaboration with Theodor Frohlich, a pediatrician interested in infantile scurvy. They fed guinea pigs on grains and other supplements, and much to their surprise, the guinea pigs came down with scurvy. At the time, there were three prevailing theories of the cause of scurvy: microbes, ptomaine poisoning, and faulty diet. When Holst and Frohlich found that the guinea pigs could be cured by the addition of fresh vegetables or fruits—while boiled vegetables lost their power to prevent scurvy— it was clear that scurvy was not the result of a germ or a toxin. Now the search was on to chemically identify the curative factor.

In 1928, the Hungarian biochemist Albert Szent-Gyorgyi was working on the energy-yielding reactions of plants and animals. In these oxidative reactions, through the action of enzymes, glucose was rapidly broken down to carbon dioxide and water. Such reactions, however, required nonprotein accessory factors in addition to enzymes. Szent-Gyorgyi found that the cortex of the adrenal gland was particularly rich in an accessory factor. When he prepared crystals of the substance, he found it was an acidic compound, hexuronic acid. (Actually, Szent-Gyorgyi wanted to call it "godnose" or "ignose" because he could not identify its exact chemical structure.) Later, he found that much larger quantities could be isolated from a common Hungarian spice, paprika, than could be obtained from the adrenal cortex. At the same time, Glen King, working in Pittsburgh, isolated ascorbic acid from lemon juice and found it to be identical to Szent-Gyorgyi's hexuronic acid. Ascorbic acid was renamed vitamin C, because McCollum had called the fat-soluble factors needed for normal growth of rats vitamin A, and the name vitamin B was given to similarly functioning water-soluble factors.

As with so many other medicinal drugs, the laboratory synthesis of vitamin C began with studies of dyes. At age 15, Walter Haworth (1883–1950) left grammar school to join his father and learn linoleum design and manufacture. This experience aroused an interest in dyestuffs and their chemistry. Through diligence and private tutoring, he was able to enroll at the University of Manchester as a student of William Perkin. At Manchester he studied terpenes and synthesized sylvestrene. He continued his studies at the University of Gottingen and received a Ph.D. from Manchester University in 1911. During World War I, Haworth studied the chemical structure of drugs at St. Andrews University (1912 to 1920) and began to investigate the composition and synthesis of sugars. He was a professor of chemistry first at Durham (1920 to 1925), and then at Birmingham (1925 to 1948); in 1933, he synthesized ascorbic acid, which allowed for cheaper commercial production. He and Szent-Gyorgyi received the Nobel Prize in 1937 for their discovery and synthesis of vitamin C.

Preventing and Curing Scurvy

Most animals make vitamin C, especially rats. Humans, however, lack this capacity, as do other primates, fruit-eating bats, and guinea pigs. Vitamin C, in the form of fresh fruits and vegetables, is required for normal health. Vitamin C is important in the manufacture of the proline-containing protein collagen, a major component of connective tissue found in tendons and ligaments and in the walls of blood vessels. (When collagen is extracted, it is called gelatin.) A deficiency in vitamin C shuts down the synthesis of collagen because the enzyme, proline hydroxylase, requires ascorbic acid as a cofactor. This role of vitamin C was dramatically shown in 1939 when Dr. John H. Crandon, a member of the research team of Dr. Charles Lund, volunteered to be the experimental subject in a test of the effect of vitamin C on wound healing. Crandon made himself vitamin C deficient and then had

Lund make a surgical incision. Biopsies were taken at weekly intervals to determine the progress of wound healing. After 6 months there was no healing; then, after intravenous administration of a gram of ascorbic acid each day for 1 week, all wounds healed. During the experiment, Crandon lost 31 pounds.

The explanation for the symptoms of scurvy—especially bleeding gums and broken blood vessels and swollen limbs—now could be explained by the capillary walls' need for collagen to retain their strength and permeability. To ensure adequate amounts of vitamin C in the diet, most foods today are fortified with it. The required daily allowance is 75 mg for an adult.

Vitamin C in lemon juice is perishable, and therefore means had to be devised for maintaining its potency. For many years, the standard method was to add 15% rum to the juice. The search for methods to preserve the vitamin C content of lemon and lime juices for long voyages, where there could be no access to fresh fruits, resulted in a very lucrative spin-off industry. In 1867, Lauchlan Rose of Edinburgh devised a method for allowing lime juice to remain active as an antiscurvy agent without the addition of alcohol as a preservative; further, by sweetening, preventing fermentation, and sealing in glass bottles, he began to produce soft drinks for public consumption. Thus, the many nonalcoholic beverages we drink today are a by-product of the 200-year-long search for the prevention of scurvy.

The Air Pollution Plague

Rickets is a disease of the bones. Persons with rickets have impaired rib development that results in a shortness of breath. The origin of the term is uncertain. It may be derived from the Greek word *rachitis* for "spine," or it may be from the Dorset County (England) dialect term *rucket* meaning "to breathe with difficulty." Or perhaps it is derived from the Saxon word *rick* meaning "hump," or the Norman word *riquets* for "hunchback," or the Welsh word *wrygates* meaning "twisted legs." No matter its precise etymological origin, rickets is an old disease that became epidemic in England after 1650, when soft coal was introduced as a fuel source. With the industrial revolution's gray pall of coal smoke and the increasing concentration of poor people in the narrow, sunless alleys of the factory towns and urban slums, the disease spread from England through the countries of northern Europe (Fig. 16.4). During the time of its high endemicity in Britain, some called it the "English sickness." William Hogarth's sketches accurately record the crippling effects of rickets in London in the late 18th century.

Rickets is the earliest example of an air pollution disease and is not, as one might believe, a disease caused by poor diet. It results from a deficiency in solar ultraviolet radiation, which is necessary for the proper calcification of the bones. With insufficient amounts of calcium, bone remains as pliable as

Figure 16.4   An urban slum in London circa 1903. (From the Getty Hulton Collection.)

plasticine, the skeleton is not sturdy, and, unable to resist stress, it bends (Fig. 16.5). Bone calcification is especially important when the long bones of the legs are lengthening and thickening; lacking rigidity, the bones bow under the weight of the body. The connection between rickets and vitamin D suggests that rickets is a dietary deficiency disease; indeed, each milk carton sold in the United States has printed on it "400 U.S.P. units vitamin D added per quart." Because of this additive, there are virtually no cases of rickets in the United States. So, is rickets a dietary deficiency disease or a deficiency of sunlight? The answer is, both.

A Look Back

In the early 19th century, a German industrial town called Wezlar (population, 8,000) was famous because house after house was occupied by children crippled by rickets. As a contemporary description put it, "The children must sit indoors . . . and most die, but of those that continue to live they develop thick joints, cease to be able to walk or have deformed legs. The head becomes large and the spine bends. Such children sit for many years without being able to move; at times they cease to grow and are merely a burden to those about them." In 1884, it was noted that in Germany and in Britain the disease was seasonal: children who were born in the fall and died in the spring had rickets, but those who were born in the spring and died in the fall had no rickets. The suggestion was that rickets occurred because children born in the fall were confined indoors during the winter months and consequently received less sunlight. In 1890, an English medical missionary, Theobald Palm,

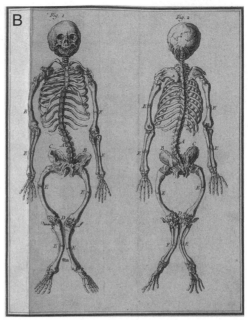

Figure 16.5   (A) A young girl with rickets standing with her younger brother and (B) the skeleton of a person with rickets. (Panel A courtesy of Corbis. Panel B courtesy of the Wellcome Library of Medicine.)

went to Japan (during the early period of Westernization [see page 363]); he was struck by the absence of rickets in that country, yet its cities were crowded with people. Palm also noted that rickets seemed to be restricted to the large industrial cities in Scotland, England, and northern Europe. He collected observations from China and Java and was impressed by the fact that European children suffering from rickets recovered from the disease after they moved to Java, yet they had received no medical treatment. He deduced that rickets was caused by the absence of sunlight and wrote, "It is in the narrow alleys, the haunts and the playgrounds of the poor, that this exclusion of sunlight is at its worst, and it is here that the victims of rickets are to be found in abundance." Palm recommended the systematic use of sunbathing as a preventive and therapy for rickets. But was the correlation of sunlight and absence of rickets a cause-and-effect relationship?

In 1908, Leonard Findlay working in Glasgow, Scotland, claimed that if he kept puppies confined in cages indoors in the dark they developed rickets, but if the puppies were unconfined and exercised out of doors they did not. He was convinced that confinement of the puppies caused the disease. Then in 1920, Sir Edward Mellanby, following on Findlay's work, kept puppies indoors on a purified diet and gave them exercise, yet they developed rickets. Adding a fatty extract of liver to the puppies' diet could reverse the rickets.

In that same year, a German pediatrician, Kurt Huldschinsky, used ultraviolet radiation on four children with advanced rickets and cured them within 2 months. Huldschinsky went on to show, using X-ray photographs, that if he exposed one arm of a child with rickets to ultraviolet light (from a mercury vapor quartz lamp), calcium salts were deposited in the irradiated arm as well as the nonirradiated one. This suggested that irradiation of the skin with ultraviolet light released in the bloodstream a substance that had the power to induce bone healing at a distance. The chemical nature of the substance, however, would not be identified for another 70 years.

In 1924, Alfred Hess in New York extended Huldschinsky's findings. He discovered that if linseed or cottonseed oil was irradiated with ultraviolet light and then fed to rats, rickets did not develop. Later, Hess was able to show that cholesterol, or a plant sterol such as ergosterol (derived from the wheat rot fungus, ergot), when irradiated and added to the diet, also prevented rickets. Today, irradiated ergosterol is added to milk and is called vitamin D_2. Hess also found that routine administration of cod liver oil in the diet—a spoonful a day—prevented rickets. These experiments on foods that were antirachitic (prevented rickets) turned medicine's attention away from the sun's ultraviolet rays and rickets came to be regarded as a dietary deficiency disease. This view was reinforced by Mellanby's experiments with puppies, from which he was able to show that the puppies became rachitic even with a diet containing yeast, milk, and orange juice, but that adding cod liver oil to the food prevented the disease. (Mellanby believed that the cod liver oil prevented rickets because it contained vitamin A; he was wrong.) If Mellanby and Hess were correct, that is, rickets was due to a deficiency in the diet, then what was the explanation for rickets developing when puppies were fed a complete diet including yeast, milk, and orange juice but kept in a confined, darkened space and Huldschinsky's ability to cure rickets in children by exposure to ultraviolet radiation?

In 1922, Elmer V. McCollum, one of the most distinguished American nutritionists, established that the antirachitic factor in cod liver oil was not vitamin A. After heating and aerating cod liver oil, it remained active, even though such treatments were known to inactivate vitamin A. Therefore, he called the fat-soluble factor in cod liver oil vitamin D. As a result of McCollum's work on vitamin D and cod liver oil, rickets came to be considered a vitamin deficiency disease—a hidden hunger. The earlier observations on the relationship between sunlight and rickets were promptly forgotten for the next 75 years.

Vitamin or Hormone?

It is now known that the ultraviolet rays of sunlight (wavelength, 290 to 310 nm) act on a sterol in the skin to produce calciferol (vitamin D_3); this travels via the blood to the liver, where the calciferol becomes calcidol, and then to the kidneys, where calcidol is chemically transformed into calcitrol, the metabolically active form of vitamin D. Calcitrol travels via the blood to the cells of the intestine and promotes the absorption of calcium from food that has been eaten; the calcium is deposited during the formation of bone. Calcitrol, as Huldschinsky clearly showed in 1919, is not a vitamin, since it is neither a micronutrient that must be obtained in the diet nor a cofactor for enzyme action; rather, it is a hormone made in the body by the action of ultraviolet light on a skin sterol that in the kidneys is converted to its active form. Kidneys thus function not only to remove metabolic ashes (urea and other nitrogenous wastes) from our blood, but they also act—along with the pituitary gland, ovary, testis, adrenal gland, and thyroid gland—as endocrine glands. Rickets is an endocrine-deficiency disease much closer to diabetes than a "hidden hunger" such as scurvy, beriberi, and pellagra. How much exposure to sunlight is necessary to make the prohormone calciferol? On a summer day in California or New York, exposure of the skin for 10 to 15 minutes for 2 to 3 times a week is sufficient. Foods, however, also contain calciferol. Cheese, egg yolk, liver, salmon, and cod liver oil all contain calciferol (when it is called vitamin D). Human breast milk and cow's milk are not very rich in vitamin D, and as a consequence, infant formulas and some weaning foods are fortified with vitamin D (40 to 50 IU/100 ml).

Rickets due to vitamin D deficiency is rare in the United States owing to the introduction of vitamin D-fortified milk and infant formula several decades ago, but even in a sunny climate, when members of the community drink little breast or cow milk, calcium deficiency can occur. In March 2001 the Centers for Disease Control and Prevention reported six cases of rickets in infants from Georgia. The infants were delayed in walking, were short for their age, showed curvature of the spine, and had a bowing of the legs and arms. The cases occurred in infants who had been breast-fed for more than 6 months without receiving vitamin D supplements. Although it is possible to obtain vitamin D through exposure of the skin to sunlight, individuals with increased skin pigmentation require longer periods of exposure because the melanin in the skin acts as a sunscreen. In the mini-outbreak of rickets in Georgia, the children with nutritional rickets were black.

The Origin of Rickets

How did humans develop rickets in the first place? There are three vitamin D zones: tropics, subtropics, and temperate. In the tropics, the dosage of ultraviolet light is high enough to synthesize vitamin D all year round; in the subtropics, the amount of ultraviolet light is sufficient for vitamin D synthesis except for one month in the year; in the temperate zone, there is insufficient ultraviolet light for vitamin D synthesis almost all of the year. Thus it is principally in the temperate zone that synthesis of vitamin D becomes critical for normal bone growth and strength.

Darker skin occurs in indigenous peoples closer to the equator, and lighter skin occurs closer to the poles. Why? Presumably, the earliest humans had lighter skin and were covered with hair, as is the case with modern-day chimpanzees. Two million years ago, about the time of *Homo erectus* (*Homo ergaster* in Africa), humans walked long distances and were very active. Some have speculated that to keep their brains cool, humans evolved to have more sweat glands and less body hair, favoring evaporative cooling. Once this occurred, these hairless humans would be faced with another problem: receiving too much ultraviolet radiation with possible damage to the skin cells. Counteracting radiation damage to the skin was the deposition of a black-brown pigment, melanin, the process of tanning. Those humans who remained in Africa and the tropics evolved a darker skin—a process that took millions of years. Were it not for melanin deposition in the skin, humans living in tropical Africa, exposed to higher levels of ultraviolet radiation, would suffer from a deficiency of folate, since it is destroyed by ultraviolet radiation. In pregnant females, folate deficiency results in birth defects such as spina bifida; in male rats, reduced spermatogenesis and infertility are the result. Therefore, a darker skin would protect not only against ultraviolet radiation damage to the skin cells, but it would also prevent folate deficiency. As humans migrated out of Africa and occupied regions closer to the poles, a darker skin would have limited the amount of ultraviolet light received, and this in turn would result in reduced calciferol production. Therefore, humans with darker skin, living in the temperate zones where ultraviolet light levels are much lower than in the tropics, would be at a selective disadvantage. Over time, however, there would be gradual selection for lighter skin through a reduction in melanin synthesis, and this would allow adequate amounts of calciferol to be made when less ultraviolet light was available. Lighter skin does have a downside: it increases the risk from skin cancer. The ability of those with lighter skins living in the temperate zones to develop a deeply bronzed skin by sunbathing prevents an oversynthesis of calciferol and lessens the risk from cancer when the amounts of ultraviolet light are great. Sun-bingeing, however, should be avoided to reduce the possibility of skin cancer; in addition, excessive sunbathing does not boost vitamin D production, because the excess calciferol is converted to an inactive form.

As humans migrated out of the tropical regions into the more temperate zones, ultraviolet light became scantier, and natural selection favored a lighter skin—as is observed in the English, Scandinavians, Germans, and some Asians. Indeed, light skin would prevent the deformities of rickets, which would certainly endanger the process of childbirth were the pelvis affected. The northern European ideal of female beauty—trim ankles, straight legs, and light skin—fits well with the hazards of rickets during childbearing.

A "fair young damsel" would be favored as a suitable mate, since she would not have suffered from rickets and probably would bear stronger children. Delicate wrists and a free-swinging walk—possible only in the absence of rickets-induced pelvic deformities—would also provide evidence that the prospective bride did not have a history of the disease. June weddings tend to bring the first baby in the spring and would favor children without rickets, as would fish on Friday and taking a baby outdoors for some fresh air and sunshine in the middle of winter. All of these became customs of those who lived in the northern climes of Europe and Great Britain. All could be viewed as adaptations to prevent rickets. And why do Inuits living in the northernmost vitamin D zone not get rickets? Because they get it in their diet of fish. Indeed, it was found that when Inuits were placed on a European diet by missionaries, they rapidly developed rickets.

It is a fact that women in all populations have lighter skin (~3 to 4%) than do males; perhaps the reason for this is that vitamin D is needed for adequate calcium absorption during their reproductive lives, especially during pregnancy and lactation. Women are also better able than men to make the most of the calcium in their food.

The perils of recent migrations of people to different vitamin D zones is that blacks may suffer more from susceptibility to rickets (e.g., the recent cases of rickets in Georgia) but have a lower incidence of skin cancers, whereas whites may suffer more from skin cancer and have a lower susceptibility of rickets save for their living in the temperate zones or in polluted air. With the increase of industrial smog, even the fair-skinned did not receive enough ultraviolet radiation, and thus rickets appeared in England and northern Europe. Rickets is not due to lack of fresh air, nor is it due to an infectious agent a defective gene, or a lack of exercise. It is due to a deficiency of sunlight, which begins the process that leads to the synthesis of the hormone calcitrol. The need for extra dietary vitamin D is the price humans must pay for their successful migration out of the tropics.

The story of dietary deficiency diseases provides us with two important lessons. Although disease may be the result of invasion of the body by a foreign agent, not all diseases are the result of germs and worms. And, on occasion, an unquestioning belief in a model (in this case, the germ theory) can diminish the power of science as a way of knowing.

Figure 17.1   *The Death Chamber* by Edvard Munch. © 2005 The Munch
Museum/The Munch-Ellingsen Group/Artists Rights Society, New York.

Chapter 17

Plagues on Order

*A creature without memory cannot discover the past; one without expectation
cannot conceive of a future.
Those who do not remember the past are condemned to repeat it.*

GEORGE SANTAYANA

Diseases, caused by infectious agents, can affect the course of human events. They have in the past, and it is certain they will do so in the future. Great plagues, such as the bubonic plague or smallpox or syphilis or influenza, can happen again. Plagues are natural and almost predictable phenomena. Although remarkable scientific advances have been made in controlling diseases through sanitation, chemotherapy, antisepsis, improved nutrition, and immunization, we continue to live in evolutionary competition with microbes, and there is no guarantee that we can always beat them at their own game. Lurking out there are germs and worms that may spread to our domestic animals, our domesticated plants, and us. These are the seeds of coming plagues (Fig. 17.1).

Many factors lead to the emergence of infectious diseases, some of which are obvious: a change in a parasite's virulence, a breakdown in public health and surveillance measures, a change in human behavior, crowding, alterations in the environment, technological advances, economic downturns, poverty, and even something as innocent as foreign travel for a vacation. Emergence of infectious disease is a complex and dynamic process, and most new infections are not caused by a pathogen that has never before been encountered. New parasites are actually old pathogens in an altered form. Some of these arise because of breaches in the species-specific barrier, but the most potent factors that drive the emergence of disease are human activities—social, economic, political, technological, environmental, and behavioral. As the cartoon character Pogo said, "We have met the enemy and he is us." Paradoxically, the current global situation favors rather than discourages the spread of epidemic disease. The reason for this is that for most of human history, populations were relatively isolated; only in recent centuries has there been extensive contact between peoples of different continents. It is estimated that by the year 2010, 50% of the world's population will be living in urban areas and there will be 24 "megacities" with populations exceeding 10 million. These areas will be able to support the persistence of some infections and allow for the emergence of others. And many of these "megacities" will be in tropical regions—places that serve as the most effective cradles for the emergence of parasitic diseases.

Movement of people from place to place has been and can be a potent force in the spread of disease. In the past, trade caravans, religious pilgrimages, and troop movements, as well as colonization and slavery, have shaped the location, nature, and spread of many diseases, including plague, cholera, typhus, smallpox, malaria, yellow fever, and river blindness. Today, because of the unprecedented volume, speed, and reach of travel, a potentially pathogenic parasite can be more easily introduced into a new geographic area by a "quick leap" using high-speed trains, planes, and ships. Pathogens that survive primarily or entirely in humans and are spread by sexual contact, droplet nuclei, and close physical contact can be readily carried to any part of the world. For example, acquired immunodeficiency syndrome (AIDS), tuberculosis, measles, pertussis, diphtheria, and hepatitis move as people move; although disease spread can be blunted by vaccination and "magic bullets," there is never a complete halt. As if this were not enough, there is an ominous new threat facing us today: bioterrorism. A nightmare scenario might be the release of an infectious agent in a "megacity," where the disease would sweep through an entire population. If the disease were smallpox, anybody born after the early 1970s when vaccination against smallpox stopped would be completely vulnerable. An outbreak in the United States would be horrific, since 120 million people were born after routine vaccination ended. In those susceptible individuals, without any immunity, the death rate could exceed 30%.

Trade, immigration, and war can set in motion the mixing of diverse genetic pools of parasitic diseases at rates and combinations never known before. As the historian William H. McNeill noted in his classic work *Plagues and Peoples*, the patterns of disease circulation have not only influenced the outcome of battles and shaped human societies; they have affected the ways in which humans behave when a novel pathogen is present. Biologically speaking, the AIDS virus isn't entirely new; it is an altered form which has moved from a host to which it is well adapted and in which it is benign to one to which it is less well adapted and in which it is lethal. Recall the situation with yellow fever, or smallpox, or syphilis and how in each of these cases the plasticity of the hereditary material has led to parasite attenuation or to virulence. Human behavior, close contact with animals, and monoculture (growing pure-bred strains) in agriculture also provide the means for the spread of infectious diseases and may contribute to a collapse in the economy and social order.

Blame It on the Rodents

A "new" disease, Argentine hemorrhagic (Junin) fever, developed because of a change in agricultural practices. When Argentine farmers turned from growing mixed crops and began clearing land to plant corn, they provided a new niche for a little mouse called *Calomys musculinis*. Herbicides were applied to increase corn production; this killed off much of the native plant life and reduced the diversity of the rodent species. Consequently, *Calomys* increased because this mouse thrives on corn, but it also carries the Junin virus. Plant more corn, get more mice, more virus, and also more human infections. Now there are more than 600 cases annually of this hemorrhagic fever.

In the summer of 1993, a severe respiratory disease appeared in the Four Corners area of Arizona, New Mexico, Colorado, and Utah. Typically, the disease was characterized by rapid pulmonary failure. A 31-year-old woman arrived at the hospital with a 4-day history of fever and sore throat followed by headache, body pains, diarrhea, and vomiting; her breathing was shallow, and she complained of chest pains when she breathed deeply. Initially, her temperature was 38.5°C, but within a day it had risen to 40.5°C. She was believed to be suffering from gastritis and dehydration, so she was placed on morphine and given intravenous fluids. Chest X rays showed her lungs to be filled with fluid. Because she did not respond to antibiotics, the preliminary diagnosis was mycoplasma or viral pneumonia. Although by the third day her blood pressure had dropped to 70 mm Hg, she made a remarkable recovery, and on her own she began to resolve the infection. Although she continued to complain of headache, cough, general achiness, and diarrhea, she was sufficiently improved in a week's time that she was discharged from the hospital. (She was lucky. Of the 160 cases in 26 states, 30% of the patients died.) When state and county health department officials conducted an investigation near the patient's home, they caught 17 rodents. Fifteen were deer mice, and one of them was positive for hantavirus, a hemorrhagic fever endemic from Western Europe to far eastern Asia. Its appearance in the United States was unique. The disease is called hantavirus pulmonary syndrome and is caused by the Sin Nombre virus and is transmitted by the urine of deer mice, whose range includes most of the 48 contiguous states. The disease is more prevalent in men than in women, probably as a result of the greater occupational exposure of men to rodents. This particular outbreak was related to a large crop of pinion nuts that was being harvested by the men on the Navajo reservation; this led to their inhalation of dried rodent urine, which was virus laden. Human-to-human transmission did not occur, but one wonders what would have been the result if the virus had acquired the capacity for pulmonary transmission, as happened with pneumonic plague.

Lassa fever is another case in point. In 1960, diamond deposits were discovered in Sierra Leone, Africa, and this encouraged human migration. More humans produced more food and garbage. Occupying the same area was the rodent *Mastomys natalensis*, a natural reservoir for the Lassa fever virus. With the increase in humans, there was an increase in virus-infected *Mastomys*, which can be considered a domesticated species because it lives in close association with human habitation. As with hantavirus, this virus is shed in the urine, the urine dries, and the virus-laden dust is carried into the air, where humans can inhale it. With an incubation period of 1 to 2 weeks, it soon causes hemorrhages and death. Outbreaks first occurred in 1969, but sporadic cases still occur today.

In 2003, there was an outbreak of monkeypox (a cousin of smallpox) in four people and possibly 30 others that was traced to a "pet swap meet" in northern Wisconsin. The source was an infected Gambian giant rat from Africa. The rat virus was transmitted to pet prairie dogs that then bit children. Clearly, the handling of exotic animals transported from their usual habitats may provide the seeds of epidemics. We were lucky in this case since human-to-human transmission did not occur, but in other cases, human infections occur that involve birds, mosquitoes, and our food.

On the Wings of Birds

West Nile virus (WNV) was first isolated in 1937 from the blood of an infected woman in the West Nile country of Uganda. The woman had a rapid onset of fever, headache, backache, muscle pains, and anorexia. She had a rash on her face and chest and swollen glands in her neck. She recovered without any treatment. Twenty-two years later, the same virus, which is antigenically (see page 219) related to St. Louis encephalitis virus and Japanese encephalitis virus, caused more severe illness—meningitis, encephalitis, and paralysis in 59 people in New York City. In 2000 to 2001, the disease spread through the South and Midwest, and by 2002, WNV was in six western states (Arizona, Utah, Nevada, Oregon, Alaska, and Hawaii), where it triggered the largest encephalitis epidemic in recent U.S. history. There were over 4,000 cases and 284 deaths. How did WNV get from Africa to America? In 1957 there was a WNV outbreak in an Israeli nursing home, and subsequent outbreaks occurred in Israel and Africa in the 1960s and 1970s. These were mild, however, and the symptoms were flu-like. But in the mid-1990s, outbreaks in Russia, Romania, and Israel had other characteristics—patients suffered from meningoencephalitis. WNV infection was now a serious neurological disorder. The most significant factor in all of the outbreaks was the involvement of the common house mosquito *Culex pipiens* as the vector. This mosquito turns out to be more than an annual transmitter, since it also has the capability of overwintering. *Culex* had never before been an important vector, but environmental conditions helped to make it so. The first epidemics reported from urban areas (i.e., in August 1999) occurred at a time when rainfall had been lower than normal; such dry conditions probably increased the number of favorable breeding sites for *Culex*, which readily lays its eggs in stagnant and polluted waters.

There also appeared to be a change in the virulence of the virus. In New York City, concurrent with human infections, there was an unusual number of bird deaths, primarily wild crows and some exotic captive birds in the Bronx Zoo. More than 14,000 cases of WNV infection occurred in horses (in 2003) with a mortality rate of 35%. The most common way humans (and horses) acquire the infection is by the bite of a *Culex* that has fed on an infected bird, especially crows and blue jays. Initially, WNV spread from Africa by migratory birds, but now person-to-person infections can occur through blood products, organ transplantation, breast milk, and intrauterine infection. WNV is not transmitted from person to person by mosquitoes—people and horses are dead-end hosts—but what would have happened if human-to-human transmission did occur?

The incidence of WNV encephalitis and death increases 20-fold among those older than 50, and the death rate is ~10% from meningitis or encephalitis due to inflammation in the brain and spinal cord with small hemorrhages and degeneration of nerve cells in the brainstem. Aside from supportive measures, there is no established treatment, and there is no FDA-approved vaccine. Prevention of WNV can be accomplished by reducing contact between humans and *Culex*; this means maintaining window and door screens, wearing long sleeved shirts and long pants when outdoors, and applying insect repellents such as DEET.

It is likely that WNV will advance through the United States within the next several years, and its path of death will be marked by the local and regional flyway patterns of virus-infected birds.

We Are What We Eat

The sign in the window of McDonald's read, "We no longer serve British beef." The year was 1996, and London was declaring itself "hamburger free." Sales of beef dropped, schools stopped serving beef, and there was widespread concern about eating beef or, for that matter, any meat product. This was not an economic boycott on British beef. Its cause was an outbreak of "mad cow disease." Fear and panic gripped the country because of the report of 10 deaths from Creutzfeldt-Jakob disease (CJD), a rare neurological disorder. Unlike most CJD cases, the victims were young, and the link to their deaths was that all had eaten British beef. The disease was called variant CJD or vCJD. The British government responded by forbidding the sale of beef, and over 250,000 cattle were slaughtered to stem the mini-epidemic. Although the numbers of vCJD deaths declined over the next several years, the cattle industry lost $40 million a year, and trade in beef with Britain was virtually nil. No one was certain that cattle slaughter and the ban on beef helped curb the spread of the disease, but the number of vCJD cases did decline. It was suspected, however, that beef infected with "mad cow disease" (technically, bovine spongiencephalopathy [BSE]) had unexpectedly jumped the species-specific barrier and had in some way managed to enter the human food supply. If the calculation of an incubation period of 25 to 30 years for vCJD is correct, then the peak of the epidemic might not occur until 2015. One scientist predicted that, if the numbers of vCJD in 1996 increased by 50% and was compounded, it would result in 200,000 cases per year. Thankfully that prediction has not been borne out!

The linkage of CJD with BSE did not begin in Britain in 1996 but in the highlands of New Guinea, with studies by an American pediatrician, Carlton Gajdusek, who had been asked to look into several cases of encephalitis among children of a hunter-gatherer tribe, the Fore. The disease was called kuru, and it was characterized by uncontrolled tremors, blurred speech, and pathological laughter. In 1957, Gajdusek described it as "a classical advancing Parkinsonism involving every age, overwhelmingly females although many boys and a few men also have had it, is a mighty strange syndrome . . . well-nourished, healthy young adults dancing about, with . . . tremors which look far more hysterical than organic, is a real sight." Gajdusek had always wanted to be a microbe hunter. As a teenager growing up in New York, he had read Paul de Kruif's book *Microbe Hunters*, and his heroes became Louis Pasteur, who devised the first laboratory-produced vaccines; Robert Koch, who identified the causative agents of anthrax, cholera, and tuberculosis; Antonie van Leeuwenhoek, who was the first human to see a parasite (*Giardia*) by looking through magnifying lenses; and Ronald Ross and Walter Reed, who flagged the *Anopheles* and *Aedes* mosquitoes as the vectors for malaria and yellow fever, respectively. Gajdusek had been well prepared for his work on kuru; he had been trained as an M.D. at Harvard Medical School, then went on to study the growth of poliovirus in kidney cells with John Enders at Harvard, immunology and hepatitis with Macfarlane Burnet in Australia, and protein chemistry with Linus Pauling at Caltech.

The disease Gajdusek investigated in New Guinea was a mysterious one. Upon autopsy, the brains of kuru victims were found to have extensive damage—there were holes and microscopic knots of protein called amyloid plaques—but remarkably, there was no sign of inflammation. The condition resembled another disease that had been described in 1913 in Breslau, Germany, by the neurologist Hans Creutzfeldt. A woman, an elderly mental patient, died 3 days after Dr. Creutzfeldt had examined her. Before her death, he noted that she had facial tics, made jerky and twitching movements of her arms and legs, and had difficulty swallowing, epileptic attacks, screaming fits, and "unmotivated outbursts of laughter." Creutzfeldt's work on this neurological disorder was interrupted by World War I, but at its conclusion he wrote up his findings and as a courtesy sent the manuscript to his colleague, Alfon Jakob of Hamburg, who had observed four other patients with the same deadly symptoms. Stained slices of brain tissue from these deceased patients were examined by light microscopy. Creutzfeldt and Jakob described the microscopic findings as "full of holes as in a sponge," and they called the condition "a spongiform change." Although the brains of these patients were severely damaged, there were no signs of an inflammatory response. The published their findings in 1920, and the disease they described was named after them, Creutzfeldt-Jakob disease (CJD).

Upon hearing of Gajdusek's findings in New Guinea, a pathologist colleague remembered the work of Creutzfeldt and Jakob and realized that the brains of kuru victims and those of CJD patients were quite similar. However, there was a critical difference in the age of the victims: the CJD patients were older, whereas those with kuru were younger women and children. Gajdusek began to search for the causes of kuru and the reasons for its pathology. In detective-like fashion, he examined the food the Fore ate; calculated the incidence of kuru in the Fore tribe; determined the contact the Fore would have with mosquitoes, flies, and ticks; and evaluated whether it was a familial disease. In his examination of the distribution of kuru among the Fore, Gajdusek carried out an epidemiological study similar to that first conducted by John Snow for cholera a century earlier (see page 160). No correlations were found for a food toxin, and vector or person-to-person contact also showed no pattern consistent with the occurrence of kuru. The disease also seemed not to be inherited. Then in 1961 to 1963, during a kuru epidemic, a husband and wife team of Australian anthropologists made a critical finding: the members of the Fore tribe engaged in ritual cannibalism. However, not all members were ancestor eaters. Only women and children partook in the practice, and the brains of the deceased relatives were considered a special delicacy. Cannibalism was not an ancient practice among the Fore. It is thought to have started in the 19th century, with the first remembered case about 1920. At its peak, kuru was estimated to have killed 1% of the population annually. With the pathology and transmission of kuru now known, Gajdusek prepared a photographic exhibit of its features, which traveled to medical museums throughout the world. One visitor to such an exhibit in London, William Hadlow, was a veterinarian. After seeing the photographs, Hadlow wrote to the medical journal *The Lancet* that the pathology seen in the kuru brains resembled a disease of sheep called scrapie, with which he was quite familiar.

Scrapie is a neurological disease causing the sheep to itch so intensely that they scrape the wool off their hindquarters by rubbing themselves against fences, trees, and walls. In addition to the scraping syndrome, the sheep stagger, have tremors, go blind, become paralyzed, and die. Where scrapie originated is unclear, but epidemics were recorded in the 19th century in parts of Europe, and by the 20th century it was endemic in Britain. In the 1930s, French and English veterinarians unintentionally demonstrated that scrapie was an infectious disease. Both groups attempted to produce a vaccine in the fashion of Louis Pasteur: brains of scrapie sheep were homogenized, formalin was added to kill the pathogen, and this vaccine was injected into the brains of healthy animals. After many months, and before the animals could be challenged, all of the "immunized" sheep came down with scrapie. Because of the long delay before symptoms appeared—5 years in sheep—a pathologist in Iceland, Bjorn Sigurdson, proposed that the infectious agent of scrapie was caused by a "slow virus." (This is in contrast to "quick" or acute virus diseases such as smallpox, polio, yellow fever, and measles.) Gajdusek, the virologist-trained pediatrician, embraced the name and used it to explain kuru's cause. When, after autopsy, the brains of the "slow virus"-infected scrapie sheep were examined, spongiform degeneration was evident. The question for Gajdusek then became, how is this "slow virus" transmitted in nature? He embarked on a series of experiments designed to satisfy Koch's postulates (see page 290). Brains from kuru victims were homogenized and injected into the brains of healthy chimpanzees. In 10 months, the monkeys came down with kuru-like symptoms, and all died shortly thereafter. It was also possible to infect other chimpanzees from these kuru-infected monkeys.

Then, using the 1878 dye methods of Paul Ehrlich, the brain tissue was stained with the aniline dye Congo red. When thin sections of the brain were examined with a light microscope, the brains were riddled with holes—they were spongiform. In Gajdusek's mind, Koch's postulates were satisfied, and now he had joined his boyhood heroes as a full-fledged microbe hunter. When the stained slides were viewed with polarizing light, the amyloid plaques (with their aggregated and knotted fibrils) stood out with a greenish glow. Again, there was no sign of inflammation in the brain tissue, indicating that there had been no immune response to the "slow virus" infection. In 1978, homogenized brain tissue from scrapie-infected sheep was concentrated and examined by Beatrice Merz using the electron microscope; she saw filaments that were twisted and tangled, but there was nothing that looked like a virus particle. As a consequence, the "slow virus" was renamed scrapie associated fibrils, SAF.

Although sheep farmers tended to believe that scrapie was a hereditary disease, there was now mounting evidence that transmission was by sheep eating the afterbirth of an infected animal or grazing in a pasture where infected animals had been kept. Scrapie and kuru seemed to follow the same transmission path: eating. Proof of kuru transmission by cannibalism was clearly shown by its decline among the Fore after the practice was stopped.

It ceased among 4- to 9-year-olds by 1968, among 10- to 14-year-olds by 1972, and among 15- to 19-year-olds by 1973. No child born after the Fore abandoned cannibalism developed kuru. Deaths from kuru among the Fore tribe dropped dramatically. In 1976, Carlton Gajdusek received the Nobel Prize "for . . . discoveries concerning new mechanisms for the origin and dissemination of infectious diseases."

In 1974, a woman who had previously received a corneal transplant came down with CJD. Examination of the donor record showed that the donor died of pneumonia but had a history of memory loss and tremors. It was later believed that the donor had suffered from CJD. Now there was the frightening possibility that CJD was transmissible by transplantation. This suspicion was tragically confirmed when other transplant recipients came down with CJD, as did those who received injections of growth hormone prepared from pituitary glands from human cadavers. Before the development of recombinant DNA methods for the synthesis of human growth hormone, pituitary pools were used as a source of hormone. Between 1966 and 1977, ~500,000 glands were used in the preparation of hormone; if only 1 in 1,000 was from an infected CJD donor, then 25 to 250 infected glands would have been included, and it would be expected that years later, several of the recipients would come down with CJD. They did.

What was the infectious agent for scrapie, kuru, and CJD? Was it the SAF in the amyloid plaques or an unusual virus? In 1976, Stanley Prusiner at the University of California, San Francisco, began carrying out biochemical analyses of SAF and found that, unlike any virus, it contained no nucleic acid but was made entirely of protein. He called the SAF "prion" for *pr*oteinaceous *in*fectious particle, and the new name stuck. Indeed, eliminating a "slow virus" as the culprit for these *t*ransmissible *s*pongiform *e*ncephalopathies (TSEs) could explain the absence of an inflammatory response in brains from victims who died from kuru, CJD, or scrapie. How can a protein-only particle, the prion, multiply without having any self-replicating genetic material such as DNA or RNA? In 1982, Prusiner and his colleagues analyzed the amino acid sequence of the scrapie protein, abbreviated PrP^{sc} (*p*rotein of *pr*ion, *sc*rapie). This 200-amino-acid sequence was then used to work backwards: amino acid sequence $\rightarrow$ RNA nucleotide sequence $\rightarrow$ DNA nucleotide sequence, to find the DNA code (or gene) that specified the prion protein (now abbreviated PrP, for *pr*otein of *p*rion). Surprisingly, the gene for PrP^{sc} from the brains of scrapie-infected animals was identical to the sequence of PrP from normal brains. The only difference between the two was that PrP could be digested with a proteolytic enzyme (such as that found in the pancreas and which is used to digest proteins in our diet) but, despite its identical amino acid composition, PrP^{sc} could not. Prusiner estimated that there are ~1,000 molecules in each SAF (or prion particle) but that in PrP^{sc} these molecules had been changed into a contorted knotted form that could not be digested with proteolytic enzymes (just as the protein in your hair is virtually indigestible by proteolytic enzymes). This of course also explains how the prion is able to resist the action of stomach enzymes and why eating can be a means for transmission. Further, Prusiner also found that if one eliminated the gene for PrP (called a "knockout") in animals susceptible to scrapie, an infection did not develop after brain inoculation of PrP^{sc}. The obvious conclusion from these experiments was that a normal protein, PrP, was changed into a disease form, PrP^{sc}, by a process of protein knotting. How could PrP be reshaped into PrP^{sc} to become infectious? The most plausible hypothesis—and it is only a hypothesis—is by nucleation. It may work, as did "ice-nine" in Kurt Vonnegut's 1963 novel *Cat's Cradle*. In the novel, a brilliant scientist invents a new kind of ice form with different properties from ordinary ice:

> "Now suppose," chortled Dr. Breed, enjoying himself, "that . . . the sort of ice we skate upon and put into highballs—what we might call ice-one—is only one of several types of ice. Suppose water always froze as ice-one on Earth because it never had a seed to teach it how to form ice-two, ice-three, ice-four . . . ? And suppose there were only one form, which we will call ice-nine— a crystal as hard as this desk—with a melting point of, let us say . . . one hundred-and-thirty degrees . . . . When the rain fell, it would freeze into hard little hobnails of ice-nine—and that would be the end of the world."

(This is what happens in *Cat's Cradle* when the nucleating ice-nine is accidentally dropped into the ocean and all the water of the Earth's surface freezes, making life impossible.) You can demonstrate the nucleation process for yourself by dissolving sugar in a glass of hot water until no more will dissolve; then cover the solution with a glass plate, turn off the heat and let it cool. In cooling, the sugar solution is supersaturated, but sugar crystals do not form because they have not been "instructed" as to the form they should take. But add a single sugar crystal, and the solution crystallizes. The crystal "seed" has supplied a template—structural information—that the solution needs to organize itself. Similarly in scrapie, kuru, and CJD, the nucleating form is PrP^{sc}; when it gets into a cell, especially nerve cells, it results in the conversion of PrP into PrP^{sc}—the abnormal form. And with more abnormal forms, there are more nucleating centers, and the cell, to use Vonnegut's words, "freezes into hard little hobnails." Stanley Prusiner received the Nobel Prize in 1996 for his work on prions.

The gene for PrP comes in two forms: M (for the amino acid methionine) and V (for the amino acid valine). If a person inherits a double dose of M, that individual is extremely susceptible to kuru or CJD. Those with one M and one V are less susceptible, and those with two V's are insusceptible. Worldwide, most of us are MV.

What is the function of the PrP, this precursor to neurological pathology? It is hypothesized that in normal cells it acts to turn off the synthesis of proteins when they are no longer needed. In effect, it is a stop sign. But when it is converted into PrP^{sc}, the aggregates lead to its depletion and the cell cannot stop making any of its proteins—the useful ones or the useless ones. The result: the cell dies. In the brain, the death of nerve cells produces holes—spongiform disease. Attractive as this hypothesis is, it doesn't completely fit with the fact that when the PrP gene is "knocked out," the animals appear to be perfectly healthy.

Thus far, we have seen how the work of John Snow, Robert Koch, and Paul Ehrlich provided the basis for Gajdusek and Prusiner to better understand the mode of transmission and the identification of the cause of kuru, but a nagging question still remained: how did a cattle disease ("mad cow") become a human plague? Beginning in the 1980s, veterinarians in Britain identified more and more cases of BSE, bovine spongiform encephalopathy. Each brain from the "mad cows" had the telltale signs of a TSE: holes and knotted protein fibrils. Initially, BSE was found to occur more frequently in dairy cows than in cows that provided meat for beef products. Could the difference once again be related to their diets? The answer seems to be yes. To boost milk production, supplemental protein was provided to dairy cows in the form of meat-and-bone meal, and that dietary supplement probably contained prions. But what is the cause of BSE in non-dairy cattle? Beef cattle are also fed, on occasion, meat-and-bone meal during the "finishing" phase, and calves are also fed this meal to maximize their growth. After dairy cows stop being good milk producers, they are slaughtered, and their meat becomes our steaks and hamburgers. No part of the cow is wasted; cattle brains are also included in the hamburger meat. In the 1980s, a time when CJD became a problem in Britain, the amount of meat-and-bone meal in animal feed was increased from 1 to 2%. Thus, BSE was probably spread by a prion-contaminated food supplement. This would have been a problem for cattle and not humans if there had been a species-specific barrier, but unfortunately (for us), the transmission barrier was not absolute. By 1993, there were reports of two British dairy farmers with CJD, and in 1994, a 16-year-old schoolgirl and an 18-year-old schoolboy in Britain died of CJD. Autopsy of their brains revealed spongiform disease. These cases of CJD in the young were unusual, since the disease is ordinarily found in the elderly; it was clearly a variant and was named vCJD. vCJD resulted in seven more cases in 1996, 77 by 2000, and over 100 by 2002. All were linked to BSE and were attributed to the eating of prion-infected beef. (Of some interest is that during the crisis there were 180,000 cases of BSE in British cattle, although because of underreporting, up to a million may have actually been infected.)

Could humans acquire vCJD by eating animal meats other than those from cattle? Possibly. Conditions for the emergence of a CJD epidemic do exist in the United States. Indeed, even if there were no BSE-infected animals, it may be that with 100 million cattle (as compared to 10 million in Great

Britain), a nucleating PrP^{sc} could spontaneously arise. If fed to other cattle, it could allow for TSEs to incubate without notice. Recycling of cattle protein occurs in the United States as it does in Britain. If this is true, then "mad cow" disease is already with us. And with 77 million Americans eating beef every day, a significant fraction of the population may be at risk for developing a fatal brain plague. Hundreds of thousands of people, but especially the young, may die in the decades ahead because of high-tech cannibalism—recycling animal protein. In this way, the kuru of New Guinea may have come to America. The irony of the TSE saga is that by ending the practice of cannibalism in the Fore tribe, kuru was eliminated, but BSE-related CJD may be the next plague because of the continuance of high-tech cannibalism.

The Rise of Virulence

Our relationship with pathogens is a dynamic one that changes with time—it is not revolutionary, it is evolutionary. As we have seen, populations of organisms contain randomly occurring diversity. When a selective pressure, such as an antibiotic or an insecticide, is applied to a population, it is possible to change the proportion of individuals in the population who are able to survive the environmental pressure, and the survivors may pass on this trait to their offspring—survival of those fittest to reproduce their kind. With time, and with the persistence of such an environment, the genetic structure of the population changes. That's how we get the protective effects of sickle cell trait and Duffy negative (see page 150).

We and all other living creatures are products of evolution—the ultimate game of chance. Bacteria, rickettsias, viruses, fungi, worms, protozoans, insects, and ticks are unaware that they are playing this game, and they know nothing about its rules. The game's rules are really quite simple: play and win, and the prize is the privilege of playing another round of the game. Biologically speaking, those participants who lose are unable to reproduce and thus they are eliminated from the game. Leaving the game is tantamount to extinction, because losers in this game of chance never get another turn; there are no second chances. Most of the players in the game eventually pay this penalty. Those players who enter the next round of play have a chance of winning proportionate to their previous success. That is, the more progeny produced, the greater the chances that some of these will be winners too. However, the game of evolution is complicated by the fact that the rules for selecting the winners can be changed from one round to the next without any predictability. In other words, the feature that allowed a win in the past need not allow a win in the future. In the final analysis, both the environment and the inherited competence of the participants determine those who will be allowed to play another round. In this sense, nature is both adversary and jury in the evolutionary game of chance.

Despite the scarcity of rules, the players' unawareness of the rules, and the lack of goals or purposes of the game, evolution proceeds in an orderly fashion. Only we humans know the rules and the goals of the game. We know how the game is played, we know what constitutes success, and yet we cannot load the dice or fix the game so that we can remain winners in the future. All we can do is try to reduce the risks. Let us consider an example, influenza and virus virulence.

"In Flew Enza"

The 1918 outbreak of influenza remains one of the world's greatest public health disasters. Some have called it the 20th century's weapon of mass destruction. It killed more people than the Nazis did and far more than did the two atomic bombs dropped on Japan. Before it faded away, it affected 500 million people worldwide; 20 to 40 million perished. This epidemic killed more people in a single year than the Black Death did in 100 years. In 24 weeks, influenza destroyed the lives of more people than AIDS has in 24 years. As with AIDS, it killed those in the prime of their life: young men and women in their 20s and 30s. During its 2-year course, more people died from the flu than from any other disease in recorded history.

The 1918 to 1920 flu epidemic brought more than human deaths: civilian populations were thrown into panic, public health measures were ineffectual or misleading, there was a government-inspired campaign of disinformation, and people began to lose faith in the medical profession. Because the flu pandemic killed more than twice the number who died on the battlefields of World War I, it hastened the armistice that ended the Great War in Europe.

Where did this global killer come from? Epidemiological evidence suggests that the outbreak was due to a novel form of the influenza virus that arose among the 60,000 soldiers billeted in the army camps of Kansas. The barracks and tents were overflowing with men, and the lack of adequate heating and of warm clothing forced the recruits to huddle together around small stoves. Under these conditions, they shared both the breathable air and the virus that it carried. By mid-1918, flu-infected soldiers were carrying the disease by rail to army and navy centers on the East Coast and in the South. Then the flu moved inland to the Midwest and onward to the Pacific states. In its transit across America, cases of influenza began to appear among the civilian population.

People in cities such as Philadelphia, New York, Boston, and New Orleans began to ask, what should I do? How long will this plague last? To minimize panic, the health authorities and newspapers claimed it was *la grippe* and there was little cause for alarm. This was an outright lie. But as the numbers of civilian cases kept increasing, and when hundreds of thousands were sick and there were hundreds of deaths each day, it was clear that this flu was serious and that the peak of the epidemic had not even been reached. In Philadelphia, undertakers had no place to put the bodies, and there was a scarcity of coffins. Gravediggers were either too sick or too frightened to bury influenza victims. Entire families were stricken with almost no one to care for them. There were no vaccines or drugs. None of the folk remedies were effective in stemming the spread of the virus, and the only effective treatment was good nursing. In cities, as the numbers of sick continued to soar, public gatherings were forbidden, and gauze masks had to be worn as a public health measure. The law in San Francisco was, if you do not wear a mask, you will be fined or jailed.

Conditions in the United States were exacerbated as the country prepared to enter the war in Europe. President Woodrow Wilson's aggressive campaign to wage total war led indirectly to the nation's becoming "a tinder box for epidemic disease." A massive army was mobilized, and millions of workers crowded into the factory towns and cities where they breathed the same air and ate and drank using common utensils. With an airborne virus such as influenza, this was a prescription for disaster. The war effort also consumed the supply of practicing physicians as well as nurses, so medical and nursing care for the civilian population deteriorated. "All this added kindling to the tinderbox."

The epidemic spread globally, moving outward in ever-enlarging waves. The hundreds of thousands of U.S. soldiers who disembarked in Brest, France, carried the virus to Europe and the British Isles. Flu then moved to Africa via Sierra Leone, where the British had a major coaling center. The dockworkers who refueled the ships contracted the infection, and they spread the highly contagious virus to other parts of Africa when they returned to their homes. In Samoa, a ship arrived from New Zealand, and within 3 months over 21% of the population had died. Similar figures for deaths occurred in Tahiti and Fiji. In a few short years, the flu was worldwide and death followed in its wake.

During the epidemic, there was a rhyme to which little girls jumped rope: "I had a little bird/And its name was Enza/I opened the window/And in flew Enza." This delightful sing-song rhyme did not describe Enza's symptoms, some of which you have probably experienced yourself: fever, chills, sore throat, lack of energy, muscular pain, headaches, nasal congestion, and a lack of appetite. It can escalate quickly to produce bronchitis and secondary infections, including pneumonia, and can lead to heart failure and, in some cases, death.

The influenza virus is highly infectious (it has an R_0 value of 10) and is spread from person to person by droplet infection through coughing and sneezing. Each droplet can contain between 50,000 and 500,000 virus particles. But the virus does not begin its "island hopping" in humans. Instead, aquatic wild birds, such as ducks and other waterfowl, maintain the flu viruses that cause human disease. Because these aquatic birds, which carry the viruses in their intestines, do not get sick and can migrate thousands of miles, the virus can be carried across the face of the earth before humans enter the picture. But the flu virus found in these waterfowl is unable to replicate itself well in humans. To become a human pathogen, it must first move to an intermediate host—usually domestic fowl (chickens, geese, or ducks) or pigs—that will drink water contaminated with the virus-containing feces of the wild waterfowl. The domestic fowl tend to be dead-end hosts, because some sicken and die, but pigs live long enough to serve as "virus mixing vessels." The flu genes of bird, pig, and human can be mixed because pig cells have receptors for both bird and human viruses. It is by this coming together of virus genes that new strains of the flu are produced.

Let us consider the eight genes of the flu virus as if they were a small deck of playing cards. If the eight different cards are shuffled, and two cards are removed from the pack at the same time, how many different combinations are there? The answer: 256, or 2^8. In the same way, if two different viruses infect the same cell and genes are exchanged randomly, there can be 256 different virus offspring. This mixing tends to occur where birds, pigs, and humans live in close proximity—predominantly in China (where the SARS [severe acute respiratory syndrome] outbreak also began). The 1997 Hong Kong "bird flu" was caused by a flu virus that became virulent by acquiring genes from geese, quail, and teal. These birds were housed together in Hong Kong poultry markets, where mixing could occur quite easily. This flu strain killed thousands of chickens before humans were infected. Eighteen people did acquire the infection from contact with the feces of infected chickens, not from other people. Fortunately, the spread to other humans was halted before person-to-person (droplet) transmission could result in a full-blown flu epidemic—health authorities enforced the slaughter of more than a million fowl in Hong Kong's markets. Had this not occurred, one-third of the human population may have sickened and died.

What are the genes that produce the new and virulent flu strains? They involve principally two surface proteins (similar to gp120 of human immunodeficiency virus; see page 92), and both are critical to virus entry into cells where the gene mixing takes place. One is called hemagglutinin (H), and the other is neuraminidase (N). The job of H is to act like grappling hooks to anchor the influenza virus to host cell receptors called sialic acid. After binding, the virus can enter the host cell, replicate its RNA, and produce new viruses. The emerging viruses are coated with sialic acid, the substance that enabled them to attach to the host cell in the first place. If the sialic acid was allowed to remain on the virus and on the host cell, then these new virus particles with H on their surface would be clumped together and trapped much like flies sticking on flypaper. The N, which resembles lollipops on the virus surface, allows the newly formed viruses to dissolve the sialic acid "glue"; this separates the viruses from the host cell and allows them to plow through the mucus between the cells in the airways along the respiratory tract, and to move from cell to cell. The entire process—from anchoring to release—takes about 10 hours, and in that time 100,000 to 1 million viruses can be produced.

There are 15 different H's and 9 different N's, all of which are found in bird flu viruses. A letter and a number designate each flu strain. For example, H1N1 is the strain that caused the 1918 pandemic, and H5N1 is the 1997 Hong Kong flu strain. Flu epidemics occur when either H or N undergoes a genetic change due to a mutation in one of the virus genes. If a virus never changed its surface antigens, as is the case with measles and mumps, then the body could react with an immune response—antibodies and cell-mediated—to the foreign antigens (see page 220) during an infection or by a vaccine, and there would be long-lasting protection. Indeed, if a person encounters the same virus, the immune system, having been primed, will swiftly eliminate that virus, and infection is prevented. But in the case of flu, there may be no immune response, because the virus changes the N and H molecules—sites where the antibodies would ordinarily bind to neutralize the virus. This mutational change of the flu virus (called antigenic drift) ensures escape from immune surveillance and allows the virus to circumvent the body's ability to defend itself. Such slight changes in antigens lead to repeated outbreaks during interpandemic years. Every 10 to 30 years, a more radical and dramatic change in the virus antigens (called antigenic shift) may occur. That means that, approximately every quarter of a century, an influenza virus emerges that the human immune system has never before encountered. When this happens, a pandemic can occur. There were flu pandemics in 1957 (Asian flu) and 1968 (Hong Kong flu) and scares in 1976 (swine flu) and 1977 (Russian flu). In 2004, the H5N1 strain resulted in 27 cases and 20 deaths in Vietnam and 16 cases and 11 deaths in Thailand. Still to be explained is why the 1918 outbreak of influenza was so pathogenic. It may have had something to do with an exaggerated innate inflammatory immune response with a release of lymphokines (see page 227), such as tumor necrosis factor; this can lead to a toxic shock-like syndrome with fever, chills, vomiting, and headache, and, possibly, death. Or perhaps the H1N1 virus that led to the 1918 to 1920 pandemic had a very different kind of H, one more closely related to that of a bird flu, not one from pigs, and for this there was little in the way of immune recognition.

Globally, influenza remains an important contagious disease, with 20% of children and 5% of adults developing symptoms. The death rate can be as high as 5%. Each fall, we are encouraged to get a "flu shot" so that when the flu season arrives in winter we will be protected. But flu shots can only protect against targeted or known strains—those whose antigenic type has been determined by scientists at the World Health Organization—not from unexpected or unidentified types. The flu vaccines used today do protect against a known type, and they do not produce disease—after the virus has been grown in chicken embryos, it is purified and inactivated. Although some vaccines consist of only a portion of the virus, such as H or N antigens, which serve to activate the immune response, weakened live-virus vaccines may give stronger and longer lasting protection. But for these vaccines, it is critical that a return to virulence not occur. Although these vaccines may be administered as a nasal spray, avoiding the pain of inoculation, they may also generate disease symptoms. Flu outbreaks will continue to plague humankind as long as there is "viral mixing," and in those at high risk (i.e., those over 50 years of age, the very young, the chronically ill, and the immunosuppressed), it may be impossible to prevent infection under any circumstances. In addition, those infected with the flu are at greater risk for pneumococcal pneumonia (a bacterial disease), which annually kills thousands of elderly people in the United States. Flu combined with pneumonia can result in skyrocketing mortality.

For a disease such as the flu, with its high R_0 value, quarantining those who show symptoms is not enough to bring the average number of new infections caused by each case to below 1—the level necessary for an epidemic to go into decline. Therefore, other measures, such as treatment and immunization, have to be employed. Flu treatments involve antiviral drugs (zanamivir and oseltamivir) that block the synthesis of N. When administered early enough, the virus dies because complete virus cannot be released. In the case of other antiflu drugs (amantidine and rimantidine), the viruses quickly acquire resistance, so these are less effective in preventing infection or in reducing the severity and duration of symptoms.

Another pandemic of influenza is inevitable. Modern means of transportation—especially jet airplanes—ensure that the virus can be spread across the globe in a matter of hours or within a day by one infected traveler. Surveillance may alert us to the possibility, and drugs may reduce the severity of illness, but neither can guarantee when or where or how lethal the next pandemic will be. What is predictable is that it will affect our lives: hospital facilities will be overwhelmed because medical personnel will also become sick; vaccine production will be slower because many of the personnel in pharmaceutical companies will be too ill to work; and reserves of vaccines and drugs will soon be depleted, leaving most people vulnerable to infection. There will be social and economic disruptions. How can we prepare for a "future shock" such as that of the 1918 to 1920 pandemic? Stockpiling anti-infective drugs, promoting vaccine developments, and increasing the methods for surveillance may all help blunt the effects, but these measures cannot eliminate them.

The Conquest of Plagues

Public perception and understanding can influence medical approaches, health policies, and sociological practices. This being so, the conquest of disease will involve both medicine and social and moral factors, and the reciprocal will also be true; that is, disease can influence the social, political, and economic structure. Epidemics have a social and political meaning as well as a biological one. Yellow fever forced Napoleon to sell Louisiana; yellow fever and malaria prevented the exploration of Africa and contributed to the presence of the United States in Panama. Disease also contributed to the distinctiveness of the southeastern states in the United States. AIDS reminds us how fear, hysteria, and stigmatization can result from the unanticipated appearance of an epidemic disease, and how public health policies can be torn apart by tensions between a person's civil rights and the population's desire for protection. Clearly, xenophobia aided and abetted by disease can act to heighten discrimination.

Here in a modified form (from Daniel Fox) are 10 generalizations about epidemic disease: (i) Never underestimate the severity of the problem when an epidemic occurs. (ii) Expect fear, anxiety, scapegoating, and attempts to segregate and quarantine. (iii) Enlist widespread support for a public health policy or program—include the medical, social, and business communities and the government. (iv) Even with enlightened and powerful science, do not expect a quick fix; it takes time, money, and lots of scientific effort to develop "magic bullets" to control the spread of disease. (v) Education can contribute to the solution, but it must be used in conjunction with other measures. (vi) Voluntary public health programs work better than compulsory ones. (vii) Recognize the importance of the cultural climate. (viii) All diseases cannot completely be eradicated—we will have many more epidemics, but we should be optimistic that it may not require hundreds of years to blunt the effects of an epidemic disease. (ix) Learn the unique natural history of the disease—the nature of the agent and how it passes through a society. (x) Know that disease has influenced our past and will influence our future, but always keep in mind these questions: How will the coming plagues be able to impact our tomorrows? Are we prepared for the next threat?

Appendix

Cells and Viruses

In the living world, the fundamental unit is the cell, and all the processes of life occur within cells. A cell may be defined as the standard of biological activity, bounded by a membrane, and able to reproduce itself independently of any other living system. All living organisms—large and small, plant and animal, human or microbe, fungus or fly, trypanosome or tapeworm—all are made up of cells. All cells are basically similar to one another, having many structural features in common. How can we recognize a cell, and what do we mean when we say it is the standard unit upon which all living systems are based? We can more easily understand the concept of a cell by making an analogy and comparing living organisms to buildings. The rooms of buildings are analogous to the cells of the organism.

Rooms and cells have boundaries with exits and entrances—rooms have walls, floors, ceilings, doors, and windows, whereas cells have walls and membranes with pores of various sizes. Rooms and cells come in a variety of shapes and sizes, and each kind of room may have its own particular kind of use, function, or specialty. Buildings may be composed of one room or many rooms. In the same way, organisms may be composed of one cell, in which case we describe them as unicellular, or of many cells, in which case they are referred to as multicellular. By combining rooms of various kinds, we can construct a variety of buildings—homes, apartments, offices, schools, and so on. Similarly, different organisms are constructed of different cell types. Just as there are no buildings without rooms, there is no life without cells.

Cells are microscopic in size and are separated from their environment by an interface or plasma membrane (Fig. A1a). In some cells, the outer surface of the plasma membrane is wrapped around with a cell wall. Everything within the plasma membrane is referred to as the cytoplasm, a jellylike material containing various structures collectively known as organelles, or "little organs." Continuing our analogy of cells with rooms, organelles might be thought of as items of furniture or equipment (such as stoves, sinks, refrigerators) that have a specialized function. Organelles are nucleus, mitochondrion,

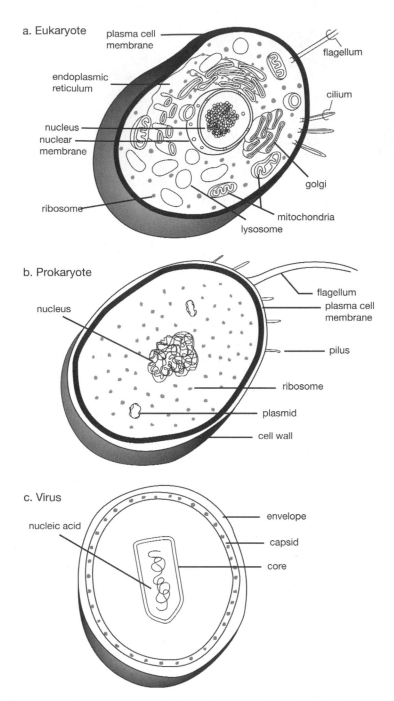

Figure A1   The appearance of cells and a virus. Although all are shown to be the same size, in actual fact they vary widely in size. The eukaryote cell (a) has been enlarged 1,000 times (actual size, 50 μm), the prokaryote cell (b) has been enlarged 25, 000 times (actual size, 2 μm), and the virus (c) has been enlarged 500,000 times (actual size, 0.1 μm).

the ribosome, endoplasmic reticulum, and Golgi apparatus (and in green plants the chloroplast).

On the basis of size, structure, and nutritional habits, we can generally divide organisms into plants (those that contain a cellulose wall and pigment, make their own food, and do not move about), fungi (those that have cell walls, are colorless, absorb their food, and do not move about), and animals (colorless, motile organisms that cannot make their own food).

Those cells with their hereditary material (nucleic acids) enclosed in a membrane (nuclear membrane) are called eukaryotes, and those with nucleic acid not enclosed in a membrane are called prokaryotes (Fig. A1b). Examples of prokaryotes are bacteria, rickettsias, and blue-green algae. The cytoplasm of prokaryotes contains no mitochondria, endoplasmic reticulum, or Golgi apparatus but is richly endowed with ribosomes; some prokaryotes can move about by means of whiplike structures called flagella. In eukaryotes, the nucleic acid is combined with protein to form chromosomes, which are enclosed by the nuclear membrane; they may move by means of cilia, flagella, or pseudopods and usually have the full array of organelles.

Viruses, smaller than bacteria, come in a variety of shapes and sizes, but they all have the same basic structure: an outer protein coat and a central core of double-stranded nucleic acid, either RNA or DNA (Fig. A1c). Viruses do not have a plasma membrane, and their genetic code is incomplete—so although the DNA or RNA contains all the information for assembling new virus particles, the information that is needed for reproduction is lacking. To reproduce, a virus must enter a living cell and use the machinery of that cell to produce its offspring. Because viruses are not completely independent, some biologists say they are not alive. However, viruses can be killed if their DNA or RNA is destroyed by irradiation or drugs, and they are able to change their genetic code by mutation. Viruses can be crystallized, just as table salt can be crystallized, and still remain infective when the crystals are dissolved and come in contact with a suitable host cell. All viruses are parasitic; however, in some cases, there is no evidence of injury to the host. Most of the viruses we are aware of cause ill effects that we call disease and therefore are regarded as pathogenic. Viruses are responsible for such maladies as acquired immuno-deficiency syndrome (AIDS), rabies, smallpox, polio, influenza, the common cold, fever blisters, genital herpes, mumps, measles, yellow fever, viral pneumonia, infectious hepatitis, severe acute respiratory syndrome (SARS), Ebola and Junin fever, and hantavirus pulmonary syndrome. Some cancers, e.g., Rous sarcoma, are caused by viruses. Viruses cannot be grown on nonliving substances but must have a living cell (either prokaryotic or eukaryotic) to supply the raw materials and enzymatic machinery necessary for their reproduction. The virus nucleic acid directs the living host cell to perform the actual synthetic operation of viral replication. Viruses, <0.1 μm in size, are too small to be seen with the light microscope but can be seen with the electron microscope.

General Works on Disease and History

A selected list of references and sources is provided for those interested in reading further. Extensive bibliographies have been avoided. In most cases, references are to more recent works since these often have extensive bibliographies of older literature.

Ackerknecht, E. H. *History and Geography of the Most Important Diseases*. New York: Hafner, 1965.

Barker, R. *And the Waters Turned to Blood*. New York: Simon & Schuster, 1997.

Berenbaum, M. *Bugs in the System: Insects and Their Impact on Human Affairs*. Boston: Addison-Wesley, 1995.

Bollet, A. J. *Plagues and Poxes. The Rise and Fall of Epidemic Disease*. New York: Demos, 1987.

Buckman, R. *Human Wildlife. The Life That Lives on Us*. Baltimore: Johns Hopkins University Press, 2003.

Cartwright, F. F. *Disease and History*. New York: Crowell, 1972.

de Kruif, P. *Microbe Hunters*. San Diego: Harcourt Brace, 1954.

Garrett, L. *The Coming Plague*. New York: Farrar, Straus & Giroux, 1994.

Goudsmit, J. *Viral Fitness: The Next SARS and West Nile in the Making*. New York: Oxford University Press, 2004.

Grob, G. *The Deadly Truth: A History of Disease in America*. Cambridge: Harvard University Press, 2002.

Hays, J. *The Burdens of Disease*. New Brunswick: Rutgers University Press, 1998.

Hudson, R. *Disease and Its Control. The Shaping of Modern Thought*. New York: Greenwood, 1983.

Karlen, A. *Man and Microbes*. New York: Putnam, 1995.

Kiple, K. F. *The Cambridge World History of Human Disease*. Cambridge: Cambridge University Press, 1993.

Knapp, V. *Disease and Its Impact on Modern European History*. Lewiston, N.Y.: Edwin Mellen, 1989.

Krause, R. M. *Emerging Infections*. San Diego: Academic Press, 2000.

Loudon, I., ed. *Western Medicine, An Illustrated History*. Oxford: Oxford University Press, 1997.

McNeill, W. H. *Plagues and Peoples*. New York: Doubleday, 1976.

Nesse, R., and G. Williams. *Why We Get Sick. The New Science of Darwinian Medicine*. New York: Random House, 1994.

Oldstone, M. *Viruses, Plagues and History*. New York: Oxford University Press, 1998.

Porter, R. *The Greatest Benefit to Mankind: A Medical History of Humanity from Antiquity to the Present*. London: Harper Collins, 1997.

Rosenberg, C. *Explaining Epidemics and Other Studies in the History of Medicine*. Cambridge: Cambridge University Press, 1992.

Sankaran, N. *Microbes and People. An A-Z of Microorganisms in Our Lives*. Phoenix, Ariz.: Oryx, 2000.

Singer, C., and E. A. Underwood. *A Short History of Medicine*, 2nd ed. Oxford: Clarendon Press, 1962.

Spielman, A., and M. D'Antonio. *Mosquito. A Natural History of Our Most Persistent and Deadly Foe*. New York: Hyperion, 2001.

Watts, S. *Epidemics and History*. New Haven: Yale University Press, 1997.

Wills, C. *Yellow Fever Black Goddess: The Co-evolution of People and Plagues*. Boston: Addison-Wesley, 1996.

Winslow, C. E. A. *The Conquest of Epidemic Disease, A Chapter in the History of Ideas*. New York: Hafner, 1967.

Notes

In researching this book, I relied on primary and secondary literature sources. These chapter notes consist of references and comments and are keyed to page numbers.

Chapter 1. The Nature of Plagues

Page 1. M. Chien, I. Morozova, S. Shi, et al., "The genomic sequence of the accidental pathogen *Legionella pneumophila*," *Science* 305 (2004): 1966–1968. Describes the genomic sequence of *Legionella* and has many references to the outbreak. In addition, there is a website on *Legionella*: http://www.cdc.gov/ncidad/diseaseinfo/legionellosis_g.html.

Page 2. R. Hajjeh et al., "Toxic shock syndrome in the United States: Surveillance update, 1979–1996," *Emerging Infectious Diseases* 5 (1999): 807–810; L. K. Altman, *Who Goes First? The Story of Self-Experimentation in Medicine* (Berkeley: University of California Press, 1998), 195–199; D. Davis et al., "Toxic shock syndrome: A case report of a postpartum female and a literature review," *Journal of Emerging Medicine* 16 (1998): 607–614.

Page 3. K. Tsang et al., "A cluster of cases of severe acute respiratory syndrome in Canada," *New England Journal of Medicine* 348 (2003): 1977–1985.

Page 4–10. L. Roberts and J. Janovy, *Foundations of Parasitology*, 6th ed. (New York: McGraw Hill, 2000); B. Buckman, *Human Wildlife: The Life That Lives on Us*. Baltimore: Johns Hopkins University Press, 2003.

Page 11–12. J. W. Leavitt, *Typhoid Mary: Captive to the Public Health* (Boston: Beacon Press, 1996).

Page 13. For an infection to persist in the population, each individual on average must transmit the infection to at least one other individual. The number of individuals each infected person infects at the beginning of an epidemic is given by R_0; this is the basic reproductive ratio of the disease or, more simply, the multiplier of the disease. The multiplier helps to predict how fast a disease will spread through the population. The simplest way of obtaining a value for R_0 is to multiply the average probability per contact per unit time (β) by the duration of infectiousness (D) by the number of contacts per unit time (c). See also *Bioscience* 46 (1996): 115; *Journal of Theoretical Biology* 169 (1994): 253.

Page 14. "SARS, lay epidemiology, and fear," *The Lancet* 361 (2003): 1739; C. Dye and N. Gray, "Modeling the SARS epidemic," *Science* 300 (2003): 1884–1885.

Page 14–15. Oldstone, *Viruses, Plagues and History*, chap. 6, p. 73–79.

Page 15–16. A. Cliff and P. Haggert, "Island epidemics," *Scientific American* 250 (1984): 138–147; M. J. Keeling, "Modeling the persistence of measles," *Trends in Microbiology* 5 (1997) 513–518.

Page 17. The calculation uses the formula $R_0 = \beta Dc$, where β is the average probability of contact, D is the duration of infectiousness, and c is the number of contacts per unit time. For HIV, $D = 0.5$, $c = 0.2$, and, if $R_0 = 1$, $\beta = 10$.

Page 17–18. R. Anderson and R. May, "Vaccination and herd immunity to infectious diseases," *Nature* 318 (1985): 323–329; R. Lenski and R. May, "The evolution of virulence in pathogens," *Journal of Theoretical Biology* 169 (1994): 253–265.

Page 18–19. McNeill, *Plagues and Peoples*, 9.

Page 19. G. Dwyer, S. Levin, and L. Puttel, "A simulation model of the population dynamics and evolution of myxomatosis," *Ecological Monographs* 60 (1990): 423–447; D. Ebert and J. Bull, "Challenging the trade-off model for the evolution of virulence: is virulence management feasible?" *Trends in Microbiology* 11 (2003):15–20.

Page 19. McNeill, *Plagues and Peoples*, 50–51.

Page 20–21. P. Ewald, *Evolution of Infectious Diseases* (New York: Oxford University Press, 1994); Nesse and Williams, *Why We Get Sick: The New Science of Darwinian Medicine* (New York: Random House, 1994).

Chapter 2. Plagues, the Price of Being Sedentary

Page 23. I. Tattersall, "Once we were not alone," *Scientific American* 282 (2003): 56–62; M. Leakey and A. Walker, "Early hominid fossils," *Scientific American* 276 (1997): 74–79; *Walking with Cavemen*, video (BBC, 2003).

Page 24. D. Johanson and B. Edgar, *From Lucy to Language* (New York: Simon & Schuster, 1996); R. Burenhalt, ed., *The Illustrated History of Human Kind* (London: Harper Collins, 1993); R. Leakey and R. Lewis, *Origins Reconsidered* (New York: Doubleday, 1992).

Page 26. R. N. Fiennes, *Zoonoses and the Origins and Ecology of Human Disease* (London: Academic Press, 1978); R. Larnick and R. Ciochon, "The African emergence and early Asian dispersal of the genus Homo," *American Scientist* 84 (1996): 538–551.

Page 26–27. S. Jones, R. Martin, and D. Pilbeam, eds., *The Cambridge Encyclopedia of Human Evolution* (Cambridge: Cambridge University Press, 1992); R. Lewin, *Human Evolution: An Illustrated Introduction*, 4th ed. (London: Blackwell, 1999).

Page 28. E. Trinkaus and P. Shipman, *The Neanderthals* (New York: Knopf, 1993); C. Stringer and C. Gamble, *In Search of the Neanderthals* (London: Thames & Hudson, 1993); I. Tattersall, *The Last Neanderthal* (New York: Macmillan, 1995).

Page 30. A. Coale, "The history of the human population," *Scientific American* 231 (1974): 41–51; E. S. Deevey, Jr., "The human population." *Scientific American* 203 (1960): 195–204.

Page 30–31. T. Malthus, *An Essay on Population* (London: Dent and Sons, 1927–1928).

Page 32–35. J. Diamond, *Guns, Germs and Steel* (New York: Norton, 1997), 108, 111, 89, 90, 208.

Page 37. McNeill, *Plagues and Peoples*, 67.

Page 37–40. Diamond, *Guns, Germs and Steel*. This discussion is based on chaps. 4, 5, and 6, p. 85–113.

Chapter 3. Six Plagues of Antiquity

Page 44. E. V. Hulse, "Joshua's curse and the abandonment of ancient Jericho." *Medical History* 15 (1971): 376–386.

Page 46–47. McNeill, *Plagues and Peoples*, 39.

Page 47–52. W. D. Foster, *A History of Parasitology* (Edinburgh: Livingstone, 1965); L. Roberts and J. Janovy, *Foundations of Parasitology*, 6th ed. (New York: McGraw Hill, 2000); J. Farley, *Bilharzia. A History of Imperial Tropical Medicine* (Cambridge: Cambridge University Press, 1992).

Page 53–55. F. Cartwright, and M. Biddis, *Disease and History*, 2nd ed. (Phoenix Mill, U.K.: Sutton, 2000) 5–7; M. Grmek, *Diseases in the Ancient Greek World* (Baltimore: Johns Hopkins University Press, 1989).

Page 55. See the website http://www.archeology.org/online/news/kerameikos.html. It has been estimated that the population in Athens was 250,000 to 300,000, and the total number of deaths was between 65,000 and 78,000. See also P. Olsen et al., "The Thucydides syndrome: Ebola deja vu," *Emerging Infectious Diseases* 2 (1996): 152–153.

Page 55–56. Thucydides, *Peloponnesian Wars*, Book 2, chaps. 47–52.

Page 56–59. A. Celli, *A History of the Roman Campagna* (London: Bale, 1933); R. Sallares, *Malaria and Rome* (Oxford: Oxford University Press, 2002). Before the Roman conquest of the Italian peninsula, the local population had a life expectancy of 28 to 42 years, and 5 to 15% of the children died during the first 10 years of life. After the conquest, life expectancy declined to ~27 years, and infant mortality increased to ~25%.

Page 60–61. Cartwright, *Disease and History*, 19–21.

Page 61–64. McNeill, *Plagues and Peoples*, 104–107, 108, 109–114; "Logic, learning and experimental medicine," *Science* 295 (2000): 800–801.

Page 64. Most scientists accept that the plague of Justinian was bubonic plague; however, a recent book—S. Scott and C. J. Duncan, *Biology of Plagues: Evidence from Historical Populations* (Cambridge: Cambridge University Press, 2001)—using epidemiological analyses contends that it was some other infectious disease, such as Ebola. See also J. Wood, R. Ferrell, and S. N. Dewitt-Avina, "The temporal dynamics of the 14th century black death: new evidence from ecclesiastical records," *Human Biology* 75 (2003): 427–448.

Bratto calculates that given an estimated population of ~290,000, about 115,000 would have contracted plague, and with a fatality rate of 20%, about 58,000 would have died. T. L. Bratto, "The identity of the Plague of Justinian (Part II)," *Transactions and Studies of the College of Physicians of Philadelphia* 3 (1981): 174–180.

Page 65. McNeill, *Plagues and Peoples*, 137.

Chapter 4. An Ancient Plague, the Black Death

General

Bray, R. S., *Armies of Pestilence. The Effects of Pandemics on History*. Cambridge: Butterworth, 1996. Chaps. 6, 7, 8, and 9.

Defoe, D. *A Journal of the Plague Year*. New York: Oxford, 1969.

Drancourt, M., and Raoult, D. "Molecular insights into the history of plague." *Microbes and Infection* 4 (2002): 105–109.

Gottfried, R. *The Black Death. Natural and Human Disaster in Medieval Europe*. New York: The Free Press, 1983.

Gregg, C. *Plague. An Ancient Disease in the 20th Century*. Albuquerque: University of New Mexico Press, 1985.

Herlihy, D. *The Black Death and the Transformation of the West*. Cambridge: Harvard University Press, 1997.

McEvedy, C. "The bubonic plague." *Scientific American* 258 (1988): 118–123.

Mee, C. L., Jr. "How a mysterious disease laid low Europe's masses." *Smithsonian* (Feb. 1990): 67–75.

Porter, R. *The Cambridge Illustrated History of Medicine*. New York: Cambridge University Press, 1996.

Slack, P. *The Impact of Plague in Tudor and Stuart England*. London: Routledge & Kegan Paul, 1985.

Tuchman, B. *A Distant Mirror: The Calamitous 14th Century*. New York: Ballantine, 1978.

Twigg, G. *The Black Death. A Biological Reappraisal*. New York: Schocken, 1985.

Ziegler, P. *The Black Death*. New York: Harper, 1969.

Specific

Page 67. Herlihy, *The Black Death*, 84–100.

Page 68–69. R. Browning, *The Pied Piper*, 1888.

Page 69–70. G. Boccaccio, *The Decameron*, quoted in Ziegler, *The Black Death*, 46.

Page 72–73. Ziegler, *The Black Death*, 58.

Page 73–74. Herlihy, 59–82.

Page 74–75. Ziegler, 84–109.

Page 75–76. Herlihy, 80.

Page 75–76. Ziegler, 84–109.

Page 76–78. Herlihy, 68–72.

Page 78–81. Herlihy, 46–52.

Page 81–82. E. Bender, "Alexandre Yersin: pursuer of plague," *Hospital Practice* (March 30, 1989): 121–148.

Page 83. L. Gross, "How the plague bacillus and its transmission through fleas were discovered. Reminiscences from my years at the Pasteur Institute in Paris," *Proceedings of the National Academy of Sciences USA* 92 (1995): 7609–7611.

Page 83–85. E. Carniel, "Plague," in *Encyclopedia of Microbiology*, 2nd ed., 3 (2000): 654–661, quoted by Ziegler, p. 18.

Page 85. M. Achtman et al., "Yersinia pestis, the cause of plague, is a recently emerged clone of Yersinia pseudotuberculosis," *Proceedings of the National Academy of Sciences USA* 96 (1999): 14043–14048.

Page 86–87. http://www.jewishgen.org/Ukraine/Photo_Album/Stamps/haffkine.htm.

Chapter 5. A Modern Plague, AIDS

General

Fan, H., R. Connor, and L. Villareal. *AIDS Science and Society*, 4th ed. Sudbury Mass.: Jones and Bartlett, 2004.

Stine, G. *AIDS Update 2005*. San Francisco: Pearson Benjamin Cummings, 2005.

Alcamo, I. E. *AIDS: The Biological Basis*, 3rd ed. Sudbury, Mass.: Jones and Bartlett, 2003.

Specific

Page 89. *Time* 157 (February 12, 2001): 37.

Page 90. Stine, *AIDS Update 2005*.

Page 90–91. R. C. Gallo, "The AIDS virus," *Scientific American* 256 (1987): 47–56.

Page 92. G. Simpson, C. Pittendrigh, and L. Tiffany, *Life* (San Francisco: Harcourt Brace, 1957), p. 18.

Page 94. "A notable career in finding out: Peyton Rous, 1879–1970," Rockefeller University *Occasional Paper* 16, 1971; H. Lechevalier, *Three Centuries of Microbiology* (New York: Dover, 1974), 282.

Page 95–96. G. M. Cooper, R. G. Temin, and B. Sugden, eds., *The DNA Provirus. Howard Temin's Scientific Legacy* (Washington, D.C.: ASM Press, 1995).

Page 96–97. Alcamo, *AIDS*.

Page 97–98. http://www.pbs.org/wnet/secrets/case-plague/interview.html.

Page 98–99. This view has been contested by A. Galvani and M. Slatkin ("Evaluating plague and smallpox as historical selective pressures for the CCR-d32 HIV-resistance allele," *Proceedings of the National Academy of Sciences USA* 100 [2003]: 15276–15279), who suggest that "smallpox alone can account for current frequencies of the HIV-1 resistance allele."

Page 99–102. Alcamo, *AIDS*, 146–147.

Page 103–104. P. Aggleton et al., "Risking everything? Risk behavior, behavior change, and AIDS," *Science* 265 (1994): 341–345; J. M. Blower and A. R. McLeary, "Prophylactic vaccines, risk behavior change, and probability of eradicating HIV in San Francisco," *Science* 265 (1994): 1451–1454.

Page 105. B. Schwartlander et al., "Resource needs for HIV/AIDS," *Science* 292 (2001): 2434–2436.

Page 105–111. J. Goudsmit, *Viral Sex. The Nature of AIDS* (New York: Oxford University Press, 1997); J. Moore, "The puzzling origins of AIDS," *American Scientist* 92 (2004): 540–549.

Page 111–112. S. Darby et al., "Mortality before and after HIV infection in the complete UK population by haemophiliacs," *Nature* 377 (1995): 79–82; P. Duesberg, *Infectious AIDS*

(Berkeley: North Atlantic Books, 1995); "The Duesberg phenomenon," *Science* 266 (1994): 1642–1648; E. Hooper, *The River* (New York: Little Brown, 1999); H. Poiner, M. Kuch, and S. Parks, "Molecular analyses of oral polio vaccine samples," *Science* 292 (2004): 743–744; *EMBO Reports* 4 (2003): S10–S14.

Page 112–115. R. A. Weiss, "HIV and AIDS in relation to other pandemics," *Science* 304 (2004): 1932–1938; J. Cohen, "Asia and Africa: on different trajectories," *Science* 301 (2003): 1650–1663; "The next frontier for AIDS," *Nature Medicine* 9 (2003); 839–842; A. S. Fauci, "HIV and AIDS: 20 years of science," *Nature Medicine* 9 (2003): 839–843.

Chapter 6. Typhus, a Fever Plague

Page 118, 124. H. Zinsser, *Rats, Lice and History* (Boston: Little Brown, 1935).

Page 119–120. Cartwright, *Disease and History*, first ed., 82–112.

Page 121. R. K. D. Peterson, "Insects, disease and military history," *American Entomologist* (Fall 1995): 147–160.

Page 122–123. L. Gross, "How Charles Nicolle of the Pasteur Institute discovered that epidemic typhus is transmitted by lice: Reminiscences from my years at the Pasteur Institute in Paris," *Proceedings of the National Academy of Sciences USA* 93 (1996): 10539–10540.

Page 123. Lechevalier, *Three Centuries of Microbiology*, 325–332.

Pages 123–125. J. W. Maunder, "The appreciation of lice," *Proceedings of the Royal Institution of Great Britain* 55 (1983): 1–31.

Page 124. Zinsser, *Rats, Lice and History*.

Page 126–127, 132. D. Raoult and V. Roux, "The body louse as a vector of reemerging human diseases," *Clinical Infectious Diseases* 29 (1999): 888–911.

Page 128–132. H. Markell, *Quarantine!* (Baltimore: Johns Hopkins University Press, 1997).

Page 132. W. Szybalski, "Maintenance of human-fed live lice in the laboratory and production of Weigl's exanthematic typhus vaccine," in *Maintenance of Human, Animal and Plant Pathogen Vectors*, eds. K. Maramorosch and F. Mahmood (Enfield, N.H.: Science Publishers, 1999), 161–180.

Page 133. Zinsser, *Rats, Lice and History*.

Chapter 7. Malaria, Another Fever Plague

Page 135. R. Kapuscinsk, *Shadow of the Sun* (New York: Vantage Books, 2002).

Page 135–136. I. Sherman, "A brief history of malaria and discovery of the parasite life cycle," in *Malaria. Parasite Biology, Pathogenesis and Protection*, ed. I. W. Sherman (Washington, D.C.: ASM Press, 1998), 3–10.

Page 137–143. R. Ross, *Memoirs* (London: John Murray, 1923); E. R. Nye and M. E. Gibson, *Ronald Ross: Malariologist and Polymath—A Biography* (London: Macmillan, 1997); W. Bynum and C. Ovary, *The Beast in the Mosquito: The Correspondence of Ronald Ross and Patrick Manson* (Amsterdam: Editions Rudolphi, 1998); P. Russell, *Man's Mastery of Malaria* (London: Oxford University Press, 1955); G. Harrison, *Mosquitoes, Malaria and Man. A History of the Hostilities Since 1880* (New York: Dutton, 1978).

Page 144. de Kruif, *Microbe Hunters*, 256–285.

Page 147–149. S. Meshnick, "From quinine to qinghaosu. Historical perspective," in Sherman, 341–354.

Page 150–152. A. Hill and D. Weatherall, "Host genetic factors in resistance," in Sherman, 445–456.

Page 153–154. W. Trager and J. Jensen, "Human malaria parasites in continuous culture," *Science* 193 (1978): 673–675.

Page 154–155. J. Barnwell and M. Galinski, "Invasion of vertebrate cells: erythrocytes," in Sherman, 93–122.

Page 156. C. Beadle and S. Hoffman, "History of malaria in the United States naval forces at war: World War I through the Vietnam Conflict," *Clinical Infectious Diseases* 16 (1993): 320–329.

Chapter 8. King Cholera

General

Grob, G. N. *The Deadly Truth. A History of Disease in America.* Cambridge: Harvard University Press, 2002.

Rosenberg, C. *The Cholera Years: The United States in 1832, 1849 and 1866.* Chicago: University of Chicago Press, 1962.

Specific

Page 159. J. Franklin and J. Sutherland, *Guinea Pig Doctors* (New York: Morrow, 1984), 181–182.

Page 160. P. Johansen et al. *Cholera, Chloroform and the Science of Medicine. A Life of John Snow* (New York: Oxford University Press, 2003); J. Snow, *On the Mode of Communication of Cholera* (London: Churchill, 1855). An excellent website is http://www.ph.ucla.edu/epi/snow.html.

Page 160. G. Bordenhave, "Louis Pasteur," *Microbes and Infection* 5 (2003): 553–555; P. Debre, *Louis Pasteur* (Baltimore: Johns Hopkins University Press, 1998); G. Geison, *The Private Science of Louis Pasteur* (Princeton, N.J.: Princeton University Press, 1995).

Page 161–162. *Essays of Robert Koch*, translated by K. C. Carter (New York: Greenwood Press, 1987); T. D. Brock and Robert Koch, *A Life in Medicine* (Washington, D.C.: ASM Press, 1999); R. Munch, "R. Koch," *Microbes and Infection* 5 (2003): 69–74.

Page 162–165. I. W. Sherman and V. Sherman, *Biology: A Human Approach* (New York: Oxford University Press, 1983); de Kruif, *Microbe Hunters*, 3–96.

Page 165. Simpson et al., *Life*.

Page 166–167. McNeill, *Plagues and Peoples*, 231.

Page 167–170. D. Barua, "History of cholera," in *Cholera*, eds. D. Barua and W. B. Greenough (New York: Plenum, 1992), 1–24; G. C. Cook, "The Asiatic cholera: an historical determinant of human genomic and social structure," in *Cholera and the Ecology of Vibrio cholerae*, eds. B. Drasar and B. D. Forest (London: Chapman and Hall, 1996), 18–53.

Page 170–171. P. Quinton, "What is good about cystic fibrosis?" *Current Biology* 4 (1994): 742–743.

Page 171. E. Ryan and S. Calderwood, "Cholera vaccines," *Clinical Infectious Diseases* 31 (2001): 561–565.

Page 172. R. Colwell, "Global climate and infectious disease: The cholera paradigm," *Science* 274 (1996): 2025–2031; C. Dold, "The cholera lesson," *Discover* (Feb. 1999), 71–75.

Page 173–175. M. Kaufman, "Blue cholera," *The Lancet* 340 (1992): 837; R. Guerrant, B. Carneiro-Filho, and R. Dilhingham, "Cholera, diarrhea and oral rehydration therapy: triumph and indictment," *Clinical Infectious Diseases* 37 (2003): 398–405.

Page 176–178. Cartwright, *Disease and History*, first ed., 157–166.

Page 178. F. Cartwright, *A Social History of Medicine* (London: Longman, 1977).

Page 178–180. G. Gill, S. Burrell, and J. Brown, "Fear and frustration—the Liverpool cholera riots of 1832," *The Lancet* 358 (2001): 233–237.

Page 180–181. Cartwright, *A Social History of Medicine*; S. Finer, *The Life and Times of Edwin Chadwick* (London: Methuen, 1952); C. Hamlin, *Public Health and Social Justice in the Age of Chadwick* (Cambridge: Cambridge University Press, 1997).

Page 181–182. J. Snow, *On the Mode of Communication of Cholera* (London: John Churchill, 1885).

Page 182–184. E. Huxley, *Florence Nightingale* (London: Weidenfeld and Nicholson, 1975). More critical views of Florence Nightingale can be found in *Eminent Victorians. Lytton*

Strachey (London: Continuum, 2002); H. Small, *Florence Nightingale, Avenging Angel* (London: Constable, 1998); and F. B. Smith, *Florence Nightingale: Reputation and Power* (New York: St. Martin's Press, 1982).

Page 184–187. http://www.entrenet.com/~groedmed/grosseisle/gi.html; http://ballinagree.freeservers.com/sumsorrow.html.

Chapter 9. Smallpox, the Spotted Plague

Page 191. J. L. Carrell, *The Speckled Monster. A Historical Tale of Battling Smallpox* (New York: Dutton, 2003); Cartwright and Biddis, *Disease and History*, 2nd ed., 63–82; Diamond, *Guns, Germs and Steel*, 157–175.

Page 191–192. J. N. Shurkin, *The Invisible Fire. The Story of Mankind's Victory Over the Ancient Scourge of Smallpox* (New York: Putnam, 1979); H. J. Parish, *Victory with Vaccines. The Story of Immunization* (Edinburgh: Livingstone, 1968).

Page 192–195. Oldstone, *Viruses, Plagues and History*, 27–44.

Page 195–202. D. R. Hopkins, *Princes and Peasants* (Chicago: University of Chicago Press, 1983); C. W. Dixon, *Smallpox* (London: Churchill, 1962); I. Glynn and J. Glynn, *The Life and Death of Smallpox* (Cambridge: Cambridge University Press, 2004); F. Fenner, D. Henderson, I. Arita, Z. Jezek, and I. Ladnyi, *Smallpox and Its Eradication* (Geneva, World Health Organization, 1988); J. Eyler, "Smallpox in history," *Journal of Laboratory and Clinical Medicine* 142 (2003): 216–220; D. Baxby, *Jenner's Smallpox Vaccine* (London: Heineman, 1981); H. Bazin, *The Eradication of Smallpox* (San Diego: Academic Press, 2000).

Page 196–197. R. Preston, *The Demon in the Freezer* (New York: Random House, 2002).

Page 202–203. S. Bernstein, "Smallpox and variolation: their historical significance in the American colonies," *Journal of Mount Sinai Hospital* 18 (1951): 228–244.

Page 204. P. J. Read, "Benjamin Jesty: new light in the dawn of vaccination," *The Lancet* 362 (2003): 2104–2109.

Page 207. P. Berche, "The threat of smallpox and bioterrorism," *Trends in Microbiology* 9 (2001): 15–18; C. McClain, "A new look at an old disease: smallpox and biotechnology," *Perspectives in Biology and Medicine* 38 (1994): 624–639; T. O'Toole, "Smallpox: an attack scenario," *Emerging Infectious Diseases* 5 (1999): 540–546.

Page 205–208. Oldstone, *Viruses, Plagues and History*, 27–44.

Page 202, 209. M. Albert, K. Ostheimer, and J. Bremen, "The last smallpox epidemic in Boston and the vaccination controversy, 1901–03," *New England Journal of Medicine* 344 (2001): 375–379; J. Farrell, *Invisible Enemies. Stories of Infectious Disease* (New York: Farrar, Straus, Giroux, 1998); J. Duffy, *Epidemics in Colonial America* (Baton Rouge: LSU Press, 1953); H. N. Simpson, *Invisible Armies: The Impact of Disease on American History* (Indianapolis: Bobbs-Merrill, 1980).

Chapter 10. Preventing Plagues

Page 211, 218–220. W. R. Clark, *At War Within: The Double-Edged Sword of Immunity* (New York: Oxford University Press, 1995).

Page 212–213. R. Goldsby et al., *Immunology*, 5th ed. (New York: Freeman, 2003).

Page 213. A. Tauber, "Metchnikoff and the phagocytosis theory," *Nature Reviews/Molecular Cell Biology* 4 (2003): 897; de Kruif, *Microbe Hunters*, 190–212.

Page 213–217. A. M. Silverstein, *A History of Immunology* (San Diego, Academic Press, 1989); A. M. Silverstein, *Milestones in Immunology. A Historical Exploration* (Madison: Science Tech Publishers, 1988).

Page 213–217. Pier, G., J. Lyczak, and L. M. Wetzler. *Immunology, Infection and Immunity* (Washington, D.C.: ASM Press, 2004).

Page 217–219, 224–226. C. Mims, *The War Within Us: Everyman's Guide to Infection and Immunity* (San Diego: Academic Press, 2000); P. J. Delver and I. Roitt, "The immune system," *New England Journal of Medicine* 343 (July 6, 2000): 37–49, and 343 (July 13, 2000): 106–115; R. Medzhitov and C. Janeway, "Innate immunity," *New England Journal of Medicine* 343 (2000): 338–343.

Page 220–223. L. O'Neill, "Immunity's early-warning system," *Scientific American* 292 (Jan. 2005); 36–43.

Page 223. G. Stollerman, "Rheumatic fever in the 21st century," *Clinical Infectious Diseases* 33 (2001): 804–812.

Chapter 11. The Plague Protectors

Page 231–232. A. Lyons and R. Petrucelli, *Medicine. An Illustrated History* (New York: Abradale, 1987).

Page 232–234. http://www.barberpole.com/art of.htm; http://www.pbs.org/kqed/demonbarber/bloodletting/the humours.html; http://www.clevelandopera.org/four/educational/ncoc/barbers.html; http://www.rcseng.ac.uk/wellcome/history_of_the_college/

Page 235–239. O. W. Holmes, *The Contagiousness of Puerperal Fever. The Harvard Classics* (New York: P. F. Collier & Son, 1909–14); M. Thompson, *The Cry and the Covenant* (New York: Doubleday, 1949); S. B. Nuland, *The Doctors' Plague: Germs, Childbed Fever and the Strange Story of Ignac Semmelweiss* (New York: Norton, 2003); F. G. Slaughter, *Immortal Magyar* (New York: Schuman, 1950).

Page 239–240. Porter, *Cambridge Illustrated History of Medicine*, 202–245.

Page 240. H. W. Haggard. *The Doctor in History* (New York, Dorset Press, 1989).

Page 240–245. V. S. Thatcher, *History of Anesthesia* (New York: Garland, 1984).

Page 246–247. J. J. Beer, *The Emergence of the German Dye Industry* (Urbana: University of Illinois Press, 1959); W. M. Gardner, ed., *The British Coal Tar Industry* (Philadelphia: Lippincott, 1915); J. Drews, "Drug discovery: a historical perspective," *Science* 287 (2000): 1960–1964; M. Marquardt, *Paul Ehrlich* (New York: Schumann. 1951).

Page 247–249. A. S. Travis, *The Rainbow Makers* (Bethlehem: Lehigh University Press, 1993).

Page 250–251. C. Walsh, *Antibiotics: Actions, Origins and Resistance* (Washington, D.C.: ASM Press, 2003); R. Hare, *The Birth of Penicillin and the Disarming of Microbes* (London: Allen & Unwin); E. Lax, *The Mold in Dr. Florey's Coat: The Story of the Penicillin Miracle* (New York: Holt, 2004).

Chapter 12. The Great Pox Syphilis

Page 255. T. G. Benedek and J. Erlen, "The scientific environment of the Tuskegee study of syphilis, 1920–1960," *Perspectives in Biology and Medicine* 43 (1999) 1–28; J. Jones, *Bad Blood* (New York: Free Press, 1981).

Page 256–259. D. S. Jones, "Syphilis, historical," in *Encyclopedia of Microbiology* 4:538–544 (San Diego: Academic Press. 2000); E. Tramont, "*Treponema pallidum* (syphilis)," in *Principles and Practice of Infectious Diseases*, 5th ed., eds. J. Mandell, J. Bennett, and R. Dolin (New York, Churchill Livingstone, 2000), 2474–2490; Cartwright, *Disease and History*, first ed., 54–81; C. Quetel, *History of Syphilis* (Baltimore: Johns Hopkins University Press, 1990).

Page 260. *Secrets of the Dead, The Syphilis Enigma*, video (PBS, 2001).

Page 261–262. Franklin and Sutherland, *Guinea Pig Doctors*, 25–57. This claim is refuted by R. Murley. R. Murley, "Guilty of perpetuating the myth that John Hunter suffered from syphilis after inoculating himself from a chancre," *World J. Surg.* 18(1994):290.

Page 262–263. Tramont, "*Treponema pallidum* (syphilis)."

Page 263–264. T. Rosebury, *Microbes and Morals* (New York: Viking, 1971).

Page 265–266. R. Desowitz, *Who Gave Pinta to the Santa Maria?* (New York: Norton, 1997); E. Tramont, "Syphilis in adults: from Christopher Columbus to Sir Alexander Fleming to

AIDS," *Clinical Infectious Diseases* 21 (1995): 1361–1371; B. Rothschild and C. Rothschild, "Treponemal disease in the New World," *Current Anthropology* 37 (1996): 555–561; A. W. Crosby, *The Columbian Exchange. Biological and Cultural Consequences of 1492* (New York: Greenwood Press, 1988), 122–164; G. Antal, S. Lukehart, and A. Meheus, "The endemic treponematoses," *Microbes and Infection* 4 (2002): 83–94; C. Fraser et al., "Complete genome sequence of the syphilis spirochete," *Science* 281 (1998): 375–388.

Page 266–267. Cartwright, *Disease and History*, first ed., 54–81.

Page 267. McNeill, *Plagues and Peoples*, 227.

Page 267–268. Tramont, "*Treponema pallidum* (syphilis)."

Page 268. F. Winau, O. Westphal, and R. Winau, "Paul Ehrlich—in search of the magic bullet," *Microbes and Infection* 6 (2004): 786–789; de Kruif, *Microbe Hunters*, 308–330.

Page 268–270. Rosebury, *Microbes and Morals*; T. Parran, *Shadow on the Land: Syphilis* (New York: Reynal and Hitchcock, 1937).

Page 268–271. Jones, "Syphilis, historical."

Page 270–271. N. Grassly, C. Fraser, and G. Garrett, "Host immunity and synchronized epidemics of syphilis across the United States," *Nature* 433 (2005): 417–425.

Chapter 13. The People's Plague: Tuberculosis

General

Cartwright and Biddis. *Disease and History*, 2nd ed., 135–147.

Daniel, T. M., *Captain of Death. The Story of Tuberculosis*. (Rochester, N.Y.: University of Rochester Press, 1997).

Dormandy, T. *The White Death. A History of Tuberculosis*. (New York: New York University Press, 2000).

Dubos, R., and J. Dubos. *The White Plague. Tuberculosis, Man and Society*. Boston: Little Brown, 1952.

Kato-Maeda, M., and P. M. Small. "User's guide to tuberculosis resources on the internet." *Clinical Infectious Diseases* 32 (2001): 1580–1588.

Reichman, L. B., and E. S. Hershfield, eds. *Tuberculosis. A Comprehensive International Approach*, 2nd ed. New York: Dekker, 2000. *Tuberculosis in America: The People's Plague*, video (PBS, 1995).

Specific

Page 275–276. L. Hutcheon and M. Hutcheon, *Opera, Desire and Death* (Lincoln: University of Nebraska Press, 1996).

Page 276–277. S. Grzybowski and E. Allen, "History and importance of scrofula," *The Lancet* 346 (1995): 1472–1474; M. Bloch, *The Royal Touch. Sacred Monarchy and Scrofula in England and France* (New York: Dorset, 1989); T. Daniel, J. Bates, and K. Dawnes, "History of tuberculosis," in *Tuberculosis: Pathogenesis, Protection and Control*, ed. B. Bloom (Washington, D.C.: ASM Press, 1984), 13–24; F. Haas and S. Haas, "The origins of *Mycobacterium tuberculosis* and the notion of its contagiousness," in *Tuberculosis*, eds. W. N. Rom and S. M. Garay (Boston: Little Brown, 1996), 3–20.

Page 277–279. V. Daniel and T. Daniel, "Old testament biblical references to tuberculosis," *Clinical Infectious Diseases* 29 (1999): 1557–1558; S. Cole et al., "Deciphering the biology of *Mycobacterium tuberculosis* from the complete genome sequence," *Nature* 393 (1998): 537–544; T. Garnier et al., "The complete genome sequence of *Mycobacterium bovis*," *Proceedings of the National Academy of Sciences USA* 100 (2003): 7877–7882.

Page 281–283. G. N. Grob, *The Deadly Truth. A History of Disease in America* (Cambridge: Harvard University Press, 2002); "The stories behind the bones," *Science* 283 (1999): 1701; M. Caldwell, *The Last Crusade. The War on Consumption, 1862–1954* (New York: Atheneum, 1988).

Page 282. Fishberg, A. M. *Pulmonary Tuberculosis* (New York: Lea & Febiger, 1932).

Page 283. "Bovine tuberculosis," *Journal of Laboratory and Clinical Medicine* 141 (2003): 1359–1360.

Page 284–285. A. M. Kraut, "Plagues and prejudice: nativism's construction of disease in the nineteenth and twentieth century in New York," in *Hives of Sickness. Public Health and Epidemics in New York City*, ed. D. Rosner (New Brunswick, N.J.: Rutgers University Press, 1995); B. Bates, *Bargaining for Life. A Social History of Tuberculosis* (Philadelphia: University of Pennsylvania Press, 1992); G. D. Feldberg, *Disease and Class. Tuberculosis and the Shaping of Modern North America* (New Brunswick, N.J.: Rutgers University Press, 1995); K. Ott., *Fevered Lives. Tuberculosis in American Culture Since 1870* (Cambridge: Harvard University Press, 1996).

Page 285–286. "Cecil John Rhodes," *The Columbia Encyclopedia*, 6th ed. (New York: Columbia University Press, 2001).

Page 286–289. T. Brock, *Robert Koch, A Life in Medicine and Bacteriology* (Washington, D.C.: ASM Press, 1999); "The aetiology of tuberculosis," in *Medicine and Western Civilization*, eds. D. Rothman, S. Marcus, and S. Kiceluk (New Brunswick, N.J.: Rutgers University Press, 1999), 319–329.

Page 289–292. P. Small and P. Fujiwara, "Management of tuberculosis in the United States," *New England Journal of Medicine* 345 (2001): 189–200.

Page 293–295. S. Blower et al., "The intrinsic transmission dynamics of tuberculosis epidemics," *Nature Medicine* 1 (1995): 815–821.

Page 295–297. F. Ryan, *The Forgotten Plague. How the Battle of Tuberculosis Was Won and Lost* (Boston: Little Brown, 1993).

Page 297–298. S. Waksman, *The Conquest of Tuberculosis* (Berkeley: University of California Press, 1964).

Page 298–301. T. Frieden et al., "Tuberculosis," *The Lancet* 362 (2003): 887–896.

Page 299–300. K. Chung and C. Biggers, "Albert Leon Charles Calmette (1863–1933) and the antituberculosis BCG vaccine," *Perspectives in Biology and Medicine* 44 (2001): 379–389; D. Barnes, "Historical perspectives on the etiology of tuberculosis," *Microbes and Infection* 2 (2000): 431–440.

Chapter 14. Leprosy, the Striking Hand of God

Page 303. D. Beckett, "The striking hand of God: leprosy and history," *New Zealand Medical Journal* 100 (1987): 494–497. It has been suggested that the biblical references to leprosy are not due to that disease but are more likely to have described smallpox.

Page 304–306. J. Troutman, "The history of leprosy," in *Leprosy*, 2nd ed., ed. R. C. Hastings (Edinburgh: Churchill Livingstone, 1994), 11–25.

Page 306. McNeill, *Plagues and Peoples*, 155–156.

Page 307–308. Z. Gussow, *Leprosy, Racism and Public Health* (Boulder, Colo.: Westview, 1989).

Page 308–309. A. Rambukkana, "How does *Mycobacterium leprae* target the peripheral nervous system?" *Trends in Microbiology* 8 (2004): 23–28.

Page 308–309. P. Brophy, "Subversion of Schwann cells and the lepers bells," *Science* 296 (2002): 862–863; S. R. Ellin, "Leprosy in history," *The Lancet* 363 (2004): 1209.

Page 309. D. Young and B. Robertson, "Leprosy—a degenerative disease of the genome," *Current Biology* 11 (2001): R381.

Chapter 15. Six Plagues of Africa

General

Azevedo, M. J. *Disease in African History: An Introductory Survey and Case Studies*. Durham, N.C.: Duke University Press, 1978.

Duffy, J. *Epidemics in Colonial America*. Baton Rouge: LSU Press, 1953.

Farmer, P. *Infections and Inequalities*. Berkeley: University of California Press, 1999.

Hibbert, C. *Africa Explored: Europeans in the Dark Continent, 1769–1889*. London: Allan Lane, 1982.

Pakenham, T. *The Scramble for Africa 1876–1912*. London: Weidenfeld & Nicholson, 1991.

Specific

Page 313–314. O. Ransford, *Bid the Sickness Cease* (London: John Murray, 1983).

Page 315–316. P. Brent, *Black Nile: Mungo Park and the Search for the Niger* (London: Gordon Cremonisi, 1977).

Page 316–317. Cartwright and Biddis, *Disease and History*, 2nd ed., chap. 8, 158–159.

Page 317–320. O. Ransford and D. Livingstone, *The Dark Interior* (New York: St. Martin's Press, 1978); M. Gelfand, *Livingstone the Doctor* (Oxford: Blackwell, 1957).

Page 320–322. H. M Stanley, *Through the Dark Continent* (New York: Greenwood Press, 1969); R. S. Hall, *Stanley. An Adventurer Explored* (London: Collins. 1974); S. Neilson-Smith, ed., *Quest. The Story of Stanley and Livingstone Told in Their Own Words* (London: Arlington, 1978); S. J. Bierman, *Dark Safari. The Life Behind the Legend of Henry Morton* (New York: Knopf, 1990).

Page 323. http://workmall.com/wfb2001/congo_democratic_republic_of_the/congo_democratic_republic_of_the_history_the_colonial_state.html.

Page 323–325. A. Hochschild, *King Leopold's Ghost* (London: Pan, 2002).

Page 325–330. T. A. M. Nash, *Africa's Bane. The Tsetse Fly* (London: Collins, 1969); M. Lyons, *The Colonial Disease. A Social History of Sleeping Sickness in Northern Zaire, 1900–1940* (Cambridge: Cambridge University Press, 1992); M. Lyons, "African sleeping sickness: an historical review," *International Journal of STD and AIDS* 2 (Suppl. 1) (1991): 20–25.

Page 325–330. K. Vickerman, "Landmarks in trypanosome research," in *Trypanosomiasis and Leishmaniasis*, eds. G. Hide, J. Mottram, G. Coombs, and P. Homes (London: CAB International, 1997), 1–37; J. J. McKelvey, Jr., *Man Against Tsetse. Struggle for Africa* (Ithaca: Cornell University Press, 1973); de Kruif, *Microbe Hunters*, 232–255.

Page 330–335. F.E.G. Cox, "A history of human parasitology," *Clinical Microbiological Reviews* 2002: 595–612; Foster, *A History of Parasitology*; R. Desowitz, *New Guinea Tapeworms and Jewish Grandmothers* (New York: Norton, 1987), 75–90.

Page 330–335. T. V. Rajun, "The eye does not see what the mind does not know. The bacterium in the worm," *Perspectives in Biology and Medicine* 48 (2005): 31–41; "Onchocerciasis," in *Illustrated History of Tropical Diseases*, ed. F.E.G. Cox (London: The Wellcome Trust, 1996), 305–309.

Page 335–337. "Dracunculiasis," in *Illustrated History of Tropical Diseases*, 287–293.

Page 337–343. S. Howard, *Yellowjack, A History* (New York: Harcourt Brace, 1934); J. H. Powell, *Bring Out Your Dead* (Philadelphia: University of Pennsylvania Press, 1993); Oldstone, *Viruses, Plagues and History*, 45–72.

Page 332–343. Franklin and Sutherland, *Guinea Pig Doctors*, 183–226; Altman, *Who Goes First?* (New York, Random House, 1987), 129–158; de Kruif, *Microbe Hunters*, 286–307.

Page 344–349. H. N. Simpson, *Invisible Armies. The Impact of Disease on American History* (Indianapolis: Bobbs-Merrill, 1980).

Page 349–352. P. Hotez and D. Prichard, "Hookworm infection," *Scientific American* 272 (June 1995): 68–74; P. Hotez et al., "Hookworm infection," *New England Journal of Medicine* 351 (2004): 799–807; E. T. Savitt and J. Young, eds., *Disease and Distinctiveness in the American South* (Knoxville: University of Tennessee Press, 1988); G. Schad and T. Nawalinsk, "Historical introduction," in *Hookworm Infections*, eds. H. Crilles and P. Ball (Elsevier: Amsterdam, 1991), 1–14.

Chapter 16. Plagues without Germs

Page 355. de Kruif, *Microbe Hunters*, 114–115.

Page 356. E. W. Etheridge, "Pellagra: an unwelcome example of Southern distinctiveness," in *Disease and Distinctiveness in the American South*, ed. T. L. Savitt and J. H. Young (Knoxville: University of Tennessee Press, 1988), p. 113, quoting D. Davidson.

Page 357–360. A. M. Kraut, *Goldberger's War: The Life and Work of a Public Health Crusader* (New York: Hill and Wang, 2004).

Page 360. C. Elvejhem et al., "Isolation and identification of the anti-black tongue factor," *Journal of Biological Chemistry* 123 (1938): 137–149.

Page 361–362. K. Bloch, *Blondes, Venetian Paintings, the Nine Banded Armadillo and Other Essays in Biochemistry* (New Haven: Yale University Press, 1994), 185–207; M. Linder, *Nutritional Biochemistry and Metabolism* (Amsterdam: Elsevier, 1985), 69–115.

Page 363–368. K. Carpenter, *Beriberi, White Rice and Vitamin B. A Disease, a Cause, a Cure* (Berkeley: University of California Press, 2000).

Page 368–374. D. Harvie, *Limeys: The Story of One Man's War Against Ignorance, the Establishment and the Deadly Scurvy* (Phoenix Mill, U.K.: Sutton, 2002); K. Carpenter, *The History of Scurvy and Vitamin C* (Cambridge: Cambridge University Press, 1988).

Page 374–375. Altman, *Who Goes First? The Story of Self-Experimentation in Medicine*, 247–248.

Page 375–378. D. R. Fraser, "Vitamin D," *The Lancet* 345 (1995): 104–107; B. Wharton and N. Bishop, "Rickets," *The Lancet* 362 (2003): 1389–1400; Z. Hochberg, ed., *Vitamin D and Rickets* (Basel: Karger, 2003), 1–13.

Page 379–381. http://www.cdc.gov.od/oc/media/pressrel/r010330.htm.

Page 379–381. W. F. Loomis, "Rickets," *Scientific American* 223 (1970): 76–91; N. Jablonski and G. Chaplin, "Skin deep," *Scientific American* 13 (2002): 72–81.

Chapter 17. Plagues on Order

Page 383–385. J. Lederberg, R. Shope, and S. Oaks, eds., *Emerging Infections. Microbial Threats to Health in the United States* (Washington, D.C.: National Academy of Sciences Press, 1992); M. E. Wilson, "Travel and the emergence of infectious diseases," *Emerging Infectious Diseases* 1 (1995): 39–46; R. Levins et al., "The emergence of new diseases," *American Scientist* 82 (1994): 52–60; L. Garrett, *The Coming Plague. Newly Emerging Diseases in a World Out of Balance* (New York: Farrar, Straus & Giroux, 1994).

Page 385–387. A. Pflieger and A Kahn, "Hantaviruses: Four years after four corners." *Hospital Practice* (June 15, 1997): 93–108; S. Wrobel, "Serendipity, science and a new hantavirus," *FASEB Journal* 9 (1995): 1247–1254.

Page 386. J. Enserink, "U.S. monkeypox outbreak traced to Wisconsin pet dealer," *Science* 300 (2003): 1639; "SARS: what we have learned," *Nature* 424 (2003): 121–126.

Page 386–387. P. Sampathkumar, "West Nile virus – epidemiology, clinical presentation, diagnosis and prevention," *Mayo Clinical Proceedings* 78 (2003): 1137–1144; "West Nile virus," *Nature Medicine* 10 (2004): 5101–5103.; D. Morse, "West Nile virus—not a passing phenomenon. *New England Journal of Medicine* 348 (2003): 2173–2174.

Page 387–394. R. R. Rhodes, *Deadly Feasts. Tracking the Secrets of a Terrifying New Plague* (New York: Simon & Schuster, 1997); S. Collins, V. Lawson, and C. Masters, "Transmissible spongiform encephalopathies," *The Lancet* 363 (2004): 51–61; C. Mushal and A. Aguzzi, "Prions," *Encyclopedia of Microbiology* 3 (2000): 809–823; C. Weismann et al., "Transmission of prions," *Journal of Infectious Diseases* 186, Suppl. 2 (2002): S157–S165; S. Mead et al., "Balancing selection at the prion protein gene consistent with prehistoric kurulike epidemics," *Science* 300 (2003): 640–643; M. Balter, "Tracking the fallout from mad cow disease," *Science* 289 (2000): 1452–1454; G. Legname et al., "Synthetic mammalian prions," *Science* 305 (2004): 673–676; S. Prusiner, "Detecting mad cow disease," *Scientific American* 291 (July 2004): 86–93; P. Brown, "Mad-cow disease in cattle and human beings," *American Scientist* 92 (2004): 334–341.

Page 395–399. J. Taubenberger et al., "Initial genetic characterization of the 1918 'Spanish' influenza virus," *Science* 275 (1997): 1793–1796; A. W. Crosby, *America's Forgotten Epidemic. The Influenza of 1918*, 2nd ed. (Cambridge: Cambridge University Press, 2003); J. M. Barry, *The Great Influenza. The Epic Story of the 1918 Pandemic* (New York: Viking, 2004);

W. J. B. Beveridge, *Influenza, the Last Great Plague* (New York: Prodist, 1978); R. G. Webster and E. J. Walker, "Influenza," *American Scientist* 91 (2003): 122–129; K. Nicholson, J. Wood, and J. Zambon, "Influenza," *The Lancet* 362 (2003): 1733–1741; "Looking the pandemic in the eye," *Science* 306 (2004): 392–397; P. Palese, "Influenza: old and new threats," *Nature Medicine* 10, Suppl. (2004): S82–S87; W. G. Laver, N. Biscofberger, and R. G. Webster, "Disarming flu viruses," *Scientific American* 280 (Jan. 1999): 78–87.

Index

Acquired immunodeficiency syndrome.
    See AIDS
Acyclovir. *See* AZT
Addison's disease, 277
Aedes mosquito, 338
Africa, AIDS epidemic in, 113–114
    as cradle of humanity, 23–27
    climate change and, 24–25
    disease and, 312f, 313
    endemic diseases of, 325
    exploration of, 314–317, 314–325
    plagues of, 313–352
Agricultural revolution, 34–35
Agriculture, early, 33
    effect of, 32–35
    lethal gifts of, 35–37
Ahrun, 194
AIDS, 4
    and syphilis, in social context, 272
    as modern plague, 89–115
    global view of, 113–115
    HIV and, 21, 99–101, 226
    incidence in Africa, 114
    magnitude of, 89–90
    protection against, 104
    sketch of child with, 88f
    social context of, 112–113
    virus causing, 8
Air pollution plague. *See* Rickets
Al-Razi, 194
Albumin, 217
Allison, Anthony, 151
Amherst, Sir Geoffrey, 194
Amodiaquine, 148
Ancylostoma, 350, 351, 352
Animals, domesticated, as source of
        disease, 39–40
    benefits of, 39

exotic, as source of viruses, 386
    founder, 38
Anopheles mosquito, 344–345, 347
Anthrax, 8
Antibiotics, development of, 250–251
Antibody(ies), 217, 221
    formation of, clonal selection theory of,
        222–224
Antimalarials, 148–149
Antisepsis, barbers, bloodletting and,
        232–240
Antoine plague, 61–62
Antoninus, Marcus Aurelius, 61–62
Aractus, 304
ARC (AIDS-related complex), 100–101
Aretus of Capodoccia, 166
Argentine hemorrhagic (Junin) fever, 385
Aristophanes, 56
Aristotle, 279
Arnold, Benedict, 202
Artemisinin, 149
Asclepios, God of Medicine, 232f, 232–233
Ashurbanipal, King, 278
Aswan High Dam of Egypt,
        schistosomiasis and, 51
Atkins, John, 327
Atovaquone, 148–149
Australopithecus, 24, 25f
Australopithecus, 2001: A Space Odyssey,
        22f, 23
Australopithecus afarensis, 24, 26
Australopithecus anamensis, 24
Australopithecus boisei, 25
Autoimmune diseases, 223
Auzias-Turenne, Joseph, 271–272
Avicenna, 44
Azidothymidine. *See* AZT
AZT, 96, 102, 245, 252

B cells, 222
β Chemokine receptor (CCR5), 98–99
Baikie, William, 317
Baltimore, David, 95
Barbarossa, Frederick, 59
Barber-surgeon, bloodletting by, 77, 78f
Barber-Surgeon Guild, 234
Barbers, bloodletting, and antisepsis, 232–240
Barlow, C., 52
Bartholin, Thomas, 200
Bastian, Henry, 335
Bauer, J., 339
BCG (bacillus Calmette-Guérin), 300
Becket, Archbishop Thomas, 124
Belisarius, General, 64
Bentley, C.A., 351
Benton, Mary, 2
Bergmann, Ingmar, 70–71, 76
Beriberi, 363f, 363–368
    curing of, 367
    prevention of, 367
    studies of, 364, 365, 366
    treatment of, 363–364
Bernstein, Leonard, 117
Biggs, Hermann, 296–297
Bilharz, Theodor, 47–48, 351
Bilharzia. *See* Schistosomiasis
Bioterrorism, 384
Black, Davidson, 27
Black Death, 67–87
    positive aspects of, 72–73
    rats and, 67–68, 72
    societal and religious changes
        associated with, 70–71
Blacklock, Donald, 333–334
Blane, Gilbert, 372
Blood, 218
Blood fluke(s), 44–45f, 47–48, 49
Blood fluke disease. *See* Schistosomiasis
Bloodletting, barbers, and antisepsis, 232–240
    basis for, 233
Blower model of tuberculosis, 294
"Blue Cholera", Margaret Kaufman, 173–175
Boccaccio, Giovanni, 69–70
Boleyn, Anne, 267
Borde, Jules, 215
Bouquet, Colonel Henry, 194
Bovine spongiform encephalopathy
    (BSE), 387–389, 393
Boylston, Zabdiel, 201–202
Braddon, Leonard, 363–364
Browning, Robert, 68–69

Bruce, David, 326–327, 328–329
Bruegel, Pieter, 70, 281
Bubo(es), of bubonic plague, 82, 82f, 83
Bubonic plague, 65, 71
    bubo(es) of, 82, 82f, 83
    control of, 86
    recent reports of, 85–86
    treatment of, 86
    vaccine for, 86–87
Bunyan, John, 282
Burnet, Macfarlane, 222–223, 388
Burton, Richard, 320

Calciferol, 378–379
Caldwell, Mark, 283
Calmette, Albert, 299–300
Camille, 1936 movie, 274f, 275–276
Camus, Albert, 70
Cannibalism, kuru and, 389, 390–391
Carbolic acid, 239–240
Caries sicca, syphilis and, 259, 259f
Caroline, Princess of Wales, 201
Cartier, Jacques, 368–369
Casimir, king of Poland, 74
Castellani, Aldo, 328–329
Caventou, Joseph, 317
Cells, and viruses, 401–403
    appearance of, 401, 402f
Centers for Disease Control and
        Prevention (CDC), 270, 271
Cercarial dermatitis, 52
Chadwick, Edwin, 180–181
Chain, Ernst, 251
Chamberland, Charles, 94
Charles VIII, 256, 266
Chemotherapy, age of, 250
Childbed fever, 234, 238
Chloroform, 243–244
Chloroquine, 148, 149
Cholera, 159–188
    and evolution, 170–171
    and "immigrant problem," 184–188
    and nursing, 182–184
    as water- or food-borne, 166
    "catching" of, 171–172
    control of, 176
    diagnosis of, current, 173
    history of, 167
    pandemics of, 167–169
    sanitation, and public health, 178–182
    susceptibility to, 171
    symptoms of, 166
    treatment of, 173, 175–176
"Cholera epidemic, street scene during,"
        Honore Daumier, 180f

Christianity, plague of Cyprian and, 63
  rise of, plagues and, 60–61
Claudius, Avidus, 61
Clinton, President William, 255–256, 308
Clinton, Sir Henry, 347–348
Cobbold, Spencer, 48
Coleridge, Samuel Taylor, description of
  scurvy, 369
Columbus, Christopher, 37, 257–258, 369
Colwell, Rita, 172
Condoms, 104–105
Consumption. *See* Tuberculosis
Contagiousness, 81–82
Cook, Captain James, 371
Copepod, cholera and, 172, 172f
Cordon sanitaires, 70, 73, 78
Corn, pellagra and, 361
Cornwallis, Major General Charles, 348
Cortes, Hernan, 191
"Court of King Cholera," in *Punch*, 179f
Cowpox, 202–203, 203f
Cox, H.R., 123
Crandon, John H., 374–375
Crapper, Thomas, 177
Creutzfeldt, Hans, 388–389
Creutzfeldt-Jakob disease, 4, 387–389
  symptoms of, 388–389
  transmission of, 391, 393–394
Cro-Magnons, 28
Cyclops, and Guinea worms, 336, 337f
Cyprian, archbishop of Carthage, 62
Cyprian plague, 62–64

da Cadamosto, Captain Alviso, 314–317
Da Rocha-Lima, Henrique, 123
da Silva, Piraja, 50
Dapsone, 311
Dark Ages, 54
Davy, Sir Humphrey, 242
DDT, 148
de Gama, Vasco, 256, 368
de Kruif, Paul, 355, 388
de Lesseps, Ferdinand, 342
de Narvaez, Panfilo, 191
de Villalobos, Francisco Lopes, 258
de Voragine, Jacobus, 63
Death Chamber, The, Edvard Munch, 382f
"Death's dispensary," from *Illustrated
  London News*, 158f
Defoe, Daniel, 71
di Tura, Agnolo, 72
Diamond, Jared, 32, 33, 38
Diamonte, Giuseppe, 37
Diarrhea, secretory, 170
Diaz, Bernal, 191

Dinesen, Isak, 267–268
Diocletian, Emperor, 63–64
Diphtheria, 214
Diphtheria, follower of Goya, 214f
Discrimination, Black Death and, 73–74
Disease(s), epidemic, generalizations
    about, 400
  human increase in, 35
  spread of, movement of people and, 384
DNA (deoxyribonucleic acid), 91–92
DNA (deoxyribonucleic acid) viruses, 95
Doctor, The, Sir L. Fildes, 230
Doctor Pestis, plague doctor, 76–77, 77f
Domagk, Gerhard, 250
Donath, W., 367
Drake, Sir Francis, 369
Dubini, Angelo, 350
Dubois, Eugene, 27
Duesberg, Peter, 111, 112
Duffy factor, 154–155
Dulbecco, Renato, 95
Dutton, Joseph Everett, 327–328
Dyes, blue, 248
  chemistry of, 246–249
  red, 247–248, 250
  yellow, 248

Echth, John, 370
Economy, and social order, Black Death
    and, 78–81
Edson, Cyrus, 130
Education, Black Death and, 78
Egypt, snail fever in, 46–47
Ehrlich, Paul, 216, 217, 246–247, 249, 250,
    268, 288, 292–293,
    390, 393
Eijkman, Christiaan, 365–366, 367
Elion, Gertrude, 96
Elvejhem, Conrad, 361
Encephalitis, due to West Nile virus,
    386–387
Encephalopathies, transmissible
    spongiform, 387–394
Enders, John, 388
Enzymes, activity of, and shape of
    molecules, 362f, 362–363
Epidemic(s), disruptive effects of, 37
  generalizations about, 400
  prediction of, 12
  projection of course of, 17
  types of, 15, 16f
Epidemiologists, 12
Erysipelas, 235
Ether, 243
Eukaryote(s), 402f, 403

Europe, and "Dominican Republic," life in, compared, 37–38
Evans, Griffith, 325–326

Factor substitution, 80–81
Fantham, H.B., 329
Faraday, Michael, 243
Farming, changes in, Black Death and, 79–80
Farming populations, life of, 34
Father Damien, 307–308
Fertilizers, 35, 36
Fever bark kina-kina, 317
Findlay, Leonard, 377
Finlay, Carlos, 343
First Operation under Ether, Robert C. Hinckley, 243, 244f
Fitness, evolutionary, 20
Flagellants, Brethren of, 75–76
Flask, swan-necked, of Pasteur, 164, 165f
Fleas, and rats, 6f, 83
    Y. pestis in, 84–85
Fleming, Alexander, 250–251, 297
Flint, Austin, 357
Florey, Howard, 251
Flu vaccines, 399
Folate deficiency, 380
Fox, Daniel, 400
Fracastoro, Giovanni, 77
Fracastoro, Girolamo, 162, 262
Frank, Anne, 127
"French disease," Felicien Rops, 257f
Froebenius, 243
Frohlich, Theodor, 373
Fujinama and Nakamura, 48
Funk, Casimir, 367

Gajdusek, Carlton, 388, 389, 390, 391, 393
Galen of Pergamum, 62, 128, 234–235, 259, 279
Gallo, Robert, 90, 96
Garnham, P.C.C., 145
Gene amplification, 251–252
Gene mutation delta-32, 98–99
Gengou, Octave, 215
Geophagy, 320
Germ theory, 81–82
Germs, as parasites, 4, 9
Giardiasis, 163
Giblin, Albert, 177
Glick, Bruce, 221–222
Globulins, 217
Glycophorin, 155
Godunov, Boris, 267
Goldberger, Joseph, 357–360
Gonorrhea, 261, 262

Gordon, Alexander, 235
Gorgas, William C., 343
Gottlieb, Michael, 90
Gram stain, 82
Grassi, Giovanni Battista, studies of malaria by, 143–145, 346, 351
Great pox. See Syphilis
Greece, city-states of, 54
    conflicts involving, 55
    democratic society in, 54
    in history of Western civilization, 53
Griesinger, Wilhelm, 351
Grijns, Gerrit, 366
Gross Clinic, The, Thomas Eakins, 236, 237f
Gruby, David, 325
Grunbeck, Joseph, 256–257
Guerin, Camille, 299–300
Guinea worm, 335–337, 336f
    eradication of, 336f, 337
    symptoms of infection with, 337
    treatment of, 336f, 337
Gutenberg, Johann, 81

Hadlow, Willliam, 389
Haffkine, Waldemar, 86–87
Haldane, J.B.S., 150
Halofantrine, 148, 149
Hancock, Elizabeth, 98
Hansen, Gerhard Armauer, 161, 309
Hansen's disease. See Leprosy
Hantavirus, 385
Harley, John, 47, 48
Hawkins, John, 314
Hawkins, Richard, 369–370
Haworth, Walter, 374
Hemagglutinin, 397–398
Hematuria, 44, 47, 49
    and parasite, 47
    endemic. See Schistosomiasis
Henry the Navigator of Portugal, 314
Henry VIII, 234, 267
Herd immunity, 18
Herlihy, David, 67
Herodotus, 46
Herrick, James, 151
Hess, Alfred, 378
Hippocrates, 56–57, 62, 128, 136, 166, 231, 233, 234, 235, 259, 278–279, 295, 350
Hissette, Jean, 334
Hitchings, George, 96
HIV, 90–91, 91f, 92f
    African connection of, 107
    and AIDS, 99–101
    as infectious disease, 101–102
    CCR5 as receptor for, 99

clinical characteristics of, 100, 100f
control of, 102–105
discovery of, 92–96
genes of, 109–110
identification of, 96
immune system and, 96–99
Kaposi's sarcoma and, 101
origins of, 105–107, 111–112
transmission of, 102
virulence of, AIDS and, 21
HIV-1, 108–109
HIV-2, 110–112
HIV virus, 8
Hoffmann, August, 246, 249
Hoffmann, Erich, 261
Hogarth, William, 281
Holmes, Oliver Wendell, 235–236
Holst, Axel, 373
Homo erectus, 26–27
Homo habilis, 25–26
Homo neanderthalensis, 28, 29f
Homo sapiens, 27–28
Hong Kong "bird flu," 397
Hookworm(s), 6f, 7, 8, 349f
    and American South, 349–352
    as worldwide problem, 351
    infections with, history of, 349–351
    life cycle of, 349
    types of, 350
Howe, General, 202
Huldschinsky, Kurt, 377–378
Human immunodeficiency virus. *See* HIV
Hunter, John, 261–262, 263
Hunter-gatherers, 28, 32–33
Hutchinson, Jonathan, 264–265
Hypersensitivity, immediate-type, 226

Immigrants
    cholera and, 184–188
    smallpox and, 208
Immune defense, 212–213
Immune system, 218–220, 219f
    HIV and, 96–99
Immunity, acquired, 214–217, 220
    antibody-mediated, 221–222
    cell-mediated, 97, 224–226
    innate, inflammation and, 220–221
Immunization, 17–18
    attenuation and, 226–228
    passive, 215
India, River Ganges, Hindu pilgrimage
    to, 167, 168f
Infection(s), carriers of, 11
    diffusion of, 16–17
    factors promoting, 383
    tropical, in Africa, 313

Infectiousness, 11
Inflammation, and innate immunity,
    220–221
Influenza, 4, 10–11
    as globally contagious disease,
        398–399
    epidemics of, 395–396
    symptoms of, 396
Influenza virus, origin of, 395
    transmission of, 396–397
Insects, as disease vectors, 39
Interstitial fluid, 218
Irrigation practices, parasite transmission
    and, 36
Isoniazid, 298, 362–363
    and rifampin, 298, 299
Ivan, Grand Duke of Muscovy,
    266–267

Jackson, Charles, 243, 245
Jail fever. *See* Typhus
Jakob, Alfon, 389
Jansen, B., 367
Jefferson, President Thomas, 207, 343
Jenner, Edward, 200, 202–206, 230
Jensen, James, 154
Jesty, Benjamin, 203
Jesus, 60–61
Jews, blamed for plague, 73–74
Job, leprosy and, 303
    stricken with plague, 17th century
        woodcut, 302f
Johanson, Donald C., 24
Johnson, Samuel, 280–281, 282
Justinian plague, 64–65

Kakke. *See* Beriberi
Kaposi's sarcoma, 90
    HIV and, 101
Kapuscinsk, Ryszard, description of
    malaria, 135
Kaufman, Margaret, 173–175
Kentish disorder. *See* Malaria
King, Glen, 374
King Léopold II, 321–323
Kitasato, Shibasaburo, 82, 215, 216
Klebs, Edwin, 214
Koch, Robert, 77, 81, 112, 161, 165,
        169–170, 182, 214, 215, 292, 355,
        365, 388, 393
    microbe for tuberculosis and, 287–289
    studies of malaria by, 143
Koch's postulates, 166, 166t, 390
Kolletschka, Jakob, 236–237
Kubrick, Stanley, 23
Kuru, 388, 389, 390–391

La Boheme, Giacomo Puccini, 275, 276
La Traviata, Giuseppe Verdi, 275, 276
Laennac, Rene, 293
Laird, Major McGregor, 316
Lander, Richard and John, 316
Langley, Joseph, 315–316
LAPDAP, 149
Larey, Baron Jean, 47
LaSalle, Robert, 340
Lassa fever, 386
Latta, Thomas, 173
Laveran, Charles Louis Alphonse,
        studies of malaria by, 136–138,
        137f, 145, 328, 348
Lazar Houses (lazarets), 305–306
Le Clerc, Victor Emanuel, 341
Legionnaires' disease, 1–2, 3, 4, 10
Leigh, Vivien, 282–283
Leiper, Robert, 48, 335
Lemons, scurvy and, 370–371, 372
Leprosarium, 308
Leprosy, 303–311
    alienation of lepers and, 305, 306f
    diagnosis of, 310
    face and hands in, 305, 305f
    facts concerning, 309
    history of, 304–309
    in Hawaiian Islands, 307–308
    lepromatous, 309
    new cases of, sites of, 309
    spread of, 311
    treatment of, 310–311
    tuberculoid type, 309
Leprovac, 311
Leuckart, Rudolph, 332
Lewis, Sinclair, 71
Lind, James, 370–371, 372, 373
Linnaeus, Carolus, 27, 335
Lister, Joseph, 239
Liston, Robert, 242
Livingstone, David, 317–320
Lock and key theory, 217
Loeffler, Friedrich, 214
Long, Crawford, 243
Looss, Arthur, 351
Louis XIV, 177
Louisiana Purchase, yellow fever and,
        340–342
Louse Hunt, The, Gerhard Ter Borch, 126f
Louse (Lice), 124f
    as parasites for humans, 125–126
    conditions for survival of, 126–127
    delousing procedures and, 127
    life cycle of, 125
    typhus and, 122–125

L'Ouverture, Francois, 341
Lowlands disease. See Malaria
Lund, Charles, 374
Lupus vulgaris, 277
Lymphocytes, 97
Lymphokines, 220

Macbeth, William Shakespeare, 280
Macroparasites, 6f, 7
Macrophages, 225
Mad cow disease. See Creutzfeldt-Jakob
        disease
Madison, James, 207, 341
Magellan, Ferdinand, 368
Major histocompatibility complex, 225, 226
Malaria, 134f, 135–156
    and American South, 344–349
    antiquity of, 56–57, 57f, 136–145
    British interest in, 317
    "catching" of, 147
    complications of, 146
    death rate from, 59
    genetic resistance to, 150–155
    in Roman Empire, 59
    mosquito transmission of, 141f,
        147–148, 346
    pathology of, 145–147
    prevention of, 149–156
    prophylaxis by Livingstone, 318–319
    resistance to, 345
    symptoms of, 135
    transmission of, 136–145
    studies of, 136–139, 137f
    vivax, 154
    worldwide distribution of, 150f, 151f
Malaria parasite, 7–8
Malaria vaccine, 155
Mallon, Mary ("Typhoid Mary"), 11–12
Malt wort, 371–372
Malthus, Thomas, 30, 31, 67
Manson, Patrick, 48, 50, 339
    studies of mosquitoes and malaria by,
        137–139, 143
Marcus Aurelius, 193
Marten, Benjamin, 281
Mather, Cotton, 201–202
Maximillian, Emperor, 258
McCollum, Elmer V., 374, 378
McGreary, Johanna, 89
McNeill, William H., 18, 19, 37, 46–47, 65,
        166, 267, 306–307, 384
MDT, in leprosy, 311
    in tuberculosis, 298
Measles, 14, 16–17
    pattern in population, 15

Medicine, effect of plague on, 76–77
  secular, 231
Mefloquine, 148, 149
Mehaffey, A.H., 339
Mellanby, Sir Edward, 377, 378
Merz, Beatrice, 390
Metschnikoff, Elie, 86, 87, 213, 225, 247
Miasmas, 130, 159, 160
Microparasites, 5f, 7, 9
Miss Evers' Boys, 254f, 255
Monkeypox, 386
Montagnier, Luc, 90, 96
Montague, Lady Mary, 200–201
Moodie, Susannah, 185
Morton, William T.G., 242, 243, 245
Mosquito(es), as vector of West Nile
    virus, 386–387
  transmission of malaria by, 137–145,
    141f, 346
Multiplier of disease (R_0), 13–14
Murray, Mungo, 318
Mussolini, Benito, 59
Musto, David, 129
Mycobacteria, 277–278
Mycobacterium leprae, 309–310
Myocobacterium tuberculosis, 290, 290f,
    291, 292–293, 301
Myxomatosis, 18–19

Nabarro, D.N., 328
Nagana, 326
Napoleon, 117, 118–121, 341
  defeat of, typhus and, 118–121
Napoleon's troops, typhus fever in 1812,
    117, 118
Napoleon's Troops in 1812, Eugene
    LeRoux, 116f
Neanderthals, 28, 29
Nelson, Lord Horatio, 119
Neuraminidase, 397–398
Neutrophils, 221
New York City, disease in 1890s,
    128–131
Ney, Marshall, 121
Niacin, 361
Nicolaier, Arthur, 215
Nicolle, Charles, 122
Nicotinamide adenine dinucleotide
    (NAD), 361
Nicotinamide adenine dinucleotide
    phosphate (NADP), 361
Nicotinic acid, 361
Nightingale, Florence, 182–184
Nitrous oxide, in dentistry, 242–243
Noguchi, Hideyo, 261

O'Brien, Stephen, 98
Onchocerca, 330, 331, 332f, 333f
O'Neill, John, 332
O'Shaughnessy, William, 173
Oswell, William, 318

Palm, Theobald, 376–377
Panama Canal, yellow fever and, 342–343
Paracelsus, 243
Parasites, 5–10
  animal, 39
  definition of, 4
  evolution of, 18–20
  germs as, 4, 9
  hematuria and, 47
  mixing of genetic pools and, 384–385
  new, 383
  plagues and, 10–12
  transmission of, 8, 36
  virulence of, 19–21
Parasitism, 4
Park, Mungo, 315–316
Parran, Thomas, 270
Pasteur, Louis, 77, 81, 82, 83, 94, 160,
    169, 170, 182, 226–228, 238, 245,
    299, 388, 390
  vaccination of sheep by, 227, 227f
Pauling, Linus, 152, 388
Pekelharing, Cornelius, 365, 366
Pellagra, 355–363
  Basic Diet 123 and, 360, 361
  clinical symptoms of, 356
  dietary experiments and, 358
  incidence of, 357
  P-P factor and, 360
  prevention of, 360
  search for cause of, 357–360
  sign of butterfly in, 354f, 355
Pelletier, Pierre, 317
Penicillin, 245, 251, 252, 268
Pepys, Samuel, 71, 124
Perkin, William, 249, 374
Perry, Commodore Mathew, 363
Pertussis, 215
Pest houses, isolation and, 73
Phagocytes, 213
Pharaoh's plague, 44–53
Phthisis. *See* Tuberculosis
Pidoux, Hermann, 287
Pied Piper of Hamelin, rhyme by Robert
    Browning, 68–69
Pizarro, Francisco, 192
Plague(s), definition of, 10
  evolution of, 18–21
  in future, 383–400

Plague(s), definition of *(Continued)*
 nature of, 1–21
 of Africa, 313–352
 parasites and, 10–12
 pneumonic, 84, 86
 prediction of, 12–18
 prevention of, 210f, 211–228
 rise and Christianity and, 60–61
 road to, 30–31
 septicemic, 83–84
 six, of antiquity, 43–65
 social and political consequences of, 400
 sylvatic, 86
Plague, The, Felix Jenewein, 66f
Plague in an Ancient City, Michael
  Sweerts, 42f
Plague microbe, transmission of, 83
Plague of Ashod, The, Nicholas Poussin, 10f
Plague of Athens, 53–56, 57f, 211
 demoralization following, 56
 funerals during, 55–56
Plague of Florence, 69–70
Plants, desirable for domestication, 38
Plasma cells, 222
Plasmablasts, 222
Plasmodium falciparum, 59, 145–147, 150,
  153–154, 155
Plasmodium species, specific for humans,
  145–147
Plato, 56
Pneumocystis carinii pneumonia, 90, 106,
  106t
Poe, Edgar Allan, 231, 282
Pope Clement VI of Avignon, 74
Population, human, growth in, 30, 31f
 levels of, in evolving populations, 34
Pott, Sir Percival, 277
Pott's disease, 277, 278, 279f, 283
Pregnancy, death rate in 1841-1846, 236,
  238t
Preston, Richard, 196–197
Priestly, Joseph, 242
Primaquine, 148
Prince Valiant, Hal Foster, 210f
Pringle, John, 239, 371
Prions, 391–392
Prokaryote(s), 402f, 403
Prontosil, 250
Prout, William, 332
PrP, 391–393
Pruner, Franz, 350–351
Prusiner, Stanley, 391–392, 393
Public health, 73
Puerperal fever, 234, 238
Pyrazinamide, 299, 362–363

Qinghaosu, 149
Quarantine, 17, 70–73, 113, 128–129, 131,
  184–187
 isolation and stigmatization by,
  128–129
Queen Elizabeth I, 177, 315, 342
Queen Victoria, 244, 321, 342
Quinine, 148, 317, 318, 319, 324–325
Quinton, Paul, 171

Ramses V (Pharoah), 193, 193f
Rat killer, Jan Georg van Vliet, 68f
Red plague. *See* Pellagra
Redi, Francesco, 163
Reed, Walter, 388
Religion, consciousness of, 74–75
Reproductive ratio of disease (R_0),
  13–14
Retrovirus(es), 91–92, 93f
Rhazes, 167
Rheumatic fever, 223–224
Rhodes, Cecil, 285–286
Rickets, 375–381
 as endocrine-deficiency disease, 378
 calcification of bones and, 375–376,
  377f
 lack of sunlight and, 375, 376f, 376–377
 origin of, 379–381
 prevention of, 378
 symptoms of, 376, 377f
 ultraviolet radiation and, 377–378
Ricketts, Howard, 122–123
Rinderpest, 330
River blindness, 330–335, 331f, 334f
 diagnosis of, 335
 economic and social consequences of,
  334
 prevention of, 335
Rivers, Eunice, 255
RNA (ribonucleic acid), 91–92
RNA (ribonucleic acid) viruses, 95
Robles, Rodolfo, 332–333
Rodents, Black Death and, 67–68, 72, 83
 viruses carried by, 385–386
Roentgen, Wilhelm, 293
Rogers, Leonard, 326
Roman Catholic Church, 74–76
Roman Empire, 58
 malaria in, 59
Roman fever. *See* Malaria
Romans, religious ceremonies and gods
  of, 60
Roosevelt, Eleanor, 282–283
Roosevelt, President Franklin D., 269–270
Rose, Lauchlan, 370

Ross, Ronald, studies of mosquitoes and
    malaria by, 137–145, 141f, 346, 388
Roundworms, 6f, 7, 8, 9
Rous, Peyton, 94–95
Rous sarcoma virus (RSV), 94–96
Roux, Emile, 214–215, 228
Ruffer, Marc Armand, 44

"Safer sex," 104
Salas, I., 356
Salmon, Edmund, 228
Salvarsan, 245, 247, 268
Sambon, Louis, 356
Sanitaria, to treat tuberculosis, 295
Santayana, George, 113, 383
SARS (severe acute respiratory
    syndrome), 3, 4, 10, 13–14
  prediction of course of, 17
Schatz, Albert, 298
Schaudinn, Fritz, 261
  studies of malaria by, 145
Schistosomes. See Blood fluke(s)
Schistosomiasis, 44, 46, 47, 50, 50f, 51
  Aswan High Dam of Egypt and, 51
  diagnosis of, 52–53
  incidence of, 51
  pathology of, 49
  treatment of, 52
Scrapie, 389–390, 391
Scrofula, 276–277, 277f, 279
  touching to cure, 279–281, 280f
Scurvy, 368–375
  ocean voyages and, 368–369
  prevention of, 370
  search for cause of, 369–373
  symptoms of, 370
  explanation for, 375
Semmelweisz, Ignaz, 236–238
Shakespeare, William, 9, 280
Sheep, raising of, 80
Shortt, H.E., 145
Sickle cell anemia, 152
Sickle cell disease, 152
Sickle cell hemoglobin, 152, 153f, 154
Sickle cell trait, 152, 153f, 345
Sigurdson, Bjorn, 390
Simian immunodeficiency viruses (SIV),
    108, 110
Simond, Paul-Louis, 83
Simpson, George Gaylord, 92
Simpson, James, 243–244, 245
Sin Nombre, 385
Skin cancer, risk of, sun-bingeing and,
    380
Slave trade, African, 314–315

Sleeping sickness, 325–330, 327f, 328f
Slim, Sir William, 156
Slim disease, 107–108, 108f
Smallpox, 62, 191–209
  acquiring of, 195
  as biohazard, 207
  cause of, 195
  contagiousness of, 198–199
  eradication of, 206
  in humans, origin of, 192–195
  in Rome, 60
  mother and child infected with, 190f
  pathologic varieties of, 198
  Pedigree of Royal British Houses and,
    198, 199f
  progression of, 195–198, 197f
  social context of, 207–209
  vaccination against, 202–207, 205f,
    384
  variolation against, 200–202, 206
Smallpox virus, 195, 196f
Smith, J.H., 242
Smith, Theobald, 228
Snail fever. See Blood fluke disease
Snow, John, 160, 177, 181, 182, 244, 389,
    393
Societal differences, accident causing,
    37–40
Sommariva, George, 259
Sontag, Susan, 275
Sooty mangabeys, HIV-2 and, 110
Sophocles, 56
Spallanzani, Lazzaro, 163–164
Speke, Hanning, 320
Spies, T.D., 361
St. Anthony's fire, 235
St. Roch, 75, 75f
St. Sebastian, 63–64, 75
Stanley, Henry Morton, 320–322
Stark, William, 371
STD (sexually transmitted disease),
    transmission
  probability of, 17
Stephens, J.W., 329
Stiles, Charles W., 351
Storms, classification of, 15
  forecasting of, 12
Strauss, Richard, 23
Street scene during cholera epidemic,
    Honore Daumier, 180f
Streptomycin, 245, 251, 252, 298
Sulfone dapsone, 308
Surgery, before anesthesia, 241f,
    241–242
Sursuta, 234

Sydenham, Thomas, 167
Sylvius, Franciscus, 281
Syphilis, 255–272
    and AIDS, in social context, 272
    and social reformers, 268–270
    chancre stage of, 262–263, 264
    clinical signs of, 258, 262
    diagnosis of, 267–268
    disseminated stage of, 262–263
    distribution and incidence of, 271
    pre-Columbian origin of, 265–266
    rapid spread of, 258
    recurrent epidemics of, 270–271
    social context of, 266–267
    spirochete(s) and, 5f, 260–261, 265–266
    tertiary stage of, 263
    transmission of, 258, 260, 264–265
    treatment of, 259, 268
    vaccines against, 271–272
Szent-Gyorgyi, Albert, 373–374

T cells, 224–225
T lymphocytes, cytotoxic, 224–225, 226
Tafenoquine, 148
Tait, Lawson, 239–240
Takaki, Kanhiro, 363
Tallyrand, 341
Temin, Howard, 95
Tennyson, Alfred Lord, 183
Tetanus, 215
Theiler, Max, 340
Thiamine, 367
Thucydides, 55–56, 57, 211
Thymus, 224
Timoni, Emanuel, 200
Tolstoy, Leo, 118
Toxic shock syndrome (TSS), 2–3, 4, 10,
    12–13
Toxoids, 216
Trager, William, 153–154
Traill, Catherine, 186
Triumph of Death, Pieter Brughel, 71f
Trotter, Captain H.D., 317
Trotter, Thomas, 372
Trudeau, Edward Livingstone, 296
Trypanosoma, 5f, 325
Trypanosomes, 325–326, 327–328, 329
Tsetse fly (*Glossina*), 326, 326f, 329
Tuberculin (PPD), 288–289, 292
Tuberculosis, 231, 275–301
    as societal disease, 301
    cell-mediated immunity and, 291–292
    control of, 295–297
    diagnosis of, 292–293
    epidemics of, 293–295

history of, 278–286
incidence of, 293–295, 294f
in United States, 300
initiation of infection in, 290–291
Jewish immigrants and, 283
microbes causing, 277–278
movement to eradicate, 297
notable victims of, 282–283, 293
organs affected by, 276–277, 277f
pulmonary, 276
search for causative agent of, 286–289
transmission of, 283–284, 290
    reduced, 293–294
treatment of, 297
    early methods of, 295–296
    vaccination against, 299–300
Turner, Peter, 1
Tuskegee Syphilis Study, 255, 262
Typhoid, 11–12
Typoid Mary, 11–12
Typhus, 56, 211
    as fever plague, 117–133
    as "immigrant problem," 128–132
    defeat of Napoleon and, 118–121
    discrimination and dissemination and,
        127–128
    in 20th century, 132
    lice and, 122–125
    origin of, 118
    symptoms of, 121–122
    transmission of, 122–125
    vaccines against, 132
    war and, 117–118

Urbanization, disease transmission and,
    43–44

Vaccination, against smallpox, 202–207,
    205f
    social context of, 207–209
Vaccine(s), development of, 228
    flu, 399
    HIV and, 103
    in plague, 86–87
    in typhus, 132
    malaria, 155
Valentin, Professor G., 325
van Leeuwenhoek, Antonie, 162–163, 388
Vibrio cholerae, 169–170, 170f
Vicars, George, 98
Villages, origin of, 35–36
Villemin, Jean-Antoine, 286–287, 288, 289
Virchow, Rudolph, 287
Virulence, intermediate level of, 19
    rise in, 394–395

Virus(es), 5f, 7, 9, 94, 95
    appearance of, 402f, 403
    as parasitic, 403
    cells and, 401–403
    diseases caused by, 403
    transmission of, 14
Vitamin B_1, 367
Vitamin C, discovery and synthesis of, 374
    importance of, 374–375
    scurvy and, 372, 373, 374
Vitamin D, synthesis of, zones of, 379–381
Vitamin D_2, 378
Vitamin D_3, 378–379
von Behring, Emil, 215, 216
von Bismarck, Otto, 288
von Pettenkofer, Max, 159–161, 162, 176, 180
Von Prowazek, Stanislaus, 123
Vonnegut, Kurt, 392
Vorderman, Adolphe, 365, 367

Waksman, Selman, 251, 297–298
Warbeck, Perkin, 256
Warren, John Collins, 243
Washington, George, 124–125, 202, 233, 347–349

Wassermann test, 270
Waterhouse, Benjamin, 207
Weigl, Rudolf, 132
Wells, Horace, 242, 245
West Nile virus, 4, 386–387
White rice plague. *See* Beriberi
Whooping cough, 217
Wilson, President Woodrow, 396
Winterbottom, Thomas, 327
Winterbottom's sign, 327, 328f
Wright, Hamilton, 363–364
Wycliff, John, 304

Yellow fever, 337–343
    and Louisiana Purchase, 340–342
    and Panama Canal, 342–343
    annual incidence of, 343
    epidemics of, 338–339
Yersin, Alexandre, 81–82, 83, 214–215
Yersinia pestis, 82–83, 84–85
    biovars of, 85

Zidovudine. *See* AZT
Zinsser, Hans, 118, 133, 211
Zoonosis, 26, 108–109
Zyklon-B, 127